Extrusion Coating Manual

4th Edition

Edited by
Thomas Bezigian

The Association assumes no liability or responsibility in connection with the use of this information or data, including, but not limited to, any liability or responsibility under patent, copyright, or trade secret laws. The user is responsible for determining that this document is the most recent edition published.

Within the context of this work, the author(s) may use as examples specific manufacturers of equipment. This does not imply that these manufacturers are the only or best sources of the equipment or that TAPPI endorses them in any way. The presentation of such material by TAPPI should not be construed as an endorsement of or suggestion for any agreed upon course of conduct or concerted action.

TAPPI PRESS
Technology Park/Atlanta
P.O. Box 105113
Atlanta, GA 30348-5113 U.S.A.
www.tappi.org

ISBN 089852-067-3

TP 0102BO65

Printed in the United States of America

Library of Congress Cataloging-in-Publication Data

The extrusion coating manual/edited by Thomas Besigian.--4th ed.
p. cm.
Includes bibliographical references.
ISBN 0-89852-067-3 (hardbound)
1. Paper coating. 2. Paper--Extrusion. I. Bezigian, Thomas, 1955-.
TS1118.F5E95 1999
676'.235--dc21 98-53265
CIP

ABOUT THE EDITOR

Mr. Bezigian received his Bachelor of Science in Plastics Engineering from the University of Lowell in 1977 and his Master of Business Administration from Bryant College in 1987. He has held numerous technical and manufacturing roles at major film and paper converters in the United States, including Cryovac Division of W.R. Grace, Mobil Chemical, Plastics Division,

Fortifiber Corporation, James River Corporation, and Schoeller Technical Papers. Since 1991, he has consulted to the film and paper converting industries around the world in technical, manufacturing, marketing, and legal capacities. He is an Adjunct Professor at the University of Massachusetts-Lowell, where he lectures on the topics of extrusion coating, cast film, and packaging.

He is a frequent speaker at TAPPI seminars and the author of numerous TAPPI test methods and publications, some of which have received Best Paper Awards by TAPPI. He is also the Polymer Processing Consulting Editor for the CONVERTING Magazine, a leading industry trade journal. He has served as an officer on TAPPI's Extrusion Coating Committee, has conducted numerous TAPPI seminars and conferences around the world, and is the recipient of the TAPPI Polymers, Laminations & Coating Division's Leadership and Service Award.

He is married and has two children, and lives in Syracuse, New York.

EXTRUSION COATING MANUAL
4TH EDITION

Thomas Bezigian
EDITOR

CONTRIBUTING AUTHORS

George L. Baker
Sharon L. Baker
David J. Bentley Jr.
Robert F. Bernacki
Maged G. Botros
Gary D. Cheney
James L. Cooper
Charles H. Crumb
Joseph D. DiGiacomo
Ray Edwards
Michael R. Fitzsimmons
Ronald H. Foster
Robert B. Gregory
Jan Gronewold
Frances M. Hammond
Paul H. Jackson
Gary D. Jerdee
David C. Kelley
William M. Kirtley
Erkki Laiho
David A. Markgraf
Frank T. Orsini
Louis W. Piffer
Michael W. Potts
Philip B. Sherman
Fred M. Singer
Joseph Turick
John Ulcej
John D. Vansant
Dan Ward
Stanton M. Weiss
Tomi Ylänen
Dean A. Zimmerman

A revision of *Extrusion Coating: Equipment and Materials,*
published by TAPPI PRESS in 1990.

CA 950205.02
A project of the Extrusion Coating Committee of TAPPI's Polymers,
Laminations and Coatings Division.

TAPPI PRESS

PREFACE

First, a word of thanks to the many people who helped make this groundbreaking book a reality. This book is the end result of dozens of authors who conducted countless hours of research, many committee members who provided untiring peer review of their fellow Extrusion Coating Committee associates' work, and the always-helpful staff at TAPPI PRESS. Together, many thousands of man-hours of work were put into this publication, which is 85% new compared to the previous edition. My personal thanks to all my friends and colleagues who not only helped make this book possible, but also who make it a fun and interesting industry within which to work.

As we enter the next century, this edition of the Extrusion Coating Manual will stand as the reference for engineers, technicians and operators alike. It includes information on classic materials and troubleshooting techniques as well as the latest materials and machine designs available anywhere in the world today. It is the sum knowledge of thousands of years of experience put to paper. The authors in this book are among the most renowned experts in the world today on the subject of extrusion coating. It is our desire that this edition of the Extrusion Coating Manual be used not only at TAPPI training seminars, but also on the job by engineers, technicians, operators, marketing personnel, management, academia, and the industry as a whole.

It is my personal hope that this manual will serve to stimulate others to get more involved with TAPPI. I can say unequivocally, that TAPPI is truly an organization in which one person can make a difference. If you see an area that needs improvement, step forward, make it known, and offer a solution. You will not be shunned, but rather only encouraged to get involved and make the positive changes needed. This is good not only for TAPPI, but also for yourself, both in terms of personal growth and career development. I can honestly say there is no other professional organization like TAPPI anywhere in the world today. Its top management fosters this culture, and its members demand it and live it, at every conference, meeting, and seminar held anywhere in the world.

We are constantly striving to fill our members' needs, and with your help, we are succeeding in that goal. I am happy to be part of this community, and to be able to provide the time required to complete this publication for my colleagues and associates. I hope and trust you will find it beneficial in your professional careers.

Thomas Bezigian
PLC Technologies Consulting Company
Syracuse, New York, USA
November 1998

TABLE OF CONTENTS

CHAPTER I

Introduction and History

Extrusion coating is a complex process involving many related variables. This book provides information on extrusion coating processes, equipment, and raw materials.

Where did it all start? We can say that it really started in 1933 with the discovery of a resinous material by Imperial Chemicals Industries Limited Laboratories in England. The material was called polythene. Basically, at that time this resin was a laboratory curiosity because it had only been produced in small quantities. Extreme interest in this material as a component in radar prompted a push on production, and in 1939 commercial production of the resin started in the United States. Two companies were producing polyethylene resin in the United States by 1940, but it was in very short supply until 1945 because of wartime allocations.

During 1946 and 1947, film samples were available on an experimental basis. Basic extrusion coating operation began in 1948, at which time the H. P. Smith Co. in Chicago developed the Loxol process for coating molten polyethylene on paper substrates. This was not truly an extrusion coating operation as we know it today since it involved the use of an extruder to feed polyethylene into a nip and two rolls to carry the substrate through the nip containing the molten polyethylene resin. One of the rolls in the nip was held stationary to create a shearing action in the hot polyethylene, resulting in a well controlled coating. This method is still used today, but it does not provide a coating that is as free of pinholes or as lightweight as those applied through conventional extrusion coating methods using a die.

Almost simultaneously in 1946 and 1947, the DuPont Co. in Arlington, NJ, experimented with extrusions involving free films and paper. The company worked with the Hartig Engine and Machine Co. to develop the equipment necessary for this extrusion coating process.

The DuPont Co. worked very closely with representatives of St. Regis in its Oswego, NY, mill to produce some of the very first polyethylene coated paper stocks for use in multiwall bags. As far as can be determined, St. Regis was the first commercial coating operation to utilize extruders, a slotted-type die, and combining rolls for coating lightweight paper stocks with polyethylene.

In 1950, the International Paper Co., at its experimental laboratories in Niagara Falls and later Glens Falls, NY, began work coating lightweight papers and paperboard stocks experimentally with polyethylene. At the same time, the Sealright Co, Inc. in Fulton, NY, was also working very closely with St. Regis to experimentally coat board stock for conversion to containers.

In 1951, the St. Regis Co. expanded its operations to Carthage, NY, with the installation of an 8-in. extruder with a 90-in.-wide die-set for its coating requirements. In 1956, the Sealright Co., working with Frank W. Egan and Co. and the DuPont Wilmington laboratories, experimentally coated paperboard stocks for use in milk container production. In 1957, the very first milk carton was produced from polyethylene coated paperboard. At that time, conventional single-service paper containers were coated with a solvent solution coating of vinyl resins. This breakthrough into the paper milk container industry resulted in the immediate phenomenal growth of polyethylene resin, and in 1955 polyethylene resin became the first billion-pound resin manufactured in the United States.

In 1960, a crash program was started by all of the single-service milk container manufacturers to convert from paraffin wax to polyethylene coating; beginning in 1960, 90 million pounds of polyethylene resins were produced. This amount skyrocketed in 1970 to 220 million pounds of paperboard coating alone and grew to almost 300 million pounds by the mid-1970s. However, the growth of the paperboard market for milk cartons was stopped in the mid-1970s with the introduction of the 1-gallon polyethylene jug. Plastic jugs rapidly penetrated the milk container market to take about 55% of the market in 1980 and about 60% in 1986, at which time the penetration slowed and the mature paperboard carton market found its equilibrium with the new plastic jug.

While the milk carton market was replaced by new technology, extrusion coating was being applied to an ever-increasing number of applications using lighter weight, more flexible papers and films. These materials were used to form a host of flexible packaging structures, which began to rival the paperboard milk carton market in size. Thin substrates such as tissues, glassines, cellophane, mylar, and aluminum foils were coated and laminated together.

The oil crisis of the mid-1970s and subsequent high energy prices gave rise to the need for lighter weight, less energy-intensive packages. This force, along with the rapid growth of convenience foods and condiments for the fast-food industry, gave rise to a very rapid growth in extrusion coating and lamination markets in the 1980s. The need for unique packages for these new markets resulted in the rapid adoption of polymers such as ethylene acrylic acid, ionomers, polyesters, polypropylene, and ethylene vinyl alcohol.

The 1980s also brought the rapid growth throughout the world of aseptic packaging in paperboard containers. Aseptic packages, along with toothpaste tube stock, advanced the technology for manufacturing paper-aluminum foil-polymer combinations by extrusion coating and laminations to new heights.

The decade from 1960 to 1970 witnessed an outstanding growth of technological advances in the extrusion coating industry. The manufacturers of the equipment required for extruding and handling the various substrates embarked on a tremendous development program involving screw designs, heating, primer application equipment, temperature control, and a myriad of developments that allowed increased speeds and wider coatings.

During the 1980s, the need for low-cost complex flexible packaging structures gave rise to the growth of coextrusion extrusion coating. By the end of the decade, almost every major extrusion coating manufacturer used coextrusion. Usually coextrusion is used to add one or two extra layers of a functional polymer such as ethylene acrylic acid for aluminum foil adhesion or heat seal layers. However, five- to seven-layer coextrusion coating and lamination equipment is coming into more frequent use.

The endless stream of technological advances has kept the process current and competitive, even though basically the process has not changed significantly except for some resin changes and new equipment handling techniques.

The continued improvement in coating technology, equipment, resins, and operating practices and procedures will enable the extrusion coating process to remain competitive, resulting in better productivity and improved efficiency with lower-cost products.

CHAPTER 2

Extrusion Coating Lines

Stanton M. Weiss

Three types of lines for extrusion coating and laminating are:

1. Thin-film or low-tension applications at operating tension levels of 8-80 lb
2. Paper and its combinations in the middle range of 20-200 lb
3. High-tension for paperboard applications at 150-1500 lb.

Extrusion coating is a process in which an extruder forces melted thermoplastic through a horizontal-slot die onto a moving web of substrate material. The rate of application controls thickness of the continuous film deposited on paper, board, film, or aluminum foil. The melt stream, extruded in one or several layers, can be used as a coating or as an adhesive to sandwich the webs together.

In extrusion laminating, the molten polymer is spread between two moving substrates in a nip of rubber and a chill roll. In a continuous operation, rolls of material are unwound, spliced on the fly, and prepared by chemical priming or other surface treatment to make the substrate receptive to the coating and to develop adhesion between the materials.

Pressure and temperature on the web and extrudate combine to produce adhesion. The substrate normally provides the mechanical strength and surface for priming. The polymer provides a gas, moisture, or grease barrier.

As today's materials, especially those used for food packaging, become more sophisticated with rigid performance standards, coating lines become more complex. Goals are high productivity, extreme flexibility, and labor-saving computerized and robotized equipment. New specialty resins offer greater adhesion, allowing line speeds to be increased.

End-use products from extrusion coating/laminating lines can be broken into six major market classifications:

1 Liquid packaging
2. Flexible packaging
3. Board packaging
4. Industrial wraps
5. Industrial products
6. Sacks.

Liquid packaging utilizes coated lightweight board, alone or combined with plastic and aluminum foil, for semirigid containers for milk, juices, water, oils, processed foods, sauces, cheese products, and aseptic liquids. The polyethylene milk carton was largely responsible for the emergence of the extrusion coating industry in the 1960s and, as more commercial uses were found for polyethylene-coated materials, the industry grew rapidly.

In the 1980s, aseptic packaging made strong inroads in replacing traditional metal and glass containers. The sterile flexible "paper bottle," which extends the shelf life of dairy products for months without refrigeration, emerged as a major alternative form of packaging.

Although aseptic packaging systems differ, most of them use paperboard/aluminum foil/plastic composite material which is formed to shape, sterilize, and filled with liquid or semi-liquid product under sterile conditions. Customized extrusion coating lines, complete with in-line laminating stations, can produce an almost unlimited variety of shapes, sizes, and printing options.

The flexible packaging classification covers the combination of plain, printed, or metalized film papers, polymers, and aluminum foil used for protection, unitizing, dispensing, or holding of commodities. These include medicine and pharmaceuticals, food, chemicals, hardware, liquids, notions, sterile products, primal meats, and condiments. Flexible packages also include take-away food wrappers and the bag for "bag-in-box" containers.

Flexible packaging lines are using progressively thinner substrates of polyester, oriented polypropylene, and metalized materials. Thin snack-food substrates require minimum tensions to assure that preprinted webs are not distorted. Machines to create the new structures are becoming increasingly complex. Thinner coating layers are more difficult to lay down on the coating machine, and often the coating head itself must be engineered to handle a variety of coating materials.

Flotation drying, using air on both sides, is widely used in flexible package manufacturing because it handles light films well. Improved drying efficiency, compared with roll support dryers, allows higher line speeds.

In board packaging, heavyweight boards are coated and laminated and then formed into rigid packages for the boxing of detergents, tobacco, liquor, frozen foods, and bakery products. Ovenable containers and ice cream cartons also fall in this category. The plastic coated containers protect against grease, moisture, and gas. Release characteristics can also be provided.

Industrial wraps cover the range of heavy or reinforced papers, films, or board used for products in which the extrusion coated material can be added to other media for composite cans, drum liners, soap wrappers, and sheet overwraps for a variety of baled materials. The coated product is not necessarily used as a unit container, but as a wrapper or part of a protective structure.

The industrial products classification takes in extrusion coated or laminated material serving various industry requirements. Products include photographic base papers, substitutes for bitumen coatings, base papers for silicone coatings, insulation backing, automotive carpet coating, and metalized film balloons. Also in this category are functional laminates such as wallpaper and disposables like tablecloths and various hospital supplies. As with industrial wraps, these products might or might not be associated with packaging.

Sacks cover materials for multiwall paper bags and plastic coated raffia. Intermediate tension lines produce coated scrim that is woven from tapes of oriented polyethylene or polypropylene. These materials are used for heavy-duty sacks, tarpaulins, and applications in building, recreational, and agricultural areas.

A final, miscellaneous classification would include coated foam used in the fast-food industry.

Obviously, no single coating line can produce all the foregoing end-use products. Different machinery is manufactured to produce everything from 10 µm film to 500 g/m^2 board, coating at weights from 7 to 100 g/m^2, at widths from 300 mm to 3100 mm, and at speeds between 20 and 600 m/min.

Equipment for extrusion coating and laminating lines can be associated with product groups, with some overlapping between groups. Substrates or web-handling characteristics distinguish the difference among thin plastic films, paper, and paperboard combinations.

The most important consideration in web processing or handling equipment is the determination of the practical range for the systems required - the maximum and minimum unwind and winder roll diameters, maximum and minimum web tension forces, splicing speeds, core diameters, and other process needs.

The unwind basically takes material in roll form and processes it in a continuous sheet over a series of idlers or driven rolls with a suitable amount of tension to minimize wrinkling yet not produce deformation. A dancer or transducer roll can be used for tension control; regenerative drives and electric or pneumatic brakes are used where applicable. Similar considerations apply to the in-feed holdback when required levels of tension differ from those of the in-line operations. These sections tend to isolate tension transients from the unwinding roll.

Electrostatic treatment is available for enhancing surface or wettability conditions of the inert substrates prior to applying aqueous solutions. The direct/reverse gravure coater can be used for priming, coating, or printing. Chemical priming is used mostly in flexible packaging lines to promote adhesion between the extrudate and substrates such as cellophane, polyester, ionomer, nylon, or polyolefin materials. Infrared preheating and vacuum rolls provide the means for polyvinylidene chloride (PVDC) coating and web handling.

Recirculation of heated air in dryers is a common energy conservation practice. Air flotation, driven roll, and idler roll or drum support dryers are selected depending on the strength, support, or tension required for the substrate. The single roll guide at the dryer exit and chill or pull rolls are needed for special web processing requirements.

The extrusion laminator along with the extruder and die system is the heart of the process. The backup chill roll, rubber roll, and large-diameter roll form a three-roll system. The two-roll laminators can be used for heavy substrates or paperboard applications. As the moving web enters the nip section, it is coated and/or laminated. Ozone in close proximity to the entering web is used for oxidation of the molten polymer for adhesion promotion.

Most of the heat is removed from the coating or laminate by the chrome-plated chill rolls. Chill rolls are normally composed of steel and are constructed with a double-shell arrangement and spirally baffled. Outer shells of aluminum have been used for high coating weights. High-velocity chilled water is circulated to maintain a temperature rise between inlet and outlet of 2-3°F.

The coated or laminated structure is normally edge trimmed at the laminator by razor, score, or shear cutting. Trim removal systems are installed in this area. Slitting can be done just prior to winding at a turret or single-drum winder.

Auxiliary unwinds are located on, near, or over the extrusion laminator to provide a secondary substrate for laminating at the nip where the extrudate acts as an adhesive. These unwinds can consist of single position or turret assemblies with flying splices for aluminum foil, oriented polypropylene, polyester, paper, or paperboard substrates.

Web turnover systems, pull rolls, coaters, infrared heating, dryers, and web processing steps after the extrusion laminator are designed according to product needs. Thickness measuring devices include infrared and clear systems.

There are two basic winding techniques. The turret winder or center wind system is used for most flexible packaging materials. Tension is controlled by a dancer or transducer roll. The same design criteria for unwinds applies to winders. The type of web, operating speed, tension range, and roll buildup must be properly controlled to wind up a satisfactory roll. Paper and paperboard products can be wound by surface methods on a single-drum winder.

Total plastic constructions require more sophisticated web controls. Many converters utilizing traditional wood cellulose substrates are specifying that their new coating lines must be able to handle all-plastic films in the future. Wider tension ranges and air flotation dryers are two principal requirements of these convertible systems.

Other features being incorporated into various lines include dc regenerative upwinds and infield holdbacks for precise and low-level tension, direct/reverse gravure for aqueous PVDC coating, infrared preheating, and vacuum rolls for web control.

In the production of photographic base papers, exacting specifications and special criteria for pigmented polymers are needed to produce coated materials that provide constantly high-quality photographs. The concept of tandem operations or coating two sides of a substrate can be applied to many flexible packaging lines that produce combinations of paper, extrusion lamination to aluminum foil, and extrusion coating a polymer for heat sealing. Higher operating line tensions can be used in producing structures with paper for granulated or powdered mixes and freezer or sugar wrap materials. Not only is polyethylene used for lamination or coating but polypropylene ionomer, acrylic acid copolymer, nylon, and ethylene vinyl acetate can be part of a converter's inventory of resins.

Single-unit pilot coating lines feature an entire coating system preassembled and prewired at the factory and mounted on a structural steel base. These lines can be completely enclosed and have applications for products such as photographic coatings, computer tape, typewriter ribbon, videotape, magnetic floppy disks, solar control film, adhesives, and many types of medical-grade coatings. Stainless steel is used when extreme cleanliness is required, as in the manufacture of magnetic tape. The "clean room" machines are designed so that any metallic particles generated by machine friction are either contained or swept away by laminar air flow. Stainless steel is also used when lines are frequently washed with solvents that could remove conventional paint.

A typical pilot coating line consists of an unwind, web cleaner, coating head, smoothing device, dryer in-field vacuum roll, air flotation dryer, dryer exit tension control, web guiding into an on-line calender, cooling station, and regrinder, all aligned on a steel frame. Pilot coating lines designed to handle narrow, 6-in. web widths can be built so that components are cantilevered. Instead of being supported by traditional sideframes, the spindles, idler rolls, force transducers, and air flotation bars are all mounted on a vertical back plate.

Extrusion coating lines are seeing increased automation. Material can now be selected from stock, delivered to the line, and handled through robotics. The entire operation can be monitored and controlled by computer.

Drives are also coming under computer process control, and there have been advances in digital drives and in energy-efficient ac inverters. While a number of different drive systems have been installed and operated, the multimotor dc system is predominantly used for extrusion coating equipment. These drives can consist of as many as 10 motors in one line with a single control to bring the web up to operating speed. The tandem follower is another drive feature whereby the extruder will increase or decrease in rate with line speed to maintain a fixed coating weight as the line speed is changed.

Fig. 1. High-speed extrusion-coating line used to produce flexible packaging-type material

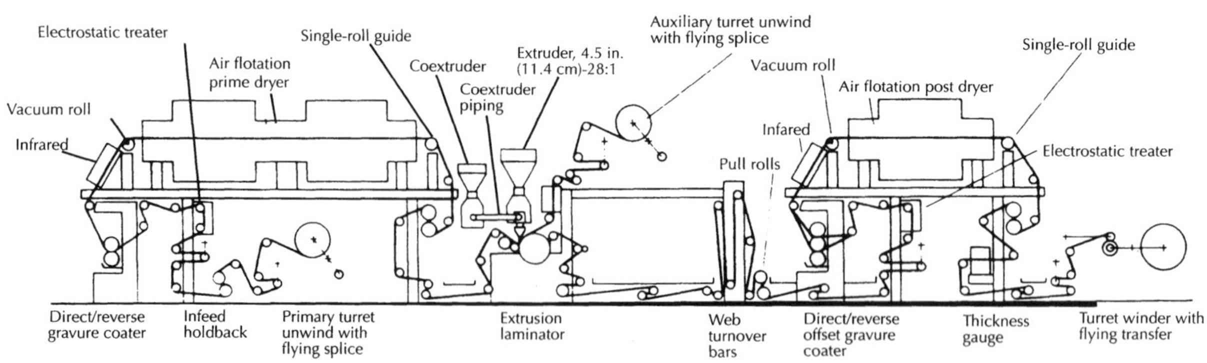

CHAPTER 3

What is Extrusion Coating?

Frank T. Orsini, Egan Davis Standard Corporation

Extrusion coating has been around for some forty years. The basic process involves the application of a melted thermoplastic onto a moving web. The thermoplastic raw material is usually in the form of very small pellets. After being fed into an extruder, it is melted and forced through a slot die. The extruder is mounted on a motorized carriage riding on rails perpendicular to the line.

The unwinds, gravure coaters, dryers, winders, and other web-handling equipment in the line are similar to those employed in other converting operations. The design of this equipment is dependent on the type of web to be processed. Typical extrusion coating operations use any of the following substrates: paper, paperboard, films, aluminum foils, and scrims.

The end products from the extrusion coating process cover many categories such as liquid packaging, flexible packaging, paperboard packaging, industrial wraps, and industrial products and sacks.

Examples of products in each category are presented below:

Liquid packaging: milk and juice gable top cartons, aseptic rectangular shaped drink boxes

Flexible packaging: snack foods, condiment packs, food, dry goods, medical packages, toothpaste type collapsible tubes, liquids, "bag-in-box"

Paperboard packaging: bakery boxes, microwaveable trays, frozen food boxes, detergent boxes, animal food boxes

Industrial wraps: drum liners, ream wrappers, composite cans, soap wrappers

Industrial products: photographic base papers, base paper for silicone coating, wallpaper, tablecloths, insulation backing, carpet coating, balloons

Sacks: cement bags, tarpaulins, swimming pool covers, grain and other agricultural or dry chemical sacks

Figure 1 shows the structure diagrams of some of the products listed.

Fig. 1. Typical product structures

Gable Top Carton
Printing
LDPE
Paperboard
LDPE

Drink Box
LDPE
Printing
Paperboard
LDPE
Foil
Coextrusion

Snack Food Bags
Polypropylene Film
Printing
LDPE
Metalized Film
LDPE

Toothpaste Tubes
Film
Printing
Coextrusion
Foil
Coextrusion
Film

Detergent Boxes
Film
Printing
LDPE
Paperboard
LDPE

Ream Wrapper
Paper
LDPE
Paper

Photo Paper
LDPE
Paper
Printing
LDPE

Fig. 1. Typical Product Structures - continued

Insulation Backing	Paper Snacks
Foil	Paper
LDPE	LDPE
SCRIM	SCRIM
Kraft Paper	Paper

Tarpaulins
LDPE
Woven Tape Fabric
LDPE

The extrusion system is made up of many key parts as listed here:

EXTRUDER

The extruder consists of a base weldment, drive motor, gear reducer, feed section, hopper, barrel with support, heaters, and flange.

Base weldment

The base weldment comes in three standard configurations. The first is called a "C" type or wraparound. This means that the gear reducer input shaft to which the drive motor is coupled will be on the same side as the extruder barrel. This configuration is desirable when the overall length of the extruder is to be minimized. The wraparound is only feasible if the diameter of the drive motor is within the limits of the space available.

The second configuration is the "Z" type. This means that the gear reducer input shaft is coupled to the drive motor on the side opposite the extruder barrel. This configuration is much longer than the wraparound. It is very desirable for maintenance for single extruder systems. The footprint of a "Z" type weldment is very narrow. This feature is quite useful when several extruders are to be mounted together to form a coextrusion system.

The third configuration is used with a small belt-driven extruder. With this configuration, you can have the advantages of small footprint and easy access, especially if the motor is mounted on top of the gear reducer. Most extrusion coating extruders use direct coupled motor/gear reducers, so this configuration is seldom used.

Drive motor

Each extruder has a dc drive to provide the rotational torque to the screw. The horsepower required varies with the size of the extruder, the output required and the resins to be processed. Typical extruder size/horsepower relationships are 6"/400 hp, 4.5"/250 hp, 3.5"/150 hp, 2.5"/75 hp. A special, field weakened drive and motor is often used when driving an extruder. The field weakening delivers maximum horsepower per rpm at lower speeds while maintaining the ability to deliver high rpm although at reduced horsepower per rpm. The need for this feature will be discussed later.

Gear reducer

Extruders operate at speeds in a range of 2-3 rpm while drooling to 300 rpm at full output. As mentioned above, the horsepower of the drive motor can be very high. The gear reducer ratio takes the motor energy at the input rpm of 1100-2300 rpm and transmits it to the screw at the reduced rpm. The melting of the thermoplastic requires both mechanical shear energy and electrical resistive heat energy. The gear reducer and drive motor provide the shear energy.

The gear reducer performs another job too. It houses the large thrust bearing which receives all the thrust produced as the shear energy is applied to the plastic via the screw. The gears and thrust bearing are lubricated by a water-cooled oil-circulating system. The oil is channeled via nozzles to the nip of the mating gears for maximum lubricating effect.

Feed section

A cast-iron water-jacketed feed section is assembled between the gear reducer and the extruder barrel. In the top of the unit is the hole into which the pelletized raw material is fed. It is water cooled to protect the resin from heat generated in the extruder barrel. The heat could cause the pellets to stick together and bridge the feed hole, interrupting the process.

Feed hopper

The feed hopper is the sheet metal device mounted on top of the feed section. It holds a small inventory of raw material to keep the extruder constantly fed. Vacuum feeding systems with low level sensors are installed as part of the hopper system. The hopper often has cooling devices installed, again to prevent bridging. Between the hopper and the feed section there is often the need to install a drawer magnet. The purpose of the magnet is to trap any metallic foreign particles that might find their way into the raw material storage device or the hopper itself.

Barrel with support

The extruder barrel is a long cylindrical shaped section bolted to the feed section usually with flanges at both ends. The barrel size is expressed as a ratio of its length over its diameter (L/D). The diameter is the nominal diameter of the extruder. The factors determining the L/D are related to screw design and will be discussed later. Since the melt temperature in extrusion coating is very high, usually around 600°F, provisions in the mounting of the barrel have to be made to compensate for the heat expansion of the metal.

A sliding barrel support is installed near the discharge end of the barrel. The support is mounted fully contracted with the extruder cold. The growth of a 4.5-in. extruder can be as much as 0.375 in.

The barrel is constructed from high-grade steel and has a bimetallic lining centrifically cast into it before final machining. The lining is designed to withstand the conditions of hard wear created by the extrusion process. The barrel is always harder than the screw flights. It is much easier to change out a worn screw than to change a barrel.

Barrel heaters

The barrel is divided into several heating zones. The size and position of these zones are usually selected to match the stages of the melting process occurring along the length of the screw. These stages will be discussed later. It will suffice to note here that both heating and cooling are required for the extrusion process. There are three popular types of heaters in use today: water-cooled cast aluminum, air-cooled cast aluminum, and ceramic band (cooled by ambient air in the off mode).

The first type, water-cooled cast aluminum, employs devices that are manufactured in two halves per set. Each half has cast into it both a series of resistive heating coils and water-cooling tubes. The resistive heat loads are connected directly to the load side of a power contactor, either mechanical or solid state. The water-cooling tubes are connected to a closed loop water-circulating system employing supply and return manifolds. The system uses distilled water with chemical treatment to protect the cooling tubes from fouling. The heat from the resultant steam is taken away via a water/water heat exchanger connected to an external water supply. In the supply line to each set of heaters, we have a solenoid valve. A temperature controller for the individual zone or for the entire extruder (e.g., computer-based system) gets feedback from a thermocouple installed deep in the wall of the barrel to sense the metal temperature. Based on the feedback, the controller will either close the heat contactor or open the cooling water solenoid.

The second type of heater, the air-cooled cast aluminum, is essentially the same as the water-cooled units. The difference is that, instead of the cooling tubes, there are fins cast in the heater. The fins provide increased surface area over which air is blown to provide cooling. The blowers and heater are shrouded in a small plenum that helps isolate the individual heat zones. The temperature controller therefore closes the heat contactor or the blower motor starter depending on whether heating or cooling is required.

The third type of heater is the ceramic band. It is the simplest of the heaters used. It employs a resistive heat load woven between pieces of ceramic material. The ceramic pieces are long and narrow so that, when fastened around the barrel, they conform to the shape of the outside diameter. The control cycle for these heaters is only on/off.

Flange

The discharge end flange is very important. It is designed to mate with the right angle adapter, which is the next component in the system. The flange is screwed onto the barrel end and then pinned in the position for perfect alignment with the mating flange. The alignment is critical since it will determine the operating position of the slot die. The joint must be rated for high operating pressure.

SCREW

The extrusion screw is the most important part of the melting process. Its design controls most of the variables that contribute to the quality of the process. The length of the screw, as mentioned above, is expressed in L/D. The sections of the screw as illustrated in Fig. 2 are the feed, transition, mixing, and metering. Also identified in the figure are the parts of the screw - namely, the root, the flights, and tip. The screw is usually constructed of high-grade steel and chrome plated in the root area. The chrome protects the root from most of the corrosive resin mixtures. The flights are either flame hardened or inlaid with a very hard material such a colmonoy. The hard material protects the screw from wear as it revolves in the barrel. The depth of the screw in the various sections is the dimension between the outside diameter of the screw and the root in the particular section. The lead of the screw is the distance along its length over which the flight makes one revolution.

Fig. 2. Single stage screw

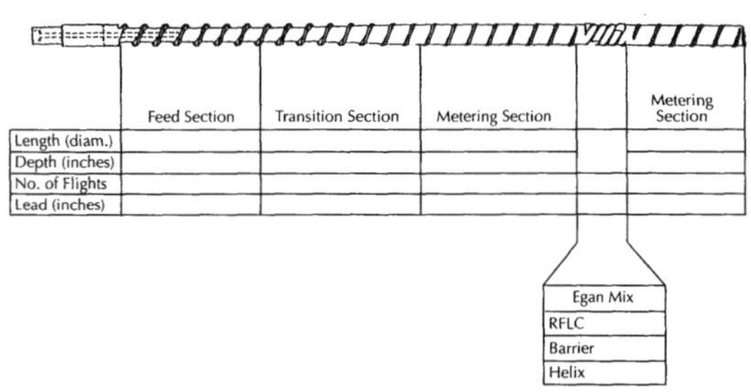

The feed area is the deepest area. It is the section of the screw that sits in the water-cooled feed section of the extruder. The depth is required to allow the pellets to flow freely into the extruder's diameter, thus filling the cavity between the root and the barrel wall.

In the transition area, where the tapered shape densifies the pellets, shear and pressure are created. The melting is achieved by the combination of the shear energy, pressure, and the heat around the barrel, provided by the barrel heaters.

The metering section is the final depth of the screw. This depth, when compared with the feed depth, determines what is known as the compression ratio of the screw. The metering depth also determines the volume of material that the screw can pump, per revolution.

Mixing sections are used to limit the amount of unmelted material in the pool. Many theories as to the design and use of mixing sections are available and beyond the scope of this text. The basic principle, however, is to introduce a barrier to the flow of the material down the barrel of the extruder, thus forcing the material being pumped to overcome the barrier at a high speed and shear which cause any unmelted material to melt. The mixing section also does as the name implies. That is, it mixes the electrically heated material riding against the barrel wall and the shear heated material riding along the root of the screw into a thermally homogeneous flow.

ADAPTER

The adapter is a device that is bolted to the discharge end of the barrel via a large flange. It has three basic functions. The first is to house a filter pack of stainless steel mesh to trap any unmelted material or any debris that slipped past the magnet in the hopper. The filter, commonly called the screenpack, can be changed by physically removing the adapter to expose the recess where the screenpack is installed. Another style uses a carrier called a breaker plate that slides into a slot in the adapter body to position the screenpack in front of the screw. The breaker plate units can be manual or use a hydraulic system which makes it a continuous operation.

The second function is to provide back pressure to the extruder by way of a tapered valve and seat. The valve is usually designed to give a linear change in pressure per length of engagement with its seat. The back pressure is needed to keep the screw filled with polymer. The last function is to direct the flow of the melt. The flow would be down in a single extruder system but can be in a number of directions in a complex coextrusion system employing several extruders.

In all cases, the adapter is connected to the die by some sort of high-temperature piping. A simple straight pipe is used in single extruder systems, while complex networks with special elbows and fittings are used with coextrusion. The piping and the adapter itself are all heated by ceramic band heaters where round surfaces exist and cast-aluminum block heaters where flat surfaces exist. The heaters are broken up in zones to match the rate of heating in each area. For instance, the narrow pipe might heat up very quickly while the large flange might heat slowly. If both were cycled by the same controller, either one will be too cold or the other will overheat. Thermocouple feedback is used in each zone.

The large body of the adapter takes a long time to heat to operating temperature. If the piping is allowed to heat up too fast, as would be the case described above, the expansion of the material in the pipe could build up pressure and explode. When microprocessor temperature controllers are used, the rate of temperature rise is determined by the slowest heating zone. In most cases, the slowest zone is the adapter body.

FEEDBLOCK

The feedblock is a device used with coextrusion systems. It is mounted above the die and receives the flows from all of the extruders in the system. The feedblock provides the preliminary shaping of the flows into flat layers and mixes the layers of the various extruders into the desired pattern. The feedblock has provisions for changing out parts of the unit, whether plates or plugs, to vary the structure to be produced. Examples of some structures are demonstrated in Fig. 3.

Fig. 3. Typical feedback structures

A/B/C/B/A

LDPE
Extrudable Adhesive
Barrier
Extrudable Adhesive
LDPE

A/B/A

Acid coploymer
LDPE
Acid Copolymer
LDPE

A/B/A/C/D

Extrudable Adhesive
Barrier Resin
Extrudable Adhesive
LDPE
Heat Seal Resin

There are several types of feedblock designs. Some employ technology of velocity matching using fins. Others use the theory of floating vanes with distribution pins. The discussion of the merits of each design is beyond the scope of this text.

DIE

The die is the last device in the extrusion system. The flow exits the die in a controlled manner and is applied directly to the moving web. The primary function of the die is to shape the flow into a melt curtain. The gap in the die through which the melt flows is adjustable, either manually or automatically. The manual dies are adjusted using wrenches and bolts that manipulate the jaw through either a sliding or flexing motion.

The automatic dies use a microprocessor and some sort of on-line gauging system. The jaw is manipulated electrically by either a resistive heater or a piezoelectric device. The gauges can be radioactive transmission, infrared reflective, infrared transmission, contact caliper, or noncontact caliper.

A fume extraction device is also incorporated in the installation of a die. The high temperature of the extrudate creates smoke which, when condensed on equipment, causes a slimy dust collecting layer. The fume extractor or hood with an exhaust blower keeps the smoke out of the area and maintains a clean operation.

Most dies are equipped with devices that adjust the width of the curtain. The devices are called deckles. There are two types of deckles, internal and external. The internal type uses a movable rod that blocks the flow by bridging the gap inside the die. The external type uses a clamping device to force a piece of Teflon-type material up against the slot to prohibit flow. Different vendors provide different variations and make claims of which type is best. That discussion is beyond the scope of this text.

The internal flow channels or manifold of the dies also fall into two types, a tapered or coat hanger shape and a straight "T" shape. The coat hanger shape requires the die to be manufactured in two halves. Each half is machined independently with half of the manifold. The two halves are bolted together with large bolts tightened to torque values of over 300 ft-lb to withstand the process pressure. The jaws or lips of the coat hanger die are usually machined as part of each half. However, some special designs employ a removable jaw on the nonadjustable side for special applications.

The "T" slot die is manufactured as three pieces. The body that is the major piece has the manifold cut into it from a direction opposite the flow. The jaws are shaped to complete the manifold and funnel the melt to the gap. Converters have varied opinions of the merits of each type. Again the discussion is beyond the scope of this text.

There are some times when resins that are to be processed in a coextrusion can have greatly different viscosities or process at different temperatures. For those applications, multi- manifold dies are used. Multi-manifold dies are actually two or three dies in one with separate feed ports from separate extruders for each manifold. Some designs allow the different melt curtains to exit the die separately and combine in the air as they approach the moving web. Others combine the layers in an area just inside the gap so that a single curtain is formed. Some single and multi-manifold dies are pictured in Fig. 4.

Fig. 4. Single and multi-manifold dies

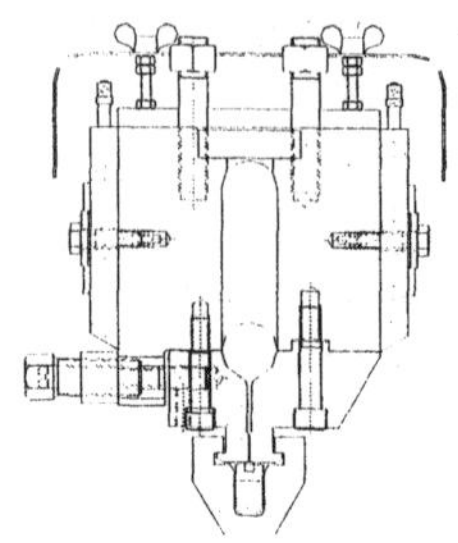

Single Manifold Die

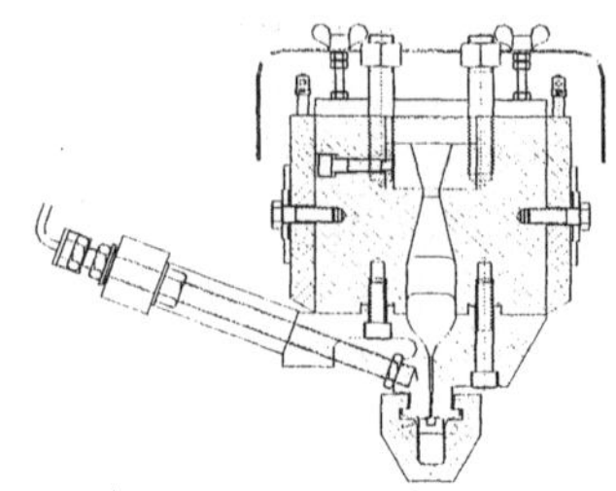

Single Manifold Automatic Die

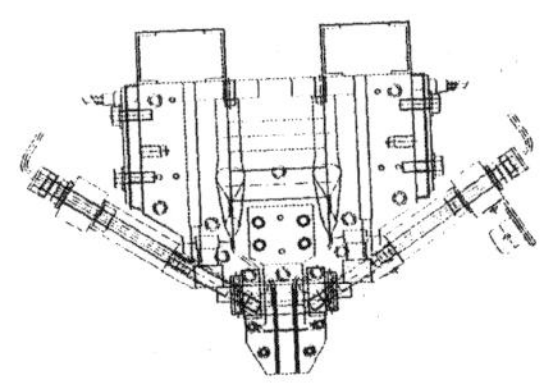

Dual Slot Dual Manifold Die

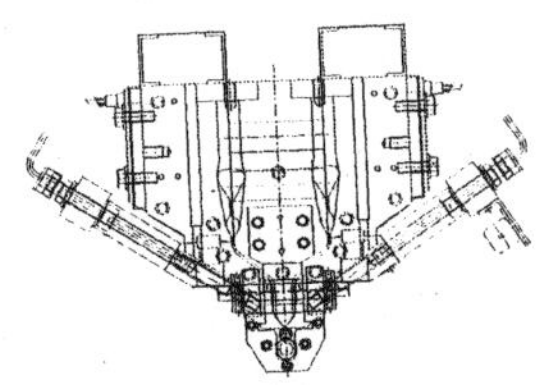

Single Slot Dual Manifold Die

Motorized carriage

The extrusion coating extruder is mounted on a motorized carriage as shown in Fig. 5. The carriage can be designed for placement of multiple extruders and is so designated as a coextrusion carriage. The motorized movement is in three directions. They are up and down, in and out of the coating position, and horizontally left and right in the line direction.

Fig. 5. Coextrusion carriage–3 extruder system

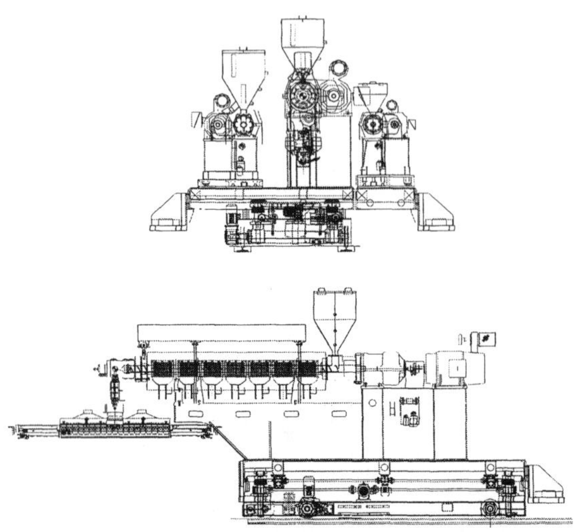

The up/down movement is very important in the extrusion coating process. The melt oxidizes as it flows in the air between the die and the laminator nip. Oxidation enhances the adhesion of the melt to the moving substrate. The higher the distance is, usually called the air gap, the longer the melt is exposed to the air and the higher the degree of oxidation that occurs. The trend for increased line speeds in extrusion coating has made very high air gaps necessary because the dwell time in the gap is reduced as speed increases.

Since the die is vertical and the position of the extrusion laminator is fixed, the carriage is provided with an in/out movement to bring the die off line. The off-line position is used to provide a place for the melt to continue to flow at low output (drool) while the line is being made ready for operation. Without the off-line movement, the drool would fill the nip of the laminator and break the web. It would be very difficult to get the process started.

EXTRUSION LAMINATOR

The extrusion laminator is horizontal nip created by a rubber-covered roll, against a large chill roll. The nip pressure is up to 125 pli. The laminator has special purpose rolls for flattening out the web (bowed or herringbone rolls) and taking out the edge sag (eccentrically adjusted skew rolls).

The chill roll diameter is determined by many factors related to its capacity to cool, e.g., the total pounds of extrudate per inch of width per hour, the heat transfer efficiency of the design, and the temperature and flow rate of the water.

The design of the chill roll is very important. The industry standard is a double shell with spirally wound baffles between the shells. This type of design keeps the cold water traveling along the outside of the diameter of the inner shell where it can be most effective. A flow through hollow design would cause the water to flow into a large puddle in the roll and not cool the outside where the cooling is needed. The chill roll surface temperature must be uniform across the face so that the cooling effect will be uniform. A common target is to maintain a change of less than 3°F per 60 in. of face width. This is achieved by pumping large volumes of water through the roll. A 24-in.-diameter roll often has a flow rate in excess of 250 gallons per minute. Another advantage of the high flow rate is that a higher water temperature can be used for a given cooling application. The desired 55°F water is fine during the winter months or in an air conditioned factory but will cause excessive condensation in the summer when the humidity and dew point are quite different. In some tropical areas, the water temperature can not go below 75°F.

The baffle design also gives strength to the roll. Because of the required heat transfer, the outside shell of the chill roll is relatively thin, about 3/8 in. The baffles spread the effects of the pressure across the whole roll. A very robust inner shell design provides the resistance to deflection.

The surface of the chill roll is very important because the chill roll determines the texture of the coating. An example can be seen in the many grades of paper used in the printing of photographs. The choices of texture (e.g., glossy, satin, matte) depend only on the surface of the chill roll used when the base paper was extrusion coated. A general rule in extrusion coating is that the rougher the surface is, the easier the material will strip from the roll because the rough roll traps more air between the web and the roll surface. When the web strips off the chill roll easily, maximum line speeds can often be achieved. A mirror surface maximizes contact between the coating and the roll. To strip the web from the mirrored roll, the speed must be selected to give time for the release without tearing the coating away from the web. There are some patented roll surfaces that are designed to give the surface appearance of the mirror with the release of the matte. Because the surface is product dependent and rolls are often changed, quick change bearings and drive components are utilized.

The nip section of the laminator can be made up of either two or three rolls. The two-roll design uses a large-diameter rubber-covered water-cooled roll against the chill roll. The diameter is sufficient to withstand the total available nip pressure. The two-roll design usually provides a large nip "footprint" which can help or hinder the process depending on the substrate being coated, etc. The large rubber roll is costly to recover.

The three-roll design uses a small-diameter water-cooled rubber-covered roll backed up by a large-diameter chrome-plated water-cooled steel roll. The centerline of the steel roll, the rubber roll, and the chill roll are kept in line. This configuration allows the use of the small-diameter rubber roll by countering the effects of deflection of the rubber roll. The rubber roll is simply squeezed between the steel roll and chill roll. This configuration allows for a small nip "footprint" and lowers the cost of recovering the rubber roll. Another advantage of using the small-diameter rubber roll is that the roll is very light in weight, thus making the changing of the roll very easy and often possible without the use of hoists.

In all laminator designs, there is an adjustable pivoted roll where the web is stripped off the chill roll. Process experience has determined that the angle at which the web leaves the chill roll and/or the degree of wrap that the web makes around the chill roll are factors that affect the quality of the product and the attainable line speed. The adjustable stripper roll provides the variability required.

Extrudate that leaves the die has a tendency to bunch up at the edges. The gap of the die can be 0.030 in., but the desired coating thickness might be less than 0.001 in. The laminator must run at a speed thirty times faster than the speed that the plastic is traveling in the air gap. The plastic is then "drawn down" to the desired thickness. As the draw down takes place, the curtain of plastic exiting the die "necks in"; that is, it is pulled thinner in width. It is the neck-in phenomenon that causes the bunch up at the edges. The bunch up is called the "edge bead" since, after the curtain is cooled, it is actually a bead attached to the edges of the web. Left in place, the edge bead prohibits the machine from winding a good roll of coated product. The roll will wind along the thick edge and be loose and uncontrolled in the center.

To overcome the problem, most extrusion coating laminators have slitting stations incorporated into the design. The number of stations depends on the number of slitting methods required to trim the various webs to be run. Some popular methods are score or crush cutters, shear cutters, and razor cutters.

A pull roll is provided as an integral part of the laminator after the slitting stations, especially when light extensible webs are to be run. The pull roll and the slitter stations are driven from a power train off of the laminator drive motor. The pull roll provides an adjustable draw of the web through the slitters while isolating the stripping of the substrate off of the chill roll from the tension required to carry the web into the next machine section.

Before we discuss line layouts, we must mention the methods used to enhance the adhesion of the extrudate to the web(s) in the extrusion coating process.

The first method is the corona treater. Corona treating in extrusion coating is the same as in other coating operations. It typically raises the surface tension to create a surface ideal for the bonding of the coating. Also, as in other coating processes, the corona treating of aluminum foil is used to remove any oil or its residue from the annealing process.

The second method is the use of a gravure-type coater to apply a chemical primer. Certain webs can only be treated for adhesion enhancement in this way. The composition of the various chemicals used is beyond the scope of this text. For some very difficult substrates or when extremely high bond strength is required, some converters use corona treatment before chemical priming.

The third method is flame treating. A flame treater uses a gas-fueled flame to "burn" the surface of webs, especially paperboard, to work much like a corona treater. The flame controller is very sophisticated and allows the ignition to occur only when a minimum line speed is reached. The exhaust from the flame treater is constantly analyzed to detect irregularities that could indicate an unsafe condition. When flame treating paper and paperboard, the flame provides an additional feature. It does indeed burn off any fibers standing on the surface of the web. These fibers can cause punctures in the coating and compromise the barrier property of the plastic. The fibers become wicks like in an oil lamp and bring moisture into the web behind the coating.

The fourth and last method we will discuss is the application of ozone into the extrusion coating nip. Ozone alone will not have any great effect, but when used in conjunction with other methods can yield very good results. As mentioned previously in our discussion of the carriage movements, oxidation is very useful in getting a good bond. The only ways to increase the oxidation level are to increase the air gap or raise the temperature of the melt. Each of these has its practical limitations without costly machine modifications or raw material changes.

The air gap is limited by the mechanical design of the machine. The temperature is limited by the allowable process temperature range for which the material being extruded was developed. Many plastics produce unpleasant odors that are retained in the web if processed at too high a temperature. The use of the web in food packaging can taint the food and therefore is not allowed. The last resort for the converter, assuming that the limits just explained cannot be overcome, is to introduce ozone into the nip. It provides the free radicals for bonding just like the other methods, although not in the quantity to be able to be the only source. Ozone treating has been used to greatly improve the bond strength at increased machine speeds. It therefore allows the converter to increase his production speed without sacrificing quality.

Line configurations in extrusion coating are developed according to the products to be run. They can be as basic as unwind, laminator, winder, and extruder or as complex as several unwinds, primers with dryers, laminators, extruders, and winders.

In recent years, extrusion coating has grown as the preferred method of lamination in some industries once dominated by dry-bond lamination. This growth has caused some converters to make the choice to either invest in extrusion coating equipment or risk losing business to some one that does.

The cost of a new extrusion coating/laminating line, with installation and commissioning, can easily cost millions of dollars. The time period between the placement of an order for a machine and the successful running of commercial product can be anywhere from 12 to 18 months. For some converters, a project of this magnitude could put them out of business. There is an alternative.

Very often, an extrusion coating station can be added to an existing line. In that way, the unwinds, gravure coaters, winder, etc., from the existing line can be utilized. The extrusion carriage installation is the bulk of the work. This work, however, can usually be done behind protective barriers with the carriage in the off-line position. The investment is much lower, and the cycle from order to production can be shortened significantly.

After the converter has had time to build a market and gain experience in the extrusion coating process, he can choose to update or replace the components of the old line or purchase additional components to move the extrusion coating station into a new line. There are many possibilities.

Now that we have discussed the equipment for extrusion coating, we need to look at the advantages and disadvantages of using this process as compared to other processes available.

ADVANTAGES

1. Speed - Extrusion coating can be performed at production speeds up to 2000 ft/min.
2. Clarity - The extrusion lamination of clear films provides a very clear web with no screen pattern often associated with roller applied adhesives.
3. Flexibility - Extrusion coating can be applied to many different webs to provide many different characteristics. The resins selected and the additives used can provide a host of solutions to product problems.
4. Profile - The extrusion coated products can be adjusted via a nuclear gauge feedback to compensate for area of the web that are not printed. In other laminations where a uniform layer of adhesive is applied, the thickness of the printing can cause distortion in the profile of the wound roll. This distortion is even more of a problem with the application of thermal stripe adhesive that is popular in the snack food industry. The flat profile in the extrusion-laminated roll means that the slit rolls will run better on form, fill and seal machines.

5. Cost - The cost of extrusion laminating was usually higher than that of other methods. The trend has reversed with the introduction of new raw materials. The coextrusion of many thin layers of materials that once had to be laminated into the product as independent films has lowered the cost, decreased the overall weight, and not sacrificed any performance. It is also very easy to bulk up the resultant web with inexpensive resin if product weight is to be maintained.
6. Sealing - The resins available today have very good heat seal strengths with superior integrity. This quality can reduce package rejects.
7. Opacity - Some laminated products require precise levels of opacity to protect the contents or provide a background color to clear reverse printed films in the structure. Blending pigment mixtures into the raw material in the extruder can provide these features.
8. Environmental - The extrusion lamination process is very friendly to the environment. The most popular chemical primers are now water based. The ozone generated by corona treaters is exhausted through ozone destruct devices. The fumes from the die area are easily cleaned if required using devices to condense the polymer from the smoke and exhaust warm air to atmosphere.
9. Expandable - There are many converters that start up their production lines with only one extruder and one carriage serving one laminator. To laminate and then coat a product, the material goes through the machine twice. The product produced is high quality and is used to build market share. As demand grows, the extrusion lamination line is easily modified to add additional extruders to the first station or additional stations to perform the lamination and coating in tandem.
10. Barrier - In the production of many different products, extrusion coating provides a practical barrier to moisture, e.g., drum liners, ream wrappers, soap wrappers, insulation backing, cement bags, and of course the gable top liquid package.
11. Strength - In products such as carpet coating, tarpaulins, swimming pool covers, and grain and other agricultural and chemical sacks, the extrusion coating becomes the material that creates the product. Particularly with tarpaulins where you have a woven fibrous cloth with no dimensional stability, the extrusion coating binds the fibers together as one unit giving strength and the needed stability.

DISADVANTAGES

1. Equipment cost - There is no doubt that the extrusion coating equipment is much more expensive than, for example, dry-bond lamination equipment. Also there are many auxiliary systems which must be purchased depending on need. Some are silos, blenders, chillers, and vacuum loaders.
2. Operating cost - Extruders consume large amounts of electrical power for both the external heating of the various sections and for the large drive motors.
3. Startup time - After a shutdown period, it takes hours for an extruder to heat up to operating temperature. Unless the machine is a dual-purpose design where it can always start up on an alternate process, the heat-up time is production time lost.
4. Rigidity - The heat history put into the web in a dry bond application provides a rigidity to the web that helps packages to stand up on the supermarket shelf. The extrusion-laminated products are usually very soft; however, resin blends have helped to overcome this problem.
5. Labor - The dry bond lamination line usually requires two men to operate it: an operator and a helper. The simple extrusion coating line requires a similar crew. However, depending on the responsibilities in the particular factory, an additional man might be needed if the line is a tandem or if there are more than two unwinds. The responsibilities referred to involve the delivery of raw material to the line and the removal of the finished product from the area near the winder. There is not enough time between rolls at production speed to do work in process material handling.

CHAPTER 4

The Extrusion Coating/Laminating Extruder

Charles H. Crumb, Black Clawson Converting Machinery LLC

INTRODUCTION

The screw-type extruder feed hopper receives thermoplastic polymeric resin materials generally in free flowing solid pellet form at approximately 20°C (70°F) from a material supply system and delivers a uniform molten flow, at approximately 330°C (625°F) in the case of low-density polyethylene, to a slot die.

A flighted auger screw rotates in a typically horizontal barrel, forming the material flow channel in which pellet conveying, compaction, melting, mixing, metering, and pumping occurs. The molten resin is then commonly passed through a screen pack filter or screen changer to remove impurities and a variable valve to create an additional pressure restriction to control or vary resin residence time in the extruder screw channel. The polymer then enters a coextrusion layering adapter or proceeds directly to a die.The material is spread the full width in a slot die and is deposited on a traveling web as a coating or an adhesive layer for laminating two webs.

This chapter discusses the equipment (Fig. 1) necessary to allow these functions to occur. Past, present, and possible future methods and designs of equipment will provide an understanding of the extrusion coating/laminating extruder.

Fig. 1. Carriage-mounted extruder

DESIGN CRITERIA

Extruder process capacity

The process capacity in kilograms (pounds) per hour required from the extruder is defined by the coating thickness and material density or coatweight, coated web width, and web speed. The extruder or extruders, in the case of coextrusion, must provide this capacity (Table 1), which

Table 1. Typical extruder size versus resin output capacity

Extruder Size	Drive Power	Typical low-density polyethylene output
65 mm (2 1/2 in.) 30:1 L/D	45 kW (60 hp)	145 kg/h (325 lb/h)
90 mm (3 1/2 in.) 30:1 L/D	75 kW (100 hp)	285 kg/h (625 lb/h)
120 mm (4 1/2 in.) 30:1 L/D	150 kW (200 hp)	545 kg/h (1200 lb/h)
150 mm (6 in.) 30:1 L/D	300 kW (400 hp)	1090 kg/h (2400 lb/h)
200 mm (8 in.) 30:1 L/D	450 kW (600 hp)	1635 kg/h (3600 lb/h)

is limited by physical size of the extruder components. The extruder size, by convention, is described by the barrel inside bore diameter and the effective screw length to diameter ratio (L/D), with the length measured from the downstream edge of the feed inlet to the discharge end of the screw. This, with the selected screw drive horsepower (hp) at some geared-in speed (or torque available), generally defines the limitations of the extruder with regard to output. Other factors such as heater power kilowatts (kW or KVA), screw design, feed port size, feed section grooving, etc., also have effects. The single-screw machine is required for normal applications; however, with some hard-to-feed materials such as hot melts or when in-line compounding is needed, a twin-screw machine is sometimes used.

It is important not only to provide a quantity of output at the desired melt temperature for adhesion to the web, but the quality or temperature and pressure uniformity and cleanliness are equally important to provide a very uniform resin thickness. Temperature nonuniformity manifests quench rate and adhesion differences or gels. Pressure nonuniformity and surging of output manifests output rate or coatweight variation.

Extruder drives

The drive motor is the prime mover for screw rotation. It must supply the torque required to convey the solid pellets and most of the energy by shearing action of the screw for resin melting and the temperature rise needed up to the required discharge temperature. The torque for mixing and the power to convey or melt the molten polymer against the screw discharge pressure created by the downstream resistance of such components as the screen pack, valve, piping, and die must also be provided. The total drive power required for extrusion coating/laminating extruders generally falls in the 3-4.7 kg (5-7.5 lb) per hour per kW (hp) range. The single-screw extruder is considered to have a constant torque drive requirement. This means that as the screw speed increases, the horsepower needed increases at a proportional rate (Fig. 2). Adiabatic or generation of resin temperature rise totally by screw frictional heat supplied by the drive power is generally desired because of the high-energy conversion efficiency. It is difficult to obtain 100% of this result with a wide range of materials or operating process requirements.

Fig. 2. Drive speed versus horsepower

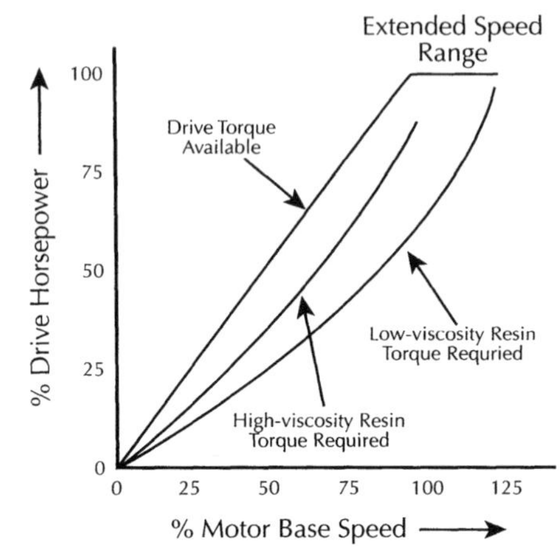

Extrusion coating and laminating applications of extruders are generally required to have a wide operating speed range. It is the variation of extruder output to web speed that defines the coat weight at a given web width. This can be accomplished by the typical 10:1 speed range of the line drive or by the typical 10:1 or greater screw speed range created by the extruder drive. It is desirable to have the web line drive interconnected with the extruder drive by a ratio follower circuit to allow the line speed to increase and the extruder to proportionately increase screw speed to maintain approximately the same coat weight.

It is also current common practice to measure or gauge the coated thickness or basis weight as well as control the extruder drive speed to maintain the average machine direction coating thickness.

A "drool" speed circuit is also desirable to allow the extruder to be purged at a very slow speed offline prior to moving on-line or during line downtime to reduce the effect of high-temperature barrel surface and long residence time degradation of the resin in a stationary screw.

The accuracy of the drive is defined by the regulation, no load to full load speed change, and drift from set speed, which is the variation of speed due to voltage, frequency, and temperature changes at the drive. Both regulation and drift are defined as a percentage of motor base or maximum speed. Speed regulations of ± 0.1% and drifts of ± 0.1% are typical in newer extrusion coating/laminating extruder drive systems. This relates to ± 1.75 rpm change due to regulation on a typical 1750 rpm drive. Load changes can be attributed to changes in head pressure due to screen pack blinding over time, screw design related surging, feed hopper bridging, resin viscosity changes due to type of polymer, processing temperature, etc. Since output is somewhat proportional to screw speed, this could account for up to ± 0.1% change in output in each case. Most drive systems will operate at ± 2-5% regulation and approximately 10% drift with standard armature feedback, in the case of a dc SCR drive for example. Tachometer feedback by actual shaft rotation measurement with an ac tach generator will improve this to ± 1% regulation and ± 1% drift. Using dc tach generators or encoders and other techniques can result in greater improvements to ± 0.1%. Zero percent regulation and drift will be possible without prohibitive cost.

Polymer viscosities at desired melt temperatures vary greatly and thereby require different screw speeds to obtain similar outputs of low and high-viscosity materials. In many applications, the variations are so great that extended constant horsepower characteristics are desired in preference to providing larger-horsepower drives geared in at faster speeds (Fig. 2). The dc drive motor, for instance, can be field weakened by a field crossover circuit in the SCR package and run at speeds above the base speed of the motor. Other types of drives also have similar capabilities of extended speed range. This additional speed is created not at constant torque but, rather, at constant horsepower. This mode of operation most generally suits the wide viscosity or melt index range of polymers used in many extrusion coating/laminating applications.

Efficiency of power conversion is becoming of greater and greater importance with increasing energy costs. Since the drive must be viewed as requiring variable speed and with the possibility of load variations, a selected drive should attempt to be as efficient as possible at all conditions of operation.

Many types of drives have been employed such as constant speed ac motors, ac motors with eddy current couplings, ac motors with variable pitched pulleys, ac motors with inverter speed control, hydraulic, hydrokinetic, dc motor generator (MG set), and dc silicone controlled rectifier (SCR).

Constant-speed ac motors were used in the very early stages of extrusion history but have disappeared due to lack of variable-speed capability.

The eddy current clutch is a drum driven by a constant-speed ac motor. Inside the drum is an output shaft-connected rotor and surrounding field assembly. When the drum rotates, an eddy current flux is created at the drum, which in conjunction with the field causes the drum and rotor to be coupled to transmit the torque and speed of the ac motor to the output shaft of the coupling. By increasing the applied field excitation voltage at a given load, the output speed can be decreased at a variable level. This operation generates heat, which is dissipated by air or water cooling. Slip or lost speed is inherent due to the inability to apply enough field to couple the two components at varying loads. Greater inefficiencies at slower speeds generally make this a dying type of drive for most extruder applications.

The constant-speed ac motor-driven belt and variable sheave drive offers low cost, high maintenance, and limited horsepower sizes. This type of drive is generally only used on the smallest and least expensive extruders.

The hydraulic hydrostatic drive is affected by use of high-pressure hydraulic oils generated by a local or remotely mounted power pack consisting of a reservoir, motor driven pump, and valving. Hoses deliver the oil to a motor in which a rotor is driven by the oil. The potential for oil leakage is the general reason for lack of support for this type of drive.

The hydrokinetic drive utilizes a constant-speed ac motor-driven impeller casing which centrifugally propels oil into a runner, thereby driving the output shaft. The change of mass of fluid allows the unit to vary speed. This drive is used very rarely due to its inefficiency at lower speeds.

The dc M.G. drive is the forerunner of the dc SCR unit. The motor generator constant-speed ac motor generates dc power from ac to drive the motor armature. The drive is now rarely used in preference of the SCR drive.

The inverter is a speed control device that allows variable frequency (hertz) and voltage to be applied to an ac motor and allows its speed to vary in relationship to its number of motor poles. This type of drive utilizes low-cost motors but higher cost controllers. Caution must be taken with regard to motor cooling because most ac motor frames are sized to affect cooling at full speed. Heat generation might not be dissipated when at reduced speed.

The dc SCR drive allows the SCR to control and rectify the dc current to the motor. This drive offers greater efficiencies than most other types of drives. The dc SCR drive is, at the present time, the most popular due to its ability to generally satisfy most of the extrusion coating/laminating extruder process requirements.

Drive connection

The drive motor is generally connected to the extruder by belts and sheaves or a flexible coupling. The connection is sized based on the drive torque it must transmit to the screw plus, typically, a 1.5 safety factor. The belt and sheaves arrangement is generally limited to lower horsepower applications due to the number of belts required, belt matching, and physical space required. Alternate sheave configurations allow the reduction ratio to be changed, thereby modifying the screw maximum speed.

The flexible coupling offers the ability to directly connect the motor to the extruder in a compact arrangement. Gear, insert, flexible steel plate, and other types of couplings must allow for slight amounts of angular and offset misalignment of the drive shafts.

In either case, the rotating parts are provided in a guard enclosure for personnel protection.

Gear reducer

Since the extruder screw in today's coating/laminating extruders rotates at 200-300 rpm maximum speeds and most types of drive systems have output base speeds of 1150 or 1750 rpm, a gear reduction unit must be provided to reduce the drive speed to the desired screw speed. The total reduction can be accomplished by single-, double-, or triple-reduction gear sets.

The gear reducer is sized based on its capacity to transmit the torque supplied from the drive. Each reducer is rated as to a theoretical mechanical horsepower capability at a given input speed and reduction ratio based on strength and durability. Published guidelines as provided by the AGMA (American Gear Manufacturers Association) are followed by most manufacturers of gear reducers. Strength is a mechanical rating of tooth load-carrying capability. Durability ratings take into account the tooth wear, radial bearing life ratings, and other factors that relate to the theoretical longevity of the unit. A comparison of the theoretical gearbox mechanical torque rating to the actual maximum applied torque from the drive is defined as the service or safety factor. This service factor for extrusion coating/laminating type extruders is recommended as 1.25 for machines operating 8 hours per day and 1.5 for 24-hour-per-day operations. These AGMA suggested service factors are guidelines and, in general, the extruder manufacturer must use standard-size gear reducers that can vary slightly above or below the numbers listed here.

The gearing in most reducers for extruder applications is either helical, double helical, herringbone, or variations thereof.

The basis of this selection is the torque and speed requirements and suitability of this type of power transmission method. This type of gearing is approximately 97% efficient, which again is an advantage over other methods of speed reduction. The parallel shaft configuration of mounting shafts and gears is also very typical. The actual gear quality is also AGMA rated. Quality levels of approximately 10 or better are used to ensure good tooth surface and form.

The gear pinions and shafts are generally supported on antifriction roller bearings which should not be, but often are, overlooked in ratings. The B10 life rating defined by AFBMA (Antifriction Bearing Manufacturers Association) is the theoretical average life in hours at which time 10% of a particular group of bearings could be expected to fail.

Bearing maximum load and speed are compared to the theoretical bearing dynamic rating capabilities. Since the average extruder operates 8000 hours per year, it is suggested that radial shaft bearings should be rated at approximately 25,000 hours as a minimum.

Noise generated by the gear reducer is of great concern with OSHA (Occupations Safety and Health Act) regulations. Present guidelines for eight hours of exposure per day are 90 dBA (decibels on the A scale) at 1 meter (3 ft) distance from the noise source. The trend for new equipment is to attempt to attain 85 dBA. The noise generated by gear mesh resonance, tooth form, surface finish, pitch line velocity, etc., can also be attenuated by the reducer housing if proper design is not followed, whether cast or fabricated construction is utilized.

As an alternate to changing belt and sheave drive connections, a change gear set can be made accessible behind a cover plate to modify the maximum screw speed. This allows the same gear reducer to be utilized for a number of applications and allows the motor horsepower of a given application to be utilized to the fullest extent to obtain greater operating efficiencies.

It might be worth noting that efficiency should be considered very important when, for instance, every 75 kW (100 hp) of drive power at 124 full load amps costs approximately US$ 76,000 per year at 10 cents per kilowatt hour to operate.

Thrust section

The screw, in its requirement to create pressure at the extruder discharge, receives an equal and opposite axial thrust force in Newtons (pounds) equal to the head pressure in Pascals (pounds /square inch or psi) times the cross-sectional barrel bore area in square meters (inches). In most coating/laminating applications, the head pressure is approximately 10,000-20,000 kPa (1500-3000 psi). This pressure is related to a reasonably large load that the thrust bearing must deal with. This force must be taken from the screw and dissipated to the base structure. The thrust bearing is typically fitted to a thrust spindle, which receives the screw shank attachment, and an external housing which is connected to the base. The thrust spindle is mounted in radial bearings for support and alignment due to the thrust bearings general lack of ability to accept radial load.

The thrust bearing housing is mounted as an integral part of the gear reducer or as a separate unit coupling connected to the gear reducer. The integral mounting arrangement offers low cost and compact mounting. The separate thrust assembly offers ease of maintenance, because the reducer does not require disassembly, and eliminates the thrust force from being applied to the reducer housing.

Different types of bearings are available to satisfy the thrust requirements. The conical roller, mounted between two tapered plates or one tapered and one flat plate, and cylindrical roller bearings are probably the most popular. In some instances where high load requirements dictate, stacked multiples of roller sections are employed. The concern for this multiple stacking centers around proper load sharing between stages. The thrust bearing is B10 rated in much the same way as previously described in the gear reducer radial bearing section. The rating is usually specified at 33,500 kPa (10,000 psi) and 100 rpm. Through evolution of bearing failures due to poor materials, design, rating methods, alignment, and lubrication, the thrust bearing usually is specified at approximately 100,000 B10 hours. This says that, based on 8000 hours per year of operation, the bearing should survive 12.5 years. This, with the recent technological advancements, produces a typically conservative life.

Lubrication system

The gear reducer housing is often utilized as the lubrication oil sump. All bearings (including the thrust bearing) and gear meshes should be lubricated to ensure long life. The splash and wipe system is a method whereby the oil is picked up, generally by the bull (large-diameter, low-speed) gear, and wiped at the top for distribution by troughs and other channeling to other parts of the box.

Another method entails a shaft-driven pump which force-feeds oil throughout the box. This method, however, at slow speeds might not provide enough volume of oil because the shaft-driven pump turns slower. The separately motor-driven pump is the most popular approach where large-speed ranges are desired and because of the higher degree of ensuring that oil flow is adequate under all conditions.

The lubrication system is quite frequently called upon to cool the oil, which is heated by the high forces of the gear meshing and bearing loads. This heat is generated from the 3% of inefficiency of the typical double reduction gear reducer and the associated radial and thrust bearings. A portion of this heat is dissipated by the reducer and thrust section housing. The remainder, if any, must be dissipated by additional cooling capacity supplied generally by a water-to-oil heat exchanger in the lubrication oil flow circuit. The lubrication oil should generally be limited to approximately 60°C (140°F) when using 4-5 EP (extreme pressure) oil to ensure reasonable oil change intervals.

Due to the high costs of gears, radial and thrust bearings, and downtime, it is generally wise to provide protection in the lube system for low flow, high oil temperature, and adequate filtration.

Feed section

The plastic pellets from bags, gaylords, silos, or railcars are delivered to the feed section material-holding hopper generally located directly above the screw. The hopper typically holds 1/2 hour of pellet usage capacity based on nominal 3 mm (0.125 in.) pellets with a bulk density of approximately 560 kg/m^3 (35 lb/ft^3). The coating/laminating extruder operates with a flood feed, meaning that the hopper is filled and the screw operates with a head or level above the screw. The material column height in the hopper has little effect on feeding performance, provided the screw flights are allowed to be filled. Hoppers filled from resin-conveying systems are automatically controlled from "low" and "filled" level sensors in the hopper. Care must be used to ensure large foreign material that might damage the internal components does not enter the machine at this point.

Magnetic traps are sometimes used for this purpose. Bridging or material hang-up in the hopper can disrupt the process and can generally be alleviated by 30° or steeper hopper sidewalls. The hopper is generally fitted with an inventory sight glass, cut-off slide, and dump chute.

The pellets flow from the hopper into a feed section in which the material enters and fills the deep screw flights in this area. For coating/laminating

Fig. 3. Extruder cross section

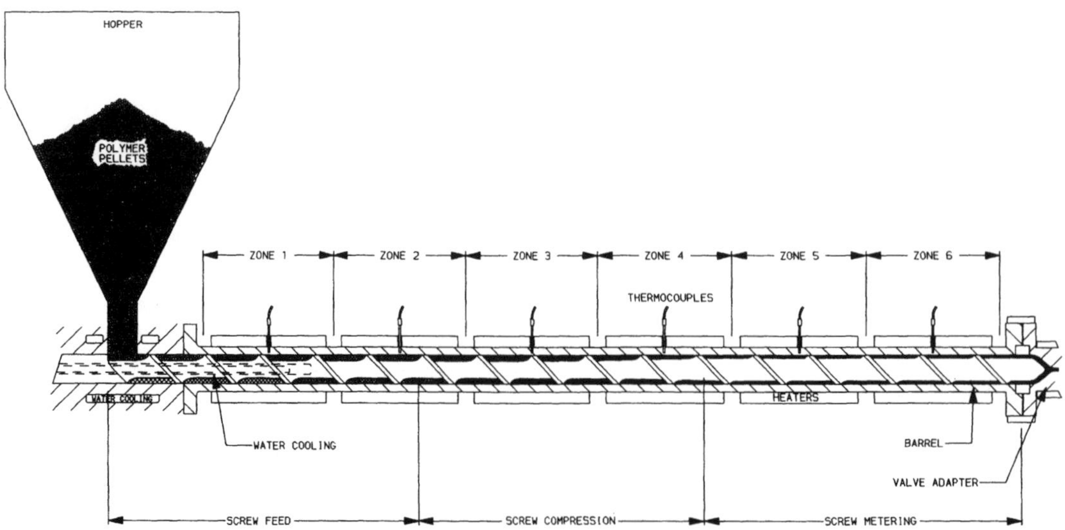

applications, the pellet feed opening is round and approximately the same diameter as the screw. The feed section usually comprises the first two to three diameters of entrained or effective screw length.

The feed section is connected to the thrust section to provide alignment of the feed section bore to the thrust sheave screw socket. Openings at the rear of the feed section allow a dropout relief to prevent polymer, generally "fluff and fines," from contacting the thrust bearing housing seals or thrust section lubricants from contacting the polymer. Lip-type seals surround the screw shank where it passes through the rear of the feed section to further lessen cross contamination on most coating/laminating extruders. The seals are skived to allow insertion without screw removal. The nature of the polymer in this area, being generally well defined full-sized pellets, make it questionable as to whether any seal at all is required. In some cases, shallow grooves are cut in the screw shank to pump any stray solids back to the feed entrance area. These grooves are sometime called "reverse pumping flights," a misnomer because they pump toward the discharge end of the extruder.

The feed section is generally bolt connected on the downstream end to the barrel and is aligned for concentricity of the screw bores. The barrel often has metal temperatures in the 100-180°C (210-360°F) range where it connects to the feed section. Therefore, it is desirable to have a means to cool the feed section to prevent premature melting of the pellet surface causing pellet bridging in the feed entrance, surging, screw blockage, etc. The feed section on most machines of a standard design are cast iron or steel to reduce manufacturing cost. Steel fabrications are common for non-typical or very-low-quantity manufacturing. In each case, water-cooling internal channels or jackets are manufactured in the feed section housing to remove heat that is conducted to it from the barrel. Plant water at 30°C (85°F) and 4-20 L/min (1-5 gal/min) is flow controlled by a hand- or automatic-valve to the jacket.

The screw in the feed section is relieved or undersized. Wear is rarely evident due to the nature of the solids conveying and non-contacting screw in this area. Some machines, however, are provided with a hardened liner. The grooved bore is a technique to increase output that is rarely used on coating/laminating extruders due to cost and the actual amount of improved output that is realized.

Extruder barrel

The barrel is the cylindrical chamber in which the main flighted effective portion of the screw rotates to create pellet compaction, melting, mixing, metering, and pumping. The barrel must contain the polymer from solid pellets at 20°C (68°F) feed temperature and 100 kPa (14.22 psi) atmospheric pressure to molten resin at approximately 325°C (617°F) and as high as 35,000 kPa (5080 psi). To endure the high temperature and pressure, the barrel body is manufactured from high-alloy, high-strength steels in appropriate wall thickness.

The bore of the barrel is required to withstand abrasive wear (in the solids compaction and melting area), occasionally corrosive materials, and wear due to the screw flights rubbing on the barrel. The screw flights, in a properly aligned machine, will actually self-center the screw in the barrel while riding on a polymer film "lubricant" between the flight and the barrel. Abrasive types of wear can be overcome by hardening the barrel's internal surface. Nitriding or case hardening the internal surface by exposing ammonia to a heated nitridable steel (usually a ferrous alloy containing amounts of aluminum and chrome), forming an absorption of nitrogen to generate 0.125 mm (0.005 inches) deep for 70 Rockwell C (Rc) hardnesses of chromium nitride and iron nitride, is an inexpensive older technique rarely used today. The bimetallic surface most commonly used is formed by centrifugally casting a lining alloy material in a steel cylinder. The cylinder is rotated (either horizontally or vertically) in a high-temperature furnace at high speeds to bond and uniformly distribute the alloy up to 5 mm (0.196 inches) thick with a finished hardness of 58-65 Rc. Certain grades of bimetallic lining alloys also can be produced with greater corrosion and abrasion resistance.

Most manufacturers of bimetallic cylinders produce three general classes of alloy materials. The first, for normal general-purpose wear characteristics, uses an iron/boron alloy producing a 58-65

Rc hardness. The second, for corrosive applications such as materials with vinyl chlorides, etc., utilizes a nickel/cobalt alloy producing a 48-54 Rc hardness. The third, for extreme wear and corrosion resistance, utilizes a matrix with numerous types of embedded carbides producing a 65-70 Rc hardness. The cost increases approximately 10% for the corrosive lining over the general-purpose lining and as much as 80-100% for the extreme wear and corrosive-resistant linings.

The bimetallic lining is bored and honed to approximately 0.05 mm (0.002 in.) concentricity, 0.082 mm/m (0.001 in./ft) of length straightness, and 8-12 rms smoothness after spinning.

The coating/laminating extruder barrel is produced with the hard surfacing cast directly into the barrel body, rather than into replaceable liners, primarily because the general purpose liners on many machines have lives of 5-20 years due to the less severe service of the process. The bimetallic lining life can be lessened by the failure of the base metal bond, cracking, spalling, or scoring from tramp metal or foreign materials. Wear of the internal surface is a gradual occurrence that does not define a clear actual life. The typical original radial screw flight to barrel bore clearance is 0.001 m/m (0.001 in./in.) of nominal screw diameter. As the barrel bore wears, this clearance increases with a coinciding decrease in extruder output due to increasing the "over the flight leakage" polymer flow. The effect of reducing the machine output very, very gradually is only realized when an evaluation of actual to theoretical output is made.

The total barrel length of coating/laminating extruders has increased over the years due to requirements for additions of mixing sections, more gradual transitions, and longer metering sections. Today it is very common to utilize 30:1 length-to-diameter machines to enable higher, more quality outputs by adding length for these screw-related elements in preference to older 24 to 28 to 1 L/Ds.

Pressure indication and protection

The barrel downstream end is fitted with a rupture tube or disc which will fail if pressure in the barrel increases to an unsafe limit. A pressure sensor with indicator is also installed to detect the actual pressure conditions at the screw discharge. An adjustable shutdown/alarm switch is sometimes also provided as an added safety device.

Barrel heating and cooling

The barrel of plastic extruders must be heated during startup to remelt polymer that remains in the machine during shutdown periods and to maintain the barrel internal surface temperature, during operation, at a level to maintain or add heat to the resin. Due to the inefficiency of adding heat to the poor thermal conducting polymer from the barrel heaters, it is preferable to add heat by mechanical screw shear heating the polymer with energy supplied by the drive.

The barrel temperature profile desired, from the feed section to the discharge, is an increasing setpoint configuration that follows the expected temperature of the resin at each point along the screw, if the screw will approximate a desired adiabatic operation. Therefore, it is desirable to divide the heating into zones capable of individual temperature setpoints. On 30:1 L/D extruders, six zones are typical.

Many types of heating methods have been successfully used in the past. The induction coil producing eddy current flux heating of the actual metal is a very efficient quick heat-up method with low maintenance, which is still available from some manufacturers and still utilized in present field operations. The typical arrangement provides a cooling jacket formed such that the cooling water is applied directly to the outside of the barrel body to provide a very efficient method of cooling. The slightly higher cost and coil removal method dictate a preference toward resistance clamped on heaters, which rely on less efficient conduction for heating. Vapor or liquid jacketed barrels are generally avoided in coating/laminating extruders due to the potential for fluid leakage in proximity to the web product. The cast aluminum or bronze, mica band, and segmented ceramic band resistance heaters are utilized in areas where no cooling is required. The barrel heat zone, in most cases, should be fitted with a means of cooling to enable the metal to be cooled in instances of temperature overshooting or quick cool shutdown. Air forced over finned heaters can be used but has a slower cooling response and efficiency due to the convective cooling method. Water cooling is an efficient, commonly utilized method whereby the water is time-proportionally flow-controlled through tubes located within the heater. Extruders having cast-aluminum resistance heaters with internal water cooling tubes on all zones are very commonplace due to stocking of one type of heater and the existing capability to water cool certain zones if desired. Cast aluminum can be utilized for extrusion barrel application zone temperatures to about 340°C (645°F). For applications above this temperature, cast bronze units can be employed at a premium price.

The water supplied for cooling to each zone must be temperature controlled and should be recirculated. It is very desirable to utilize a small reservoir containing mineral-free water for this purpose. Failure to use mineral free water will lead to premature fouling of the system and a loss of cooling efficiency. A pump draws water from the reservoir and directs it to a supply manifold containing individual zone flow metering valves and on-off solenoid valves to time proportion water, as required, through the heater. After the water passes through the heater, it is collected in a drain manifold to be cooled in a water-to-water heat exchanger prior to returning to the tank. Flexible reinforced hose with screwed fittings has replaced the older copper pipe with compression fittings, virtually eliminating leakage from thermal shocking, on the feed and drain connections to the heater.

The barrel zone desired temperature is accomplished by utilizing a temperature sensor, usually a type "J" thermocouple, installed about midway in the zone length and so that the junction is immersed approximately 50% of the way into the barrel wall. The sensor signal is detected by a controller that is capable of receiving an operator-set desired setpoint. The controller must compare the setpoint value to the actual value and provide heating, cooling, or neither. The controller accuracy and capability to maintain actual zone temperatures as close as possible to the setpoint are of utmost consideration to attain steady-state zone temperature to help avoid fluctuation in the resin temperature. Time proportioning heat/cool type of individual zone controllers or system microprocessors are typically used for this purpose. Wiring in the heater area must be suitable for the high temperatures generated.

Barrel covers

Sheet metal covers with underside insulation and access to the heater zone are desired for operator protection from heater terminals and high-temperature surfaces. The insulation also reduces the convective heat losses into the machine room. The covers should be short in length to reduce weight for easy removal by the operator.

Feed screw

The feed screw is the rotating auger device designed to provide the resin at the desired process conditions. The screw is most often an alloy steel which has been machined, stress relieved, and heat treated for maximum dimensional uniformity and strength. The screw flight tips are flame, spray, or weld-applied hardened to reduce wear. It is appropriate to have the barrel lining harder or more wear-resistant than the screw flight tip hardness. This allows the flight tip to be the sacrificial element because it's easier to rebuild the outside diameter of the screw than the inside diameter of the barrel. Proper selection of compatible flight tip and barrel lining materials should be made to ensure long life.

The alloy steel screws are generally protected with 0.025 mm (0.001 in.) chrome or nickel to prevent rusting and to resist corrosion due to polymers with components such as vinyl chloride and fire retardants.

Screw water cooling

The screw is usually bored to the depth of the feed section to promote feeding characteristics and prevent bridging due to premature melting. The assembly consists of a dual-passage rotary union with the water supply pipe connected to the screw and a return tube to drain the water and possible steam generated. A flow-metering valve allows the amount of cooling to be controlled.

Screw removal device

A manual or hydraulic jacking tool for pushing and disengaging the screw shank from its drive socket is required for screw changes, inspection, or cleaning.

Carriage base

The carriage base provides the support of the individual components such as the drive, gear reducer, barrel supports, barrel cooling systems, and die support. The barrel supports must be designed to accept thermal expansion, generally toward the die end, due to the normally fixed drive and reducer components.

These supports should maintain both horizontal and vertical alignment. The axial thermal expansion can account for as much as 13 mm (0.5 in.) of movement on a 150-mm (6 in.) 30:1 L/D as an example.

The slot die and the extruder barrel are oriented perpendicular to the path of web travel. This configuration can have a fixed-position coater/ laminator or fixed extruders. Today most coater/laminator arrangements have the ability to move up or down to adjust resin "draw-down" distance from the die to cooling roll, right/left to adjust point of contact of resin on the web, and in/out to move the die off-line for inspection, web threading, drooling, etc. If the extruder carriage base is incorporated with all three movement directions it is usually referred to as an xyz arrangement or design. Since it is very difficult to fix the extruder and just move the die on- and off-line, most extruders are mounted on motorized pushbutton-controlled carriages with wheels mounted on aligning rails. A safety horn should be provided to indicate carriage movement. End stops and limit switches are often used to properly position the die over the web. The draw-down adjustment can aid in product adhesion, oxidation, quench rate, and other characteristics and can be accomplished by moving the laminator rolls or extruder up and down. The same is true of the right and left or upstream and downstream web direction movement, which can allow the resin material to have a longer residence time on the chill roll or desired contact point and can be accomplished by moving the laminator rolls or extruder.

When the extruder is movable, plant service (electricity, cooling water, and air) must be flexible-connected, generally through a protecting power track or festoon to the extruder carriage.

In the case of coextrusion (use of additional extruders for multiple resin layering applications), the carriage base must have accommodation for the additional extruders and must also move with the main extruder. In the case of future capability for coextrusion, the extruder base with all three motions is generally much more difficult and expensive to add to the base structure. If present space is provided, then valuable plant floor space can be utilized. The in/out only extruder carriage base with movable coater/laminator does not have this inherent problem.

Piping and wiring

The piping from the barrel cooling system heat exchanger, lubrication cooling heat exchanger, feed cylinder jacket, and screw cooling assembly must be interconnected to common feed and drain manifolds to allow single lines to be run through the power track.

Wiring for items such as the heaters, thermocouple sensors, barrel cooling systems, and lubrication systems are typically connected to intermediate junction boxes for easy maintenance prior to the power track. The extruder drive is usually wired directly to the power source through the power track due to the large wires involved.

Screen pack filter

The resin must be filtered after leaving the barrel and screw area to ensure that clean polymer is deposited on the web. The filter also offers the ability to not only remove foreign matter, but it can reduce fish eyes, gels or unmelted material in the flow. The most common type of filtration utilized in coating/laminating extruders consists of a mesh screen or multiple screens of the same or alternating meshes forming a screen pack. Since the polymer viscosity's and pressures are much greater than the screen pack could support by itself, a flat backup breaker plate is employed to support the screen material that is then mounted on the upstream side. The breaker plate consists of numerous holes drilled through a plate to allow polymer flow through the plate and proper screen support but without the breaker plate contributing excessively to the system pressure drop. The screen material is placed in a restraining groove in the breaker plate to prevent it from moving out of position during installation. Screen meshes utilized vary with the amount of filtration desired. Typical screen packs are made up of two or three composites of 60 to 120 mesh screen with the heaviest mesh against the breaker plate to provide support, from pressing through the holes, for the first or generally finer meshes and to increase the effective open area of polymer flow.

In operation, the screens retain foreign material that builds at a somewhat constant rate to blind or block off the polymer flow. As this occurs, the block-off creates an increase in head pressure. The extruder, when started with new screens, might have a clean pressure drop of 7000-14,000 kPa (1000-2000 psi). The pressure then increases with time until levels of approximately 17,500-35,000 kPa (2500-3000 psi) are attained that, for this process, begin to reduce throughput while increasing melt temperature from long residence shear heating to undesirable levels. The screen pack should then be replaced. If the screen pack is not replaced, the increasing excessive pressure will cause an unsafe condition that will exceed the mechanical design of the equipment. Serious personal injury and/or machine damage could result.

Many types of mounting arrangements have been used for mounting the breaker plate. On older machines, the breaker plate was mounted between two separable flanges. This is accomplished with rigid bolted, swing bolted, "C" clamp, breach lock, or other methods of clamping, and by actually utilizing the breaker plate as the metal-to-metal polymer seal ring in the flange female counterbores. These methods rely on the clamping force to prevent polymer leakage and the type for its ease or quickness to minimize the time required for plate changing. The mounting in the counterbores requires that the downstream adapter and die be moved to provide access for removal. The side entry breaker plate is an alternative that allows the breaker plate to be installed into the side of a housing at right angles to the polymer flow. This allows downstream equipment to remain in position during breaker plate changes. The side entry unit is manufactured as a housing with bolting or clamped doors, which provide the resin sealing while allowing the breaker plate insertion into the flow only while the extruder flow is stopped. Very limited attempts have been made to utilize slide plate or cylindrical plug configurations due to the high precision parts required to prevent leakage of the very-low-viscosity resin at the process conditions. The infrequent change periods (usually one per couple of weeks) dictate that the operation be simplistic and inexpensive. This can be typically scheduled so that it coincides with other line downtime periods.

The breaker plate area and equipment placed downstream of the screw tip should be heated to maintain the metal temperature at the same temperature as the molten resin inside. It should be considered the screw's job to attain the desired melt temperature and the downstream components' job to maintain it. It is also desirable to break downstream equipment, with varying masses into separate heating zones instead of trying to balance heater capacities to prevent hot spots.

A pressure-sensing device is placed directly after the screen pack to allow, in conjunction with the pressure sensor at the downstream end of the screw, the operator the ability to detect the pressure drop and thus the pressure increase due to the screen pack blinding.

Valve assembly

After passing through the screen pack, but prior to the die, a polymer pressure-restricting device is desirable to alter the total head pressure desired for good mixing in the screw. A fixed orifice is a possibility rarely used. The more common variable-needle valve is conveniently placed in the intersection of the right angled turn from the horizontal extruder screw and the vertical connection to the top of the die. The needle and seat are cylindrical sections that allow pressure to vary nearly linearly with each turn of the thread system provided for needle movement. The valve can be utilized to maintain a constant head pressure in opposition to the increasing breaker plate pressure drop in an attempt to maintain constant screw mixing and shear heating. In practice, the valve is usually set at a "ball park" condition and the pressure buildup varies over the preset point.

The valve stem is fitted with a hex nut, handwheel, control side extended handwheel, or motorized adjustment to turn the needle. Access to this adjustment is an important consideration.

A pressure sensor downstream of the valve is a valuable tool to determine the contribution of the valve to total pressure drop by the differential pressure obtained by subtracting the value from the pressure sensor value located after the screen pack, for example.

The valve body is resistance-heated with clamp-on band heaters and is usually one zone of temperature control.

A flow channel connects the valve body to the tip of the slot die. This channel can contain a static mixing device which is placed in the flow path to reduce cross-channel temperature stratification of the polymer melt prior to entering the die. Alternately a dynamic mixer placed on the end of the screw or as a separate motorized device consists of a rotating element and stationary seat, which is gap controlled to define intensity of mixing could be implemented. Although a post-mixing device can aid in improving uniformity by as much as plus or minus a couple of degrees, many lines are run without such a device. It is also in this pipe where the immersion melt thermocouple is generally mounted to determine polymer temperature without causing flow interruptions in the die. The thermocouple is an immersion style with the tip junction located at or as close as possible to the center of the melt flow cross section.

A variable-depth sensor is valuable to detect the cross-channel temperature variation. The variable-depth unit is prone to freezing in position, bending due to the high resin forces on a relatively small cross section, and other problems that make the device suitable as a diagnostic tool that would be substituted for the more substantial fixed-depth unit used in production.

An additional melt pressure sensor is sometimes used to determine pressure drop across the static mixing device.

Where coextrusion is employed, the polymer connection piping might need to be longer to allow the polymer flow to be directed to a combining adapter, dual manifold or dual-slot die for either internal or external sandwiching prior to being deposited on the web. It is important to design these feed pipes to ensure temperature uniformity and pressure restraining capability.

Exhaust hoods

As the molten resin emerges from the slot die at the elevated temperatures involved, it gives off a smoke that should be removed from the extrusion room. Pickup hoods and suction bower entrain and discharge the smoke to an exhaust area. Since the extruder is movable, the ductwork must allow for the movement.

BIBLIOGRAPHY

1. Scott Collins (Ed.), Plastics Machinery Equipment Mag. pp. 26-28, Sept., 1982.
2. Bernhardt, E.C. (Ed.), Processing of Thermoplastic Materials, Reinhold, New York, 1959.
3. Tadmor, Z. and Klein, I., Engineering Principles of Plasticating Extrusion, Van Nostrand Reinhold, New York, 1970.
4. Mainstone, K.A., "Extrusion," in Web Processing and Converting Technology and Equipment (D. Satas, Ed.) Van Nostrand Reinhold, New York, 1984, pp. 122-146.
5. Wrede, H.G., "Chapter 2: Extrusion Coating Extruder"; Hoagland, J.M., "Chapter 4: Adapters"; and Palmer, R.W., "Chapter 8: Extruder Drives" in Extrusion Coating Monograph No. 48, TAPPI PRESS, Atlanta, 1973.
6. Extrusion Coating of Paper and Paperboard: Equipment and Material, TAPPI PRESS, Atlanta, 1990.

CHAPTER 5

Screw Design Considerations for Extrusion Coating

Robert B. Gregory

Extrusion coating applications create a particularly difficult set of operating conditions to the extruder screw designer. He or she must include not only a wide range of polymers in the design criterion, but also a wide range of output rates and temperatures for many of these plastics. This paper will provide the converter with an insight into the screw designer's realm, along with some suggestions for establishing a reasonable performance specification for new extruder or screws being purchased.

Extruder performance must be quantified as much as possible for each product being produced. For some parameters, it can only be qualified in a more general manner since quantification exceeds the present state of the art. To amplify on these statements, examine the quality variables that establish the acceptability of the performance of the line.

Many of the quality variables are quantifiable and can easily be codified. These are considered objective in nature since numbers are easily assigned to describe acceptable value ranges. A typical list of these would include:

Output rate

No number is typically considered as important as the production rate of a machine. A specified rate is usually established for a specific coating weight, width, and line speed based on process economics. Therefore, it is a rare specification that does not first and foremost spell out minimum and maximum productivity in pounds per hour or some equivalent units.

Mean melt temperature

Extrusion coating is particularly sensitive to a narrow window of acceptable melt temperatures. Too low and substrate adhesion is affected. If the adhesion is too high, the effects of oxidative degradation to the polymer exceed acceptable limits.

Variability of polymer melt temperature

Excessive variations in melt temperature cause product defects such as applesauce appearance in the coating or bands of differential adhesion. This variable has two dimensions since the melt temperature can change with either time or the position in the melt stream.

Rate stability

Variations in rate, usually called surging, can cause coating "machine direction" weight variations that injure the output quality. Surging can usually be detected by rationing the minor variation in pressure, which modulates the average value of discharge pressure to the mean pressure.

Resin degradation

The shearing caused by the stresses of the extrusion process, coupled with the high melt temperature involved in coating, can cause molecular weight changes due to chain scission and oxidation.

A more difficult list of quality variables to identify and quantify are those which are subjective in nature. These include, but are not limited to:

Odor

A very important factor to the food packaging industry. How would you specify an acceptable odor level?

Barrier quality

Many coatings are expected to provide barrier qualities to meet MVTR, CO_2, or O_2 specifications. In some cases, the extrusion process manner is an important contributor to the problems of barrier quality.

Optical properties

Coatings can be evaluated by their haze, gloss, or color acceptability. Gels can also import a negative quality to the product appearance.

Physical properties

Such variables as heat sealability, printability, strength, or impact resistance are, in some coated structures, an important aspect of processing quality.

Putting the extrusion coating line into its proper perspective, the extruder can, and often does, influence all the subjective as well as objective factors listed. Thus, an intelligent buyer will research a pending screw purchase in light of the designer's experience in the science of screw design and also, in some cases more importantly, in the art form of screw design.

EXTRUDER PERFORMANCE

Most extruders must perform a number of separate functions in series to meet the converters' quality satisfaction. These are listed in the order in which they occur from the feed end of the extruder to the discharge of the screw. While some are optional, many are absolute requirements for successful extrusion coating applications. Included are:

1. Solids conveying
2. Compaction of the solid bed
3. Melting
4. Venting (optical)
5. Dispersive mixing (optional)
6. Distributive mixing
7. Pumping.

Every screw used in extrusion coating will be required to perform these variables in their daily use. It is therefore important to assign an operating problem to the appropriate lateral location in the extrusion process. Thus, output rate and surging are typically upstream problems while mixing related variables are centered downstream.

SUPPORTING THE EXTRUSION PROCESS

An extruder or screw cannot make the proverbial "silk purse from a sows ear." Two factors other than the screw design contribute heavily to the success of each extrusion coating process. It is imperative that the feedstock in the form of the plastic and any additives be commensurate with the final specifications. In addition, the operator must input the proper zone settings, screw speed, and discharge pressure setting to finalize successful extrusion coating.

Put into the proper perspective, a finely turned process is the combination of the best of screw design and processing equipment coupled with the best feedstock and optimized operational variables.

RESIN PROCESSING PROPERTIES

Resin properties that must be factored into the extrusion coating equation include the following:

Solid feed properties

Includes such factors as coefficient of friction, angle of repose, internal angle of friction, bulk density, and compressibility.

Thermodynamic

Specific thermodynamic properties which control the processability of plastic include specific heat, heat of fusion, melt density, and thermal diffusivity.

Rheological

The molten plastic flow properties related to viscosity, shear rate, and shear stress must be known and utilized to optimize the process equipment and control settings. The storage and loss modulus of the melt will couple in the complex modulus to predict the melt stability at a specific rate and temperature. This viscoelastic measurement is important to identifying such problems as edge weave.

Chemical and additive properties

Resin chemistry keying to average molecular weight, molecular weight distribution, internal and terminal saturation, and thermal stability will affect extrudability as well as end product quality.

Be prepared to provide the screw designer with these data when purchasing new equipment. The development of new homo- and copolymers for extrusion coating applications has broadened the data bank of information required for intelligent design applications.

COEXTRUSION COATING

Coextrusion coating has added yet a new dimension to the problems of extrusion coating. Most applications involving the Dow patented combination adapter technology require that the various layers being coextruded be evenly distributed in the proper ratio and with a minimum of interfacial instability.

The demands placed on extruder performance involving melt temperature, temperature homogeneity, and rate stability are increased by the coextrusion process. The demonstrated need to match viscosities at the polymer interfaces of the structure has been documented in the literature[1] as well as in practice.

Some of the barrier resins used in coextrusion coating are difficult to extrude and require special handling in terms of equipment design, screw design, and operating conditions. Some polymers such as EVOH and PVDC are relatively new to the extrusion coater and will likely require thorough operator retraining to handle the problems of startup, optimization of operating hardware and software, and shutdown or purge procedure.

ON-LINE MEASUREMENTS

There are significant advantages to collecting qualitative data from the extruder to quickly and properly optimize the coating process. It is an obvious fact to experienced extrusion coaters that the coating process can add a very inexpensive laminate to an expensive substrate, which in turn leverages the loss cost of producing an hour or more of scrap or out-of-specification product to many times that of the extrusion cost alone. At times of extreme resin shortage, just the loss of the coating plastic can be considered a severe loss.

On-line instrumentation will feed data to either the operator or a microprocessor for analysis of process quality. Interestingly, only the objective parameters of quality are usually on-line "measurable." The subjective items often require evaluation time in a quality control laboratory of as much as 24 hours. A great deal of bad product can be created in 24 hours while waiting for test results.

The use of a melt thermocouple and pressure transducer properly located after the screw, coupled with data output to rapid response indicators, recorders, or microprocessors will allow an appropriately trained operator to view some of the important operating conditions. In turn, it allows an evaluation of the critical objective quality variables to be conducted quickly and corrective action taken when necessary to repair the process and minimize the losses occurred by "out-of-spec" performance.

EDUCATION

Extruder operators need to have a basic understanding of the extrusion process in addition to the substrate processes of let-off, priming, treating, printing, cooling, and windup. Few extrusion coating line operators are trained sufficiently to either recognize the physics involved in the plastics extruder or to know the proper response to correct extrusion-related problems. Most operators have learned their jobs by "trial and error."

In-plant seminars offer a relatively inexpensive way to educate these personnel in a nontechnical

1. Han, C. D., Shetty, R., "Studies on Multilayer Film Extrusion Interfacial Instability in Flat Film Coextrusion," Polymer Engineering and Science, vol. 18, no. 3, February 1978.

way. An educational program properly presented at a high school graduate level should pay back to the employer who supplies the appropriate in-plant training handsome financial rewards in addition to the increased morale of operators who feel confident of their knowledge of the extrusion process and its control.

Extrusion coating is a complex process utilizing large and expensive lines to create a costly web structure. The extruder is often buried in the line to the point of being virtually out of sight to the casual observer. Most operators know that the extruder can create problems that are difficult and expensive to solve.

New technology is introducing even more problems to the processing of plastics. Coextrusion has given the coater a new tool and process to increase process flexibility. At the same time, coextrusion has narrowed the window of optimum performance of the extruder. The operator skill involved in successful coextrusion is, or should be, higher than conventional extrusion coating.

Intelligent equipment selection coupled with the proper instrumentation will allow a skillful operator, utilizing the proper plastics, to produce a high-quality, profitable product. Operator training is recommended to improve or hone the skills required to optimize the process.

TYPES OF SCREW DESIGN

All screws are not "created equal" for each application. One need only trace the history of extruder screw design evolution to recognize the virtual explosion of design concepts which are now available to produce molten plastic for extrusion coating, blown film, and cast film.

Although still around, the day of the "standard meeting screw" to do all jobs is disappearing fast. A number of factors are reshaping the demand for better screw designs. These factors include:

More output rate for an equivalent extruder

- Improved quality of the melt created by the extruder
- Better wear characteristics of the screw in service yielding longer screw life
- The ability to handle more difficult to extrude polymers
- A requirement for greater latitude of operation by handling a greater variety of plastics with broad latitude of output conditions.

Figure 1 illustrates the flight and screw root configuration of eight designs that are traditionally considered single-stage screws. Screws A through F are metering screws with a definite feed, transition, and metering section length. Screws G and H are not metering screws, but have been used with various plastics and rubber compounds for more than 100 years. It is interesting that Miller *(1)* suggests a design similar to Fig. 1G as being proper for the extrusion of LLDPE.

Fig. 1. Variations of standard single-stage extruder screws

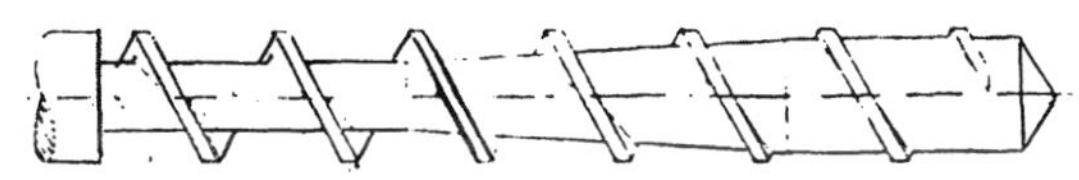

A. Standard Single Stage Screw

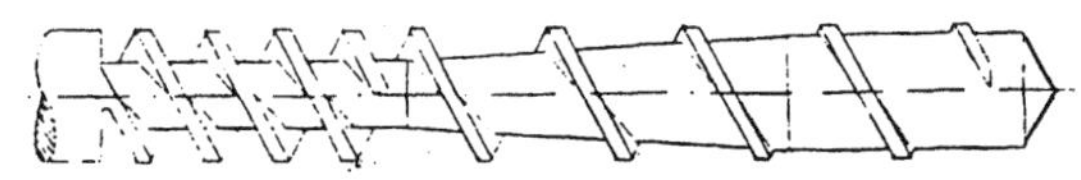

B. Standard Single Stage Screw with Additional Flight in the Feed

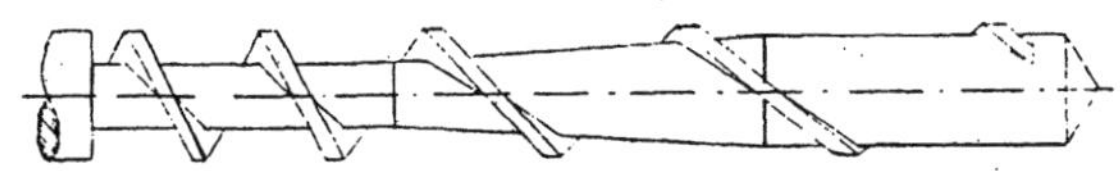

C. Variable Pitch Screw–Increasing Helix Angle

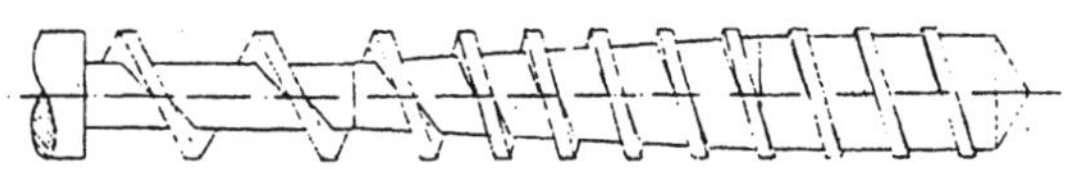

D. Variable Pitch Screw–Decreasing Helix Angle

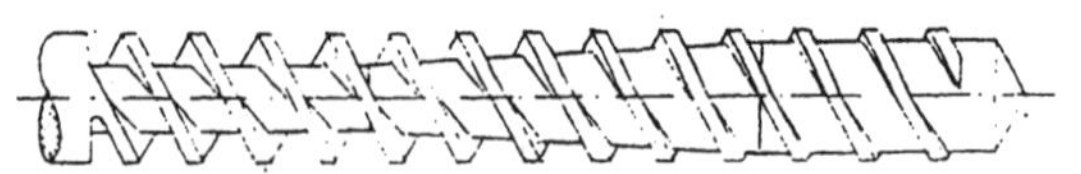

E. Standard Double Flighted Single Stage Screw

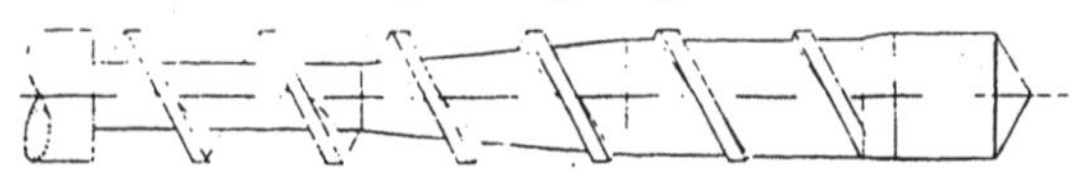

F. Square Pitch Screw with Torpedo or Mixing Head

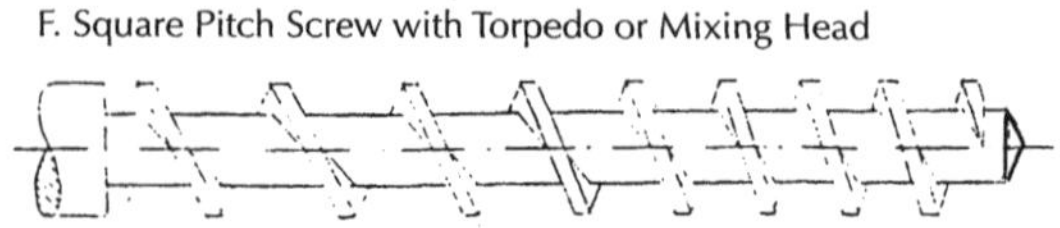

G. Variable Pitch Screw with Constant Root Diameter

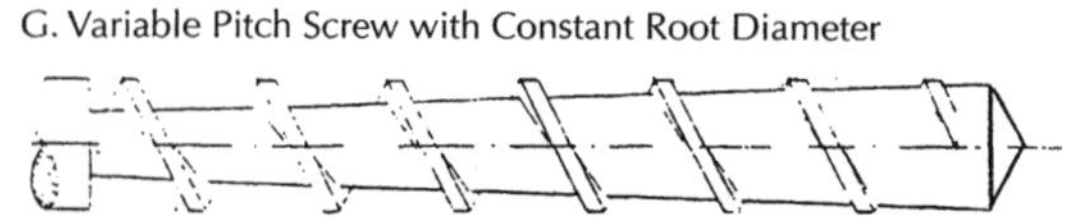

H. Screw with Tapered Foot Diameter Over Full Length

An illustration of the diversity of screw design technology, Rauwendaal *(2)* recommends a design similar to Fig. 1C with a narrowed flight width to reduce the energy dissipation involved in this portion of the screw. It is hard to imagine two more different screws than Fig. 1C and G, both of which are targeted for the same application of LLDPE extrusion.

Screw design evolution took a dramatic, though rather unplanned, forward step when the two-stage screw (Fig. 2) was introduced to allow for gas removal from the polymer melt. In addition to the advantages claimed for an extruder with a vent, three design factors were explored which changed screw and extruder designs significantly.

Fig. 2. Two-stage screw—for venting

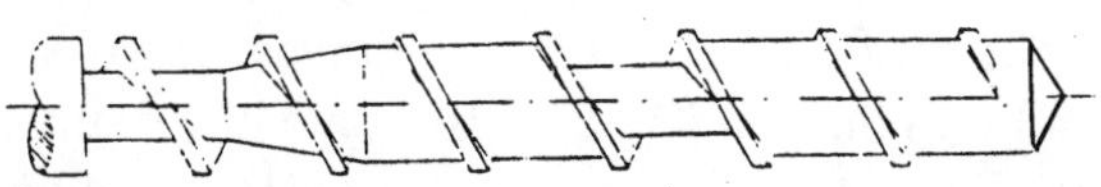

Because the vented extruder is essentially two pumping screws in series, it was necessary to lengthen the extruder to accommodate the process. Vented extruders pushed the length to diameter ratios of extruders from 12:1 or 16:1 L/D to 24:1 or 28:1 in a matter of a few years. The advantages of a longer L/D ratio in nonvented applications became readily apparent.

Two staging of screws also opened the door to the concept of separating the screw into various sections specifically designed to do one and only one quality improvement operation. Again, it was discovered that two-stage screws operating nonvented, if properly designed, could significantly improve extrudate quality over its single-stage counterpart. Two-stage screw designs without a vent have led to many significant changes in feed screw technology since the middle of this century.

The third advancement to the state of the art of screw design evolved around the insertion of the blister (Fig. 3) at the end of the first-stage metering section in a vented screw. The blister became a barrier to limit the flow of plastic in the second-stage feed section. It was originally installed to limit or eliminate the problem of vent flooding.

Fig. 3. Blister rings

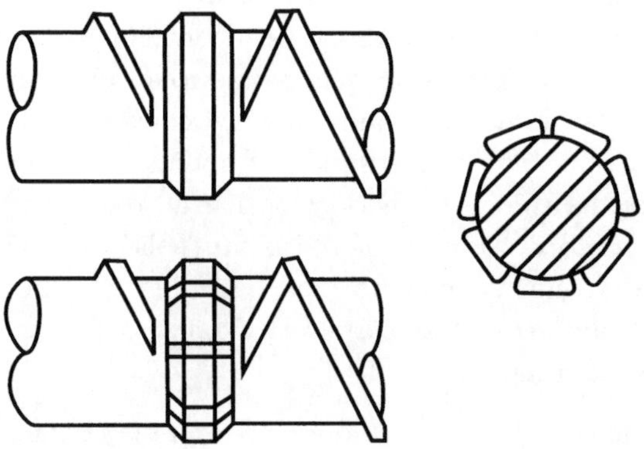

A barrier such as the blister forced all of the melt to be exposed to a higher shear rate and shear stress than would be experienced in a normal metering screw. The results involving extrudate quality were often dramatic. This reduced the problem of unmelted feed pellets seen in the discharge melt. In addition, some improvement in dispersive mixing was noted. Again, the advantages of the barrier were quickly transferred to nonvented screw designs.

The mention of barrier screws opens the door to a large number of designs. In general, two basic concepts of barriers are currently marketed. The short barrier screw *(3–5)* is shown in Fig. 4. A number of designs vary the axial location of the mixing section as well as the number of mixing sections.

Fig. 4. Short barrier section screw

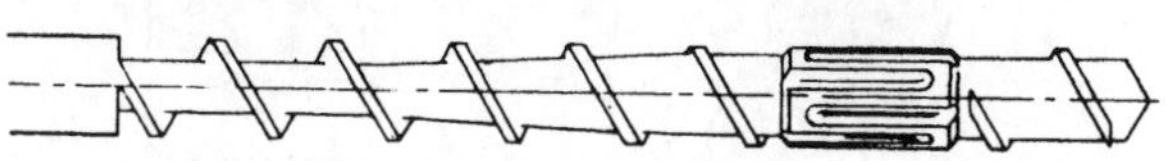

The long barrier screws appear to have been an invention of Charles Maillefer in Europe in the mid-1950s. Paul Geyer *(6)* received a U.S. patent which paralleled the Maillefer design in 1968. Figure 5 illustrates the Maillefer or Geyer design. Note that to reach the die, the solid plastic in the channel on the left must be able to flow from the solids channel across the barrier flight to the melt channel.

Long barrier screws of the melt draining concept are claimed to overcome the problem found in poorly designed nonbarrier screws, mainly that of solid bed breakup. Solid bead breakup is known to reduce extrusion quality due to air entrapment in the unmelted solids that are conveyed in iceberg fashion through most, if not all, of the length of the screw. In theory, the barrier dam restrains the forward movement of the solid bed while allowing the melt formed over the solid bed to "drain" over the barrier clearance to the downstream flight.

Many variations on the long barrier theme are now marketed under a variety of trade names. The variations include adjusting the helix angle of the barrier or principle flights, adjusting the width and depths of the solid and melt channels, and in some cases making the barrier flight less than the full length required to totally cross the channel as shown in Fig. 5.

Fig. 5. Long barrier of Maillefer or Geyer design

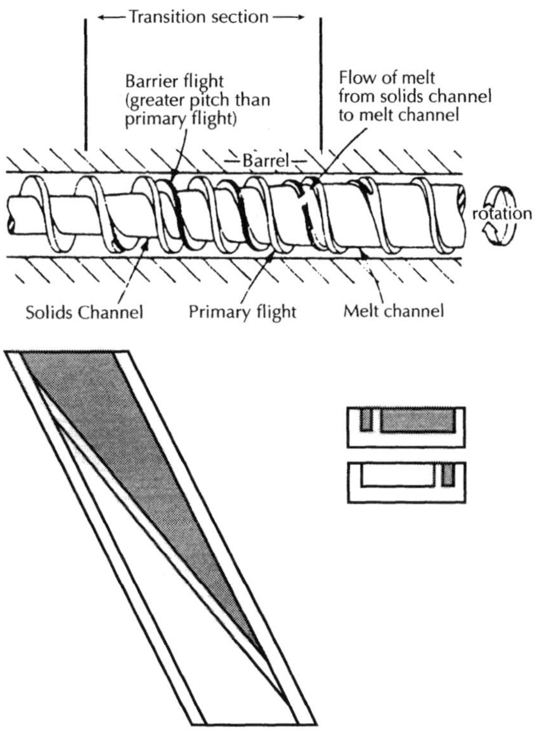

It is important to recognize that the barrier location in relation to the rate of melting is important using the Maillefer/Geyer approach. Since the rate of melting is dependent upon many variables such as screw speed, barrel zone set temperatures, screw root temperatures, and resin characteristics such as molecular weight and slip additive ratio, the long barrier screw can be less forgiving to operation at conditions other than the original design specification.

In summary, it is safe to say that no single screw design provides the best answer to all of the specifications required to perfect an extrusion process. Additionally, variations in operating variables such as screw speed or barrel zone temperature settings can cause an otherwise acceptable screw to perform poorly. Experience has shown that there are definitely poor screws, good screws, and better screws. The best screw design is still hiding in the wings for future generations to discover. At this point in time, we need to be content with the melt quality obtained from the better screw designs.

REFERENCES

1. Miller, J.C., Tappi J. 67(6): (1984).
2. Rauwendaal, C., Plastics Tech. Mag. 29(9):61 (1983).
3. Gregory, R.B. and Street, L.F., U.S. pat. 3,411,179 (1968).
4. LeRoy, G., U.S. pat. 3,486,192 (1969).
5. Gregory, R.B., U.S. pat. 3,788,614 (1974).
6. Geyer, P., U.S. pat. 3,375,549 (1968).

CHAPTER 6

Feed Screw Design

Robert B. Gregory

FEED SCREW TERMINOLOGY

The typical plastics extruder feed screw will probably have the general appearance of the root profiles of Fig. 1 or 2. The screw profile of Fig. 1 is representative of the single-stage screw, which has been an industry standard in the form shown for at least the past decade, while the screw profile shown in Fig. 2 is one of a more modern design.

Fig. 1. Typical single-stage extruder feed screw diagram

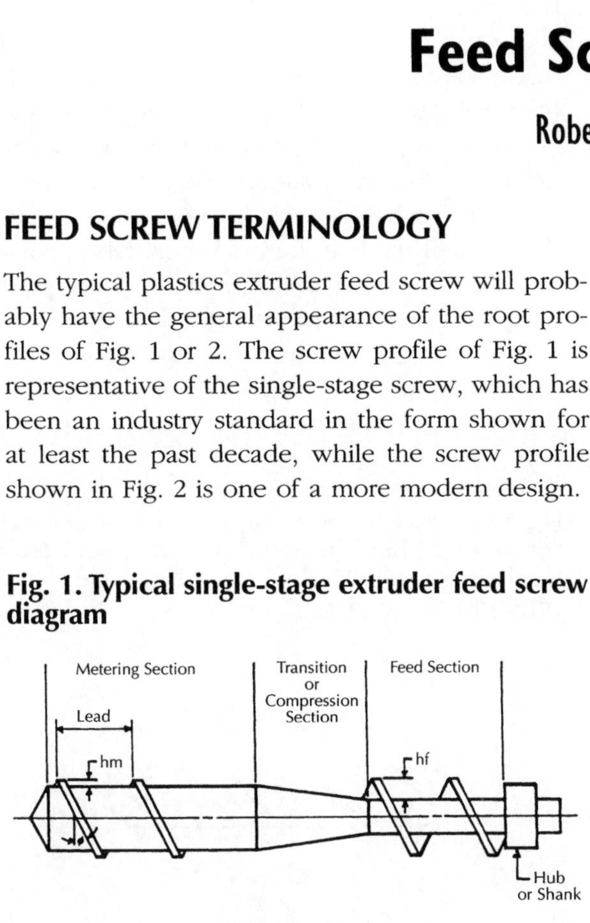

Fig. 2. Typical multiple-stage extruder feed screw diagram

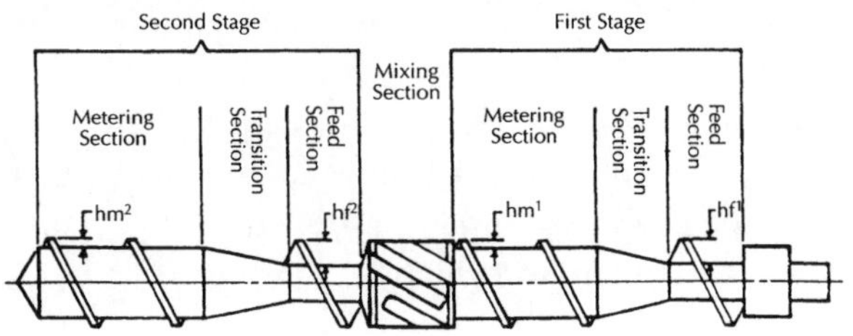

Section terminology for screws is consistent with the deeper channel h_f or h_f1 and h_f2 being called feed sections and the shallow channel depth h_M or h_M1 and h_M2 being called metering sections. The conical or tapered section is called either a transition or a compression section with a preference favoring the use of transition section since the word "compression" does not necessarily accurately describe the section function.

Screws are characterized by the ratio of their length to diameter. The extrusion industry has established certain ratios of lengths to diameter as standard.

The current standard ratio is between 28:1 and 32:1 on extrusion coating lines in the field. The L/D ratio of an extruder is one of the important factors in determining the extruder's ability to produce the desired quantity, quality, and temperature of plastic to the laminating die.

MATERIAL OF CONSTRUCTION

A feed screw is normally driven from the shank end with the total applied screw torque passing through the first increment of the feed flight and dissipating gradually to zero at the screw tip. It is a typical engineering paradox that the maximum screw torque must be transmitted through the smallest-diameter portion of the screw root diameter.

It is not surprising to find that a screw root in the feed section can be highly stressed depending upon the process torque requirements. The materials of construction, therefore, tend to vary from a conventional AISI1020 steel for low-torque applications to AISI4140 heat-treated steel for higher-torque applications to special alloys for very-high-torque applications.

FLIGHT LAND PREPARATION

Typically, extruder feed screws are so long in relation to their section moment of inertia (i.e., deflect due to lack of stiffness) that a major portion of the screw flight in an empty extruder lies in contact with the bottom of the barrel inside diameter (ID) as shown in Fig. 3. The melted thermoplastic polymer acts as a lubricant between the feed screw flight land and the barrel ID at startup to cause the screw to lift and essentially re-center itself in the barrel during operation.

Thus, if the screw is rotated at a high flight land surface velocity in a dry barrel, the possibility exists that the land will gall on the barrel hard surface material. The recommended startup procedure for any extruder is to start the screw rotating very slowly (5-10 rpm) and to feed plastic to the screw quickly after initiating rotation.

As soon as melted plastic begins to discharge from the downstream end of the process, the screw's rotational velocity can be increased without undue fear of galling.

Fig. 3. Extruder feed screw laying on bottom of the barrel ID prior to being filled with thermoplastics

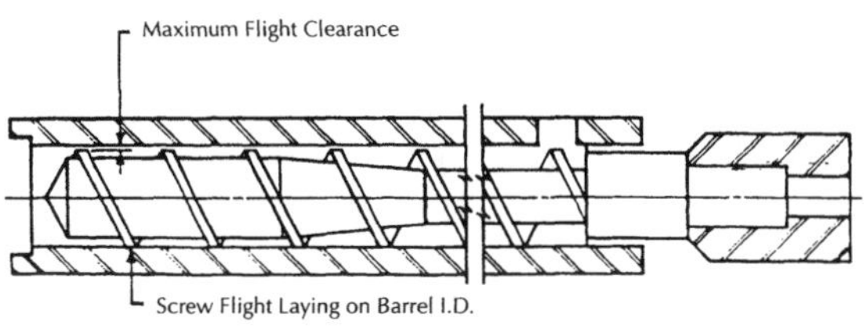

Several methods of screw flight land preparation exist to prevent excessive wear due to erosion, corrosion, or galling. The earliest standards involved case hardening. Of these, flame hardening was the most popular in the United States. A flame-hardened 4140 or 4340 flight surface might initially have a surface hardness as high as Rockwell C60 to C70; but, if the screw is to be used at elevated temperatures such as the typical 600-650°F for extrusion coating, the surface hardness will be reduced by the drawing process to a lower level, possibly as low as C45 to C50. There is some evidence to indicate that these lower levels of hardness will still be satisfactory for some extrusion processes, providing the polymer being processed maintains its film strength under the flight land during extrusion.

Recent applications of various hard surfacing alloys such as Stellite 1, 6, or 12, Colmonoy 56, or Xaloy 800 to the screw flights as shown in Fig. 4 have had increasing degrees of field acceptance. These alloys, by virtue of being hard, are also brittle so that the hard surface alloy on the ductile core will probably crack in one or more places due to their thermal stresses during fabrication or mechanical stress during operation. The process of applying hard surfacing alloys is critical to prevent base metal dilution of the hard alloy while at the same time obtaining good base metal penetration to achieve the required metallurgical bond.

Fig. 4. Cross section of hard surfaced extruder screw flight land (method A). Cross section of hard surfaced extruder screw flight land (method B).

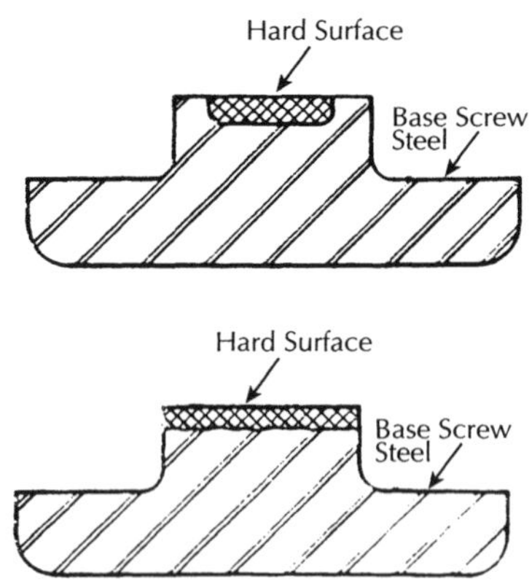

Reliable extruder feed screw suppliers are competent to explain the advantages and disadvantages of the flame or nitride hardened flight versus the alloy hard surfaced flight.

Chrome plating

Many extruder feed screws are supplied with a thin layer of chrome plate over the surface in contact with the plastic. Such a surface has been thought to lower the coefficient of friction between the solid feed pellets and the screw. Such an honor for a chrome-plated surface appears to be undeserved.

In any case, a good industrial-grade chrome-plated surface is very hard and will protect the softer base metal from scratching when the screw is cleaned with abrasive materials such as a steel wire brush. This advantage alone probably warrants the cost of plating.

Chrome plating cannot be considered as an absolute barrier to chemical corrosion since chrome in the presence of free HCl, such as is liberated with the decomposition of PVC, will combine chemically with the HCl leaving that part of the screw free of plating.

The purchaser of a feed screw should honor the judgment and history of his equipment supplier in the selection of screw alloys, surface preparation, etc., or be willing to accept the responsibility of mechanical failure should such occur. In any case, an extruder screw can still fail in service with improper treatment by operating personnel.

Extrusion process analysis

The history of extruder design reveals a gradual trend toward the recognition by the process designer of the subjective as well as the objective factors by which the process must be evaluated.

Objectively the user requires that his extruder provide the following:

1. The desired output rate
2. The desired extrudate temperature throughout the range of processing rates of the extruder
3. A minimum of rate surging, usually measured by discharge pressure variations or machine direction gauge variations
4. A uniform melt temperature with a minimum of time-dependent variations of temperature.
5. A minimum process residence time, consistent with meeting the other objective requirements.

The list of subjective process requirements is usually a little more difficult to describe since these requirements are established by judgment and not easily measurable. They might require the extruder to do the following:

1. Provide the desired level of pigment dispersion and distribution
2. Produce a final product having haze and gloss levels within acceptable limits
3. Minimize or eliminate fisheyes, gels, and other imperfections possibly originating in the plastic feed
4. Minimize polymer degradation from any processing source.

Some of these specifications can be met by incorporating the design features obtained by relatively simple engineering design calculations. Others are much more difficult to confidently engineer if their solutions are at all predictable. For example, the rate performance of an extruder feed screw with predetermined barrel zone settings and an intimate knowledge of the physical properties of the polymer can usually be computed with a high degree of accuracy. At the same time, the discharge mean melt temperature and the required power can be confidently calculated. By contrast, there are no available mathematical tools with which to accurately prejudge an extruder feed screw's ability to eliminate gels, produce a low haze, high gloss film, or to minimize polymer degradation.

With this introduction to the limits of the combined art and science of extruder feed screw design, we should examine the known physics of the extrusion process. The single-screw extruder is considerably more complex in its functioning than the "meat grinder" with which it is so often compared.

TYPES OF FLOW

An extruder is usually fed a granulated solid plastic pellet. The form varies from cubes to cylinders and even possibly to a lens shape. The screw must convey these solids, in some fashion, to the discharge end of the machine. During the transit, it is mandatory that the energy level of the plastic feed be raised to the point that the nature of the "solids" feed is lost and the semi-liquid state of a "melt" is obtained. Two basic forms of conveying are thus employed in the process, with a different set of mathematical tools required to predict the performance of each.

Solids conveying

The first type of conveying is solid in nature. Our definition of solids conveying is limited to the state where all adjacent molecules maintain their relative position during the conveying process. There are two forms of solids conveying in the extruder as illustrated in Fig. 5.

Fig. 5. The two forms of solids conveying in the extruder

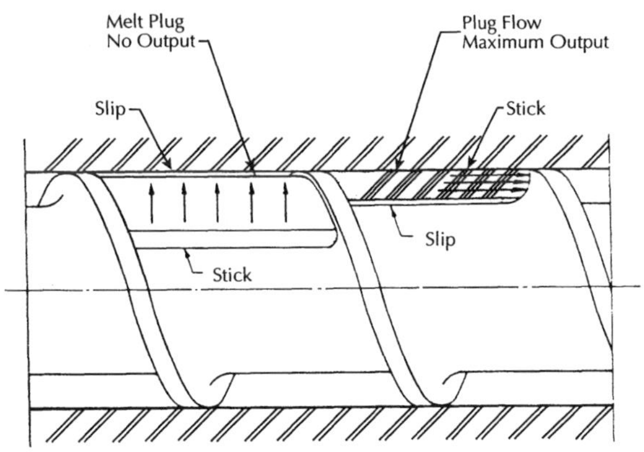

"Plug flow" is diagrammed on the right of Fig. 5 and occurs when the solid plastic material adheres to the inside wall of the extruder barrel while slipping on the wetted perimeter of the screw channel. This type of movement is analogous to holding a nut on a screw and turning the screw. The nut will advance on the screw in relation to the volume of the thread the screw displaced by one turn.

The contrasting flow to "plug flow" is illustrated on the left of Figure 5 and is often called "melt plug." In this state, the solid plastic adheres to the wetted perimeter of the screw while slipping against the barrel inside wall.

This type of flow obviously produces no axial movement to the plastic since the feed is spinning with the screw. This state is analogous to turning the screw with the nut on the thread but not holding the nut to keep it from turning. The nut in this case does not advance with respect to the screw. At first glance, this condition might be considered to be disastrous to the success of the extrusion operation. Yet we will see later that the success of the melting process is generally dependent upon the occurrence of some "melt plug" flow during the solids conveying phase of extrusion.

What condition of the polymer, inside wall of the barrel, and screw wetted perimeter will promote adhesion of the feed to the barrel and allow it to slip on the screw for at least part of a turn of the screw? Two factors are immediately apparent to control the slipstick condition.

The first is the ratio between the area at the screw channel and the area at the barrel wall for a unit lead of screw. This ratio should be less than one for the same coefficient of friction between the plastic feed and the two surfaces in question to be assured that some plug flow will occur during a revolution of the screw.

A solution of the computation of the area of a screw channel shows that for the common helix angle of 17.7°, there is decay in area with increasing channel depth. On the other hand, it is obvious that when the channel depth is very shallow, the ratio of the area of the screw to the barrel approaches unity.

The area relationship and its importance to forwarding is one valid reason for eliminating the term "compression ratio" from the screw designer's vocabulary. The use of a fixed value of "compression ratio" or channel depth ratio independent of the diameter tends to make the feed section depth more shallow in relation to its diameter of a large-diameter extruder; this then leads to a less favorable area ratio and possibly a feed or solid conveying problem.

The other important factor in controlling whether the polymer will slip or stick to the appropriate surface is, of course, the value of friction coefficient that exists between the two surfaces. It has been shown *(1, 2)* that the coefficient of friction of a polymer, such as low-density polyethylene, against a steel surface varies with the temperature of the steel.

Figure 6 shows more recent data collected for two polymer temperatures paralleling the work Grant published. There is a band of steel temperatures where the friction coefficient is maximized and several where it is minimized. Thus, if one wishes to promote a high coefficient of friction between the solid polymer and the inside wall of the barrel, the steel temperature should be maintained in the range of 200°F to 400°F. Conversely, if it is desirable to promote a low coefficient of friction between the feed and the screw, the data plotted in Fig. 6 shows that it is desirable to maintain the screw surface temperature below 150°F or above 500°F.

It is obvious that the screw design is dependent *not only* upon the screw geometry to promote a successful operation but also the proper environmental temperatures. The extruder operator has control of steel temperatures though the extruder zone settings he or she selects.

Fig. 6. Coefficient of friction low-density polyethylene at 70°F and 150°F vs. steel at 70°F to 550°F polyethylene-Union Carbide DYNH

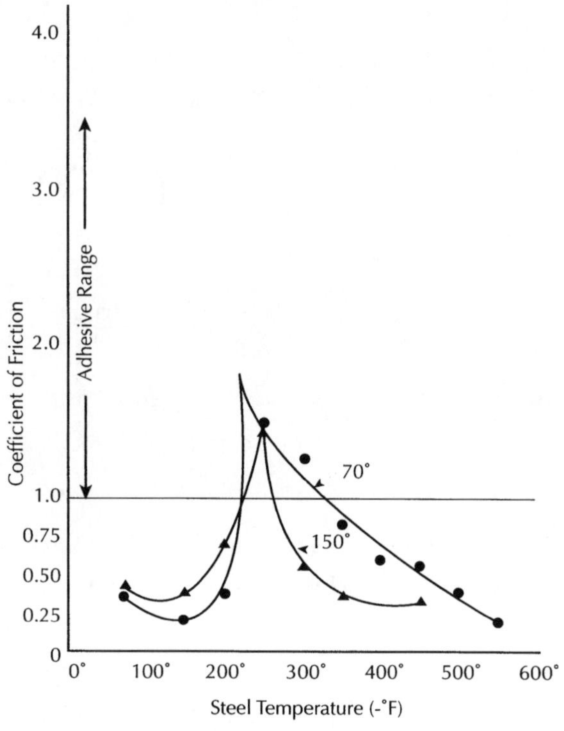

Plug flow equation

If the coefficient of friction between the screw and the resin is very low while the coefficient between the plastic and the barrel is high, an equation *(3)* for computing the "plug flow" volumetric conveying rate follows:

$$Q_s = \pi^2 D^2 N h \sin \varnothing \cos \varnothing \left(1 - \frac{ne}{t}\right) \quad (1)$$

Where

Q_s = Volumetric conveying rate, in.^{3}/s

D = Screw outside diameter, in.

N = Rotational velocity, rps

h = Channel depth, in.

ø = Flight helix angle

n = Number of flight starts

e = Flight width, in.

t = Lead, in.

MELT FLOW EQUATION

The third form of conveying that the extruder screw must perform is illustrated in Fig. 7. This is called melt flow and occurs when the plastic adheres to all of the wetted surfaces, and adjacent molecules separate permanently by the application of a shear stress.

Fig. 7. Cross section of an extruder barrel showing melt flow and pressure flow

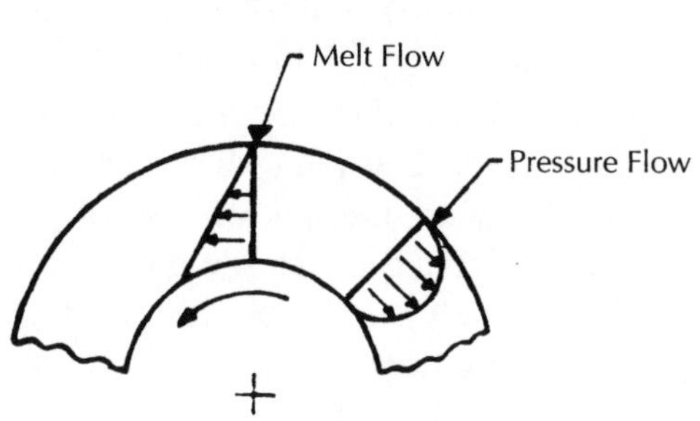

This type of flow has been called "drag flow" in the literature and can be conveniently computed by the following Eq. 2.

$$Q_D = \pi^2 D^2 N\, h \sin \varnothing \cos \varnothing \left(1 - \frac{ne}{t}\right) F \qquad (2)$$

Where

Q_D = Volumetric melt conveying rate, in.^{3}/s

F = Factors to account for curvature and the channel shape.

Comparison of plug vs. melt flow in a screw

Note the similarity of Eqs. 1 and 2. In broad terms, a screw channel of fixed depth can convey two times the solids volume rate using plug flow than it can convey melt by drag flow. Considering the feed section channel depth of the average feed screw is typically much deeper than the metering section (melt conveying section) by a factor ranging from two to six times, it can be seen that the solids conveying rate of the screw in plug flow can be four to twelve times the melt flow capacity of the metering section. Since the usual density of the solid plastic fed to the extruder varies from approximately 40-80% of the melt density, such differences in volumetric conveying capacity are not justified by the difference of densities alone to maintain a rate balance in the extruder.

Why should the feed section be so deep if its rated volumetric capacity is so much greater than needed? How can rate equilibrium be established with such a rate discrepancy? The feed section must be deep to optimize the ratio of wetted perimeters as previously described. Rate equilibrium, therefore, is determined by the vectorial combination of "plug flow" and "melt plug" in the feed section of the extruder with the percentage of each self-adjusting to account for density differences. Thus, even though the environment has been optimized to promote plug flow, polymer slip with the barrel inside wall must be forced.

Since this slip occurs with a maximum coefficient of friction at the barrel wall during the period of "melt plug" flow in the feed section of the screw, it is a logical conclusion that a significant amount of energy is introduced into the process in a manner similar to a "brake shoe." This increment of energy, combined with that transferred by conduction from a usually hotter barrel wall to the cooler plastic, provides the necessary total energy to melt the plastic.

Output rate formula

One method of predicting the output rate of an extruder based upon melt conveying is available in the literature *(3, 4)*. The general form of simplest expression follows:

$$Q_T = Q_D \pm Q_P \qquad (3)$$

Where:

Q_T = Total volumetric output rate, in.^{3}/s

Q_D = Drag flow rate, in.^{3}/s

Q_P = Pressure flow rate, in.^{3}/s.

Equation 2 describes the method of putting the drag flow rate Q_D. The pressure flow equation is:

$$Q_P = \frac{\pi D h^3 \Delta P \sin^2 \varnothing F_P}{12 \mu L}\left(1 - \frac{ne}{t}\right) \qquad (4)$$

Where:

ΔP = Incremental pressure difference, psi

L = Section length being evaluated, in.

μ = Apparent viscosity of melt, (lb - s)/(in.^{2})

F_P = Factors to account for curvature and channel shape.

The algebraic sign for the pressure flow component (Q_P) of Eq. 3 is dependent upon the direction of the pressure gradient along the screw. Figures 8 and 9 diagram two possible pressure profiles, with Fig. 9 showing a higher pressure at the beginning of the metering section than at the discharge.

If the metering section indicated by length L is evaluated for rate performance as indicated in Eq. 3, the pressure flow rate component (Q_p) will be positive or additive to the drag flow rate.

Conversely, a pressure profile as seen in Fig. 9 will produce a negative value to the pressure flow rate. The total rate will be less than the drag flow rate.

Usually the pressure profile witnessed by a number of sensitive pressure transducers along the barrel will not show a single value of pressure at any one point as plotted in Figs. 8 and 9 but will produce an envelope surrounding the mean value. This variability of pressure has been observed to typically increase with distance downstream from the feed opening to a maximum value at about half the screw length. The variability then tends to decrease to a more desirable level at the discharge end of the extruder. The value of ΔP as illustrated in Fig. 8 will actually be modulated by ΔP^1, thus producing an unstable or surging rate.

Fig. 8. Pressure profile of a feed screw showing a positive pressure flow

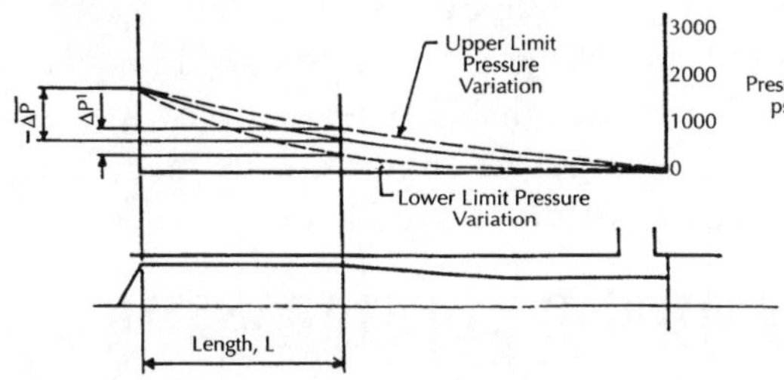

Fig. 9. Pressure profile of a feed screw showing a negative pressure flow

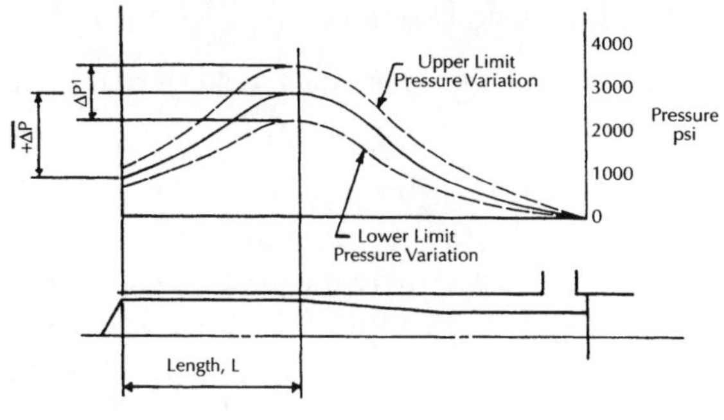

The value of the pressure flow component of rate for the condition found in Fig. 9 is considerably more stable since ΔP^1 is smaller. The resulting output rate surge is less than the condition described in Fig. 8.

EXTRUSION ENERGY REQUIREMENTS

The energy balance of the extrusion process had been briefly summarized in the literature *(6)* as follows:

$$Z_T = Z_H + Z_P + HL \qquad (5)$$

Where

Z_T = Total energy delivered to the process

Z_H = Energy required to heat the polymer

Z_P = Energy required to generate the necessary discharge pressure

HL = Heat loss from the process.

Expanding the individual of this balance one finds that:

$$Z_T = Z_M + q \qquad (6)$$

Where

Z_M = Input mechanical power, in.·lb/s

q = Input heat by condition, in.·lb/s.

Further

$$Z_H = C_P\, Q\, \Delta T \text{ or } \Delta H\, Q \qquad (7)$$

Where

C_P = Specific heat (average), in.·lb/in.^{3}/°F

Q = Volumetric extrusion rate, in.^{3}/s

ΔT = Temperature rise in plastic, °F

ΔH = Enthalpy change in plastic, in.·lb/in.^{3}

Finally

$$Z_P = Q\Delta P$$

Where

ΔP = Pressure rise in plastic, lb/in.^{2}

Figure 10 shows a typical enthalpy of low-density polyethylene. Note that the energy change required to heat a pound of this plastic from a 70°F ambient temperature to 600°F for extrusion coating will be approximately 350 Btu per pound. This amount of energy must be available from the two energy sources at the desired rate if the plastic is to be heated to the desired temperature level.

The rate of mechanical energy input to the thermoplastic in the melted state has been developed *(3)* as follows:

$$Z_M = \frac{\pi^3 D^3 N^2 \mu}{h}\left(1 - \frac{ne}{t}\right) \qquad (8)$$

From this expression, it is apparent that the mechanical energy input varies in an exponential relationship with the screw speed (N). At first glance, it might appear that mechanical energy (Z_M) is truly proportional to N^2 but, due to the non-Newtonian characteristic of most plastic polymers, the effect of the exponent is reduced. This reduction is caused by the reduction in viscosity caused by an increase in shear rate within an increasing value of screw speed (N).

In any case, it is not difficult to understand that at very low screw speeds with their resulting low output rates, the bulk of the energy to heat the plastic must be available from the barrel zone heaters, for there is little mechanical energy input to the process.

Conversely, at high screw speeds with their resulting high output rates, little if any energy is required from the barrel zones to provide the energy to extrude since the mechanical energy input is large due to the speeds' exponential effect on the available mechanical energy.

Fig. 10. Enthalpy of low-density polyethylene

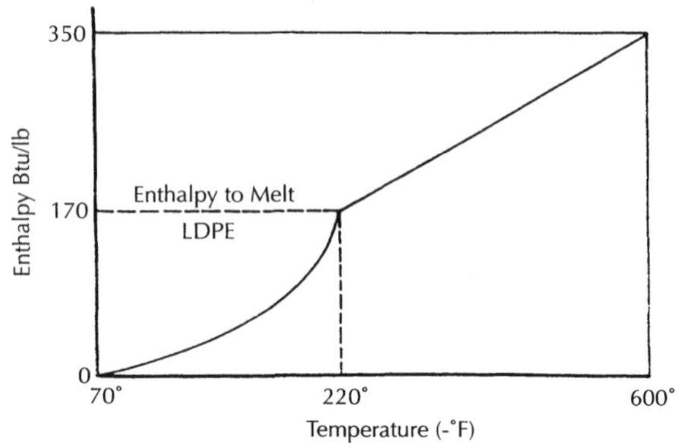

No ironclad rule of thumb concerning the output rate per unit of mechanical power can be stated to cover the entire range of an extruder's potential output rate. Nevertheless, most users of extruders are only interested in their operation at rather high output rates where the available heat from conduction is quite small and can often be ignored.

An interesting point is reached with increasing screw speed and, of course, increasing rate for any particular extruder. The point in question occurs where the mechanical energy input exceeds the desired sum of the Z_H and Z_P energy in Eq. 5. Any speed beyond that point requires that some process energy be removed to maintain the enthalpy Z_H at the desired level.

CONDUCTIVE HEAT TRANSFER

Jepson *(6)* computed a mathematical relationship for coefficient of heat transfer between the plastic in the screw channel and inside barrel wall of an extruder. His work described the coefficient as being proportional to the screw speed (N) and the radial flight land clearance.

Figure 11 shows some data collected from two extruder sizes that correlate with Jepson's work.

The importance of this data is apparent when one wishes to establish a mathematical model of the extrusion process where the heat flow to and from the process must be evaluated. It can be seen that the barrel of an extruder operating at high speed can be a fairly efficient heat exchanger.

An efficient cooling system on an extruder is often found to be more important that the heating system to the economics of the process to be performed. The heat loss variable (HL) in Eq. 5, therefore, is found to be a significantly large number and can represent 30-50% of the total energy into the process at high speeds.

Fig. 11. Film coefficient of heat transfer vs. rpm

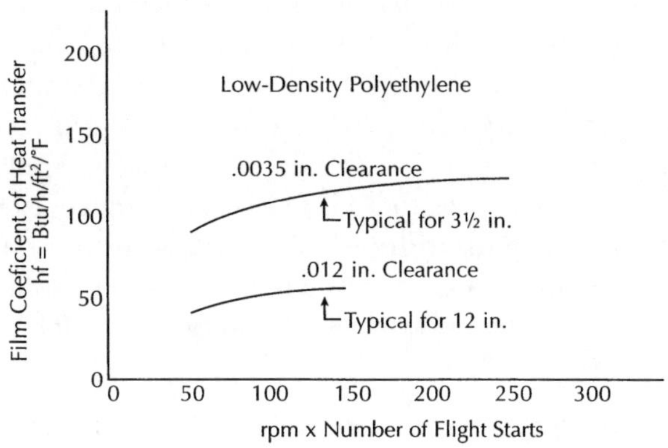

Is it any wonder then that the modern extrusion coating extruder is being supplied with water-cooled barrel zones? With this feature incorporated, maximum rates have been significantly increased over recent years *(7)*.

MIXING SCREW DESIGNS

The use of a screw design of the type shown in Fig. 2 is also responsible for upgrading the performance of extrusion equipment. The principal reason for the enhanced qualitative performance of such a screw is the improved level of distributive as well as dispersive mixing caused by the mixing section. A barrier mixing section is usually combined in the multiple-stage feed screw profile.

How is the performance of an extruder quantitative? Two measurable variables are important indicators of the quality of the extrudate. A short-time constant immersion thermocouple, located in the melt stream in the adapter or downspout of the process and feeding a signal to a rapid response recorder is capable of measuring variations in the melt temperature with respect to time *(8, 9)*. A short-time constant-strain gauge pressure transducer can measure variations in pressure with time on a similar recorder

It is recognized that the variation of melt temperature is in part a measure of the level of mixing in the extruder. Pressure fluctuations partially measure the rate surging present in the process. Maddock *(10)* stated that "problems of mixing or product homogeneity and those of rate uniformity or surging are essentially independent phenomena."

The barrier-type mixing section is available in several forms *(7, 9, 10)*. Each is similar in that all of the polymer must pass through a restrictive gap that produces a level of shear stress higher then that obtainable in the normal screw channel of the type shown in Fig. 1.

MIXING IN AN EXTRUDER

McKelvey *(11)* and Mohr *(12)* have divided the scope of mixing into two categories. They are "dispersion" and "distribution."

Dispersive mixing is an operation which reduces the agglomerate size of minor constituent to the possible limit of the ultimate particle size suitable for either concurrent or future distributive mixing in a major matrix.

Distributive mixing is an operation employed to increase the randomness of the spatial distribution of one or more minor constituents in a major matrix with no change in the ultimate particle size of the minor constituents.

Mixing covers the following typical extrusion functions:

1. Dispersion (i.e., deagglomeration) of a pigment in a polymer followed obviously by a distributive mixing operation
2. Dispersion and distribution of two different resins together to make a new "alloy"
3. Dispersion and distribution of a high-molecular-weight constituent in a virgin polymer to reduce the end product "gel count"
4. Dispersion and distribution of low-temperature resin in a matrix of higher temperature melt in order to approach a homogeneous melt temperature
5. Dispersion, if necessary, and distribution of additives such as fillers, reinforcers, stabilizers, plasticizers, and lubricants into a polymer

matrix in order to manufacture a desired compound.

The list is only representative and not the total scope of the term "mixing" in the extrusion process.

How can a particle of pigment, as an example, be deagglomerated or dispersed? The only reliable process of dispersing in an extruder requires the use of the shear stress of a major constituent to rupture the agglomerates of the major constituent to produce the desired level of particle size.

It is well known *(13)* that the level of shear stress is dependent upon the viscosity as well as the shear rate in the extrusion process:

$$\tau = \nu \dot{\gamma} \tag{9}$$

Where

τ = Shear stress, lb./in.^{2}

ν = Viscosity, lb. · s/in.^{2}

$\dot{\gamma}$ = Shear rate, lb/s.

The logical conclusion to the use of this mathematical relationship is that if dispersion is a function of the shear stress applied to minor constituent. In the case of an extruder, the shear stress will be increased by either a high level of primary fluid viscosity, a high level of applied shear rate, or preferably both.

A high level of viscosity is the most easily obtained from a low melt temperature at the point of dispersion. A high shear rate can be easily accomplished in an extruder by vectorially combining the rotational shear rate with the shear rate caused by pressure flow in tight clearance orifices.

If dispersion is not followed by distributive mixing, the end product might not be satisfactory. The reason, using pigment as an example, is that the process of deagglomeration by shear stress will streak the ultimate particle size pigment along the line of applied stress. The concentration of pigment will then vary from spot to spot in the melt. Conversely, if the first operation of mixing is not dispersive in nature where dispersion must be accomplished, the process of only randomly distributing undesirably large agglomerates will produce a meaningless quality to the final product.

Recognizing that redistribution of the minor constituent is one of the primary functions of the screw extruder, let us examine those aspects of the process that will improve distributive mixing. The literature *(11, 12, 14)* reveals that the way to spatially reorient two molecules starting in close proximity in the extrusion process is to apply a high shear rate to the mixture or produce a high ratio of pressure flow to drag flow in the screw. In either case, it is desirable to have the viscosity at a relatively low level to promote distributive mixing.

The shear rate in a process is, by definition, the slope of the flow velocity vector profile. Thus the higher the shear rate, the greater the separation which will occur between adjacent molecules per unit of travel in the extruder, i.e., improved distributive mixing.

Pressure flow in the feed screw *(4)* is shown to be inversely proportional to the viscosity of the melt and proportional to the pressure across the screw section and the cube of the channel depth being analyzed *(4)*.

From this relationship, it can be seen that to increase the ratio of $Q_P : Q_D$, the resin viscosity should be low (i.e., high temperature), the pressure increment must be high, and the screw channel significantly deep.

A high ratio of pressure flow to drag flow, in fact, increases the apparent shear rate to a value considerably above the level predicted by rotational shear rate alone. In addition, as indicated by past work *(14)* with glass barrel extruders, the direction of the velocity vector fans within the screw channel will change. The improved distributive mixing results from the summation of these two phenomena.

PROCESS EVALUATION

To repeat, the evaluation of the screw extruder's ability to perform to the user's satisfaction of output rate and quality can usually be categorized by two type variables, one objective in nature and the other subjective. The objective more easily measure criterion of extruder performance. Some of the typical objective requirements are:

1. Produce maximum rate per unit of capital investment
2. Rate to be constant or with an acceptable minimum of variation with respect to time
3. Polymer to be at one desired temperature (i.e., isothermal) or with a minimum of variations
4. Resign degradation caused by oxidation thermal reactions, or chain incision due to high shear stress, is to be minimized.

The subjective variable list is much more abstract and might appear as follows:

1. The extrusion process is to produce the desired level of pigment dispersion and distribution.
2. The finished product is to have a haze, gloss, or transparency level acceptable to the customer's judgment.
3. Fisheyes, gels, and other nondescript imperfections possibly occurring in the original feed are to be minimized or desirably eliminated.
4. Product imperfections caused by a bimodal molecular weight distribution of feed, such as might be obtained from a mixture of virgin and scrap polymer, should be eliminated or minimized.

Each of the subjective, as well as most of the measurable of objective process parameters, are dependent upon the ability of the screw extruder to do a satisfactory job of *mixing*.

SUMMARY

The theory behind the operation of an extruder feed screw is extremely complex. The information contained in this chapter is but a scratch on the surface of the available technology. Some of the flow equations (Eqs. 2, 3, and 4) are under attack *(15)* due to their "many shortcomings."

The accuracy of the simple flow equations presented in this chapter is surprising in light of the complexity of the flow within the channel of the screw. There can be no quarrel with the fact that some "shortcomings" do exist in their use, but for a basic understanding of extrusion philosophy their use seems to be adequate.

It is important to realize that the performance of a feed screw design within an extruder is dependent upon the selection of the proper barrel zone settings coupled with an adequate barrel heating and cooling system. In addition, the drive must be capable of supplying the necessary screw torque at the screw speed selected. Finally, the quality of the plastic being fed to the extruder should be good, with a minimum of gel producing oxidized or high-molecular weight particles. The lot-to-lot variations of the rheological properties of the resin should be an acceptable minimum.

It is impossible to divorce the qualitative and quantitative performances of a feed screw from the extruder in which it is installed, the operator, and the resin used in the extruder. Nevertheless, the feed screw is still the heart of the extrusion process.

REFERENCES

1. Grant, D., Rubber and Plastics Age 38(4):328 (1957).
2. Gregory, R.B., Friction Coefficients of Plastics and Steel, ANTEC, Society of Plastics Engineers, Stanford, CT, 1969, p.114.
3. Paton, J.B., Squires, P.H., et al., "Chapter 4" in Processing of Thermo-plastic Materials (E.C. Bernhardt, Ed.) Reinhold, 1959.
4. McKelvey, J.M., Polymer Processing, Wiley, 1962, Chap. 10, p. 250.
5. Maddock, B.H., Kabelitems 105(3):3 (1958).
6. Jepson, C.H., "Future extrusion studies," Industrial Engineering Chemistry, 1952, p. 45.
7. Bunte, W.S. and Gregory, R.B., New Techniques in Extrusion Coating Screw Design, TAPPI PRESS, Atlanta, 1967, p. 80.
8. Gregory, R.B., "Extrudate temperature measurement another dimension," RETEC, The Society of Plastic Engineers, Inc., Japan Section, Kyoto, Japan, 1968.

9. Street L.F. and Gregory, R.B., "Upgrade your PP extrusion," Plastics Technology 15(8):47 (1969).

10. Maddock, B.H., "An improved mixing screw design," ANTEC, Society of Plastics Engineers, Stanford, CT, 1967, p. 835.

11. McKelvey, J.M., Chapter 12 in Polymer Proceeding, Wiley, 1962, pp. 299-300.

12. Mohr, W.D., "Chapter 3" in Processing Of Thermo-plastic Materials (E.C. Bernhardt, Ed.) Reinhold, 1959, p.117.

13. Metzer, A.B., "Chapter 1" in Processing Of Thermo-plastic Materials (E.C. Bernhardt, Ed.) Reinhold, 1959, p.6.

14. Clapp, J.B., Mohr, W.D., and Starr, F.C., "Flow patterns in a non-Newtonian fluid in a single-screw extruder," ANTEC, Society of Plastics Engineers, Stanford, CT, 1961, Sect. 2-3, p. 4.

15. Klein, I. and Tadmor, Z., Modern Plastics (1968).

CHAPTER 7

Extrusion Dies for Flat Sheet, Film, and Coatings

John Ulcej, Extrusion Dies, Inc.,

INTRODUCTION

Extrusion dies are used in the production of flat, continuous webs of various materials. The die's most common application is found in the extrusion of thermoplastics, either as webs by themselves, or as laminates to other materials. Extrusion dies are also used to apply aqueous and solvent-based solutions to webs. The main function of the die is to provide the means to uniformly spread the material being processed to the desired width and thickness, in a controllable, uniform manner.

DESIGN CRITERIA

Types of dies

The selection of an extrusion die can be classified into three broad categories, based on the product that is to be produced: film dies to process thin materials, sheet dies to process thick materials, and dies used to apply low viscosity coatings.

Dies used in film production can be manually or automatically adjusted (Fig. 1). Uniform flow distribution is accomplished by the manifold design, and a flexible lip is used for fine tuning the material thickness. In some applications, a restrictor bar is used to aid in tuning the die. Coating dies that are used to laminate a polymer to another material are similar in design to film dies. Coating dies are also used to apply low viscosity material to a substrate (Fig. 2). Coating dies used to proximity coat, or wipe on low viscosity media, are able to have the die lip faces offset, creating a precisely controlled gap that establishes the coating thickness. These dies are usually provided with a support system to accurately position the die.

Fig. 1. Typical autoflex die

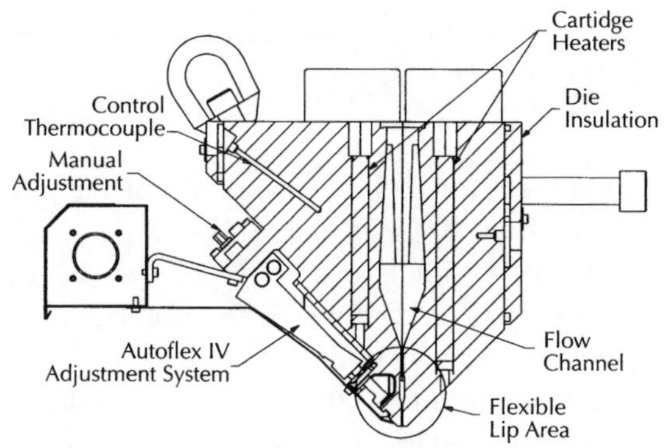

Fig. 2. Coating die

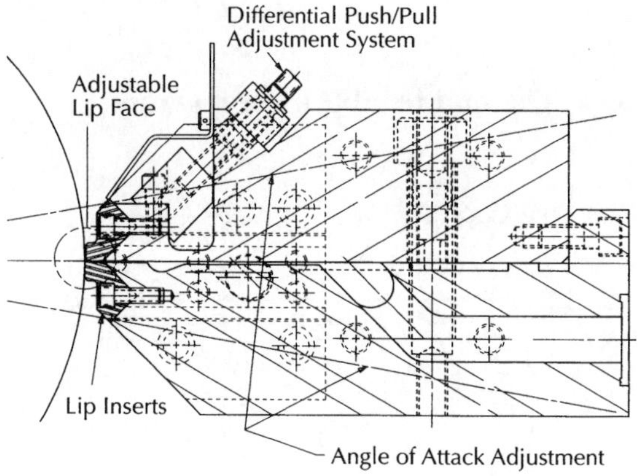

Sheet dies, used to produce heavier gauge materials, have a sliding lip and a restrictor bar (Fig. 3). The flow distribution is controlled within the die by the adjustment of the restrictor bar, and the sliding lip is used to control the thickness of the sheet.

Fig. 3. Typical sheet die

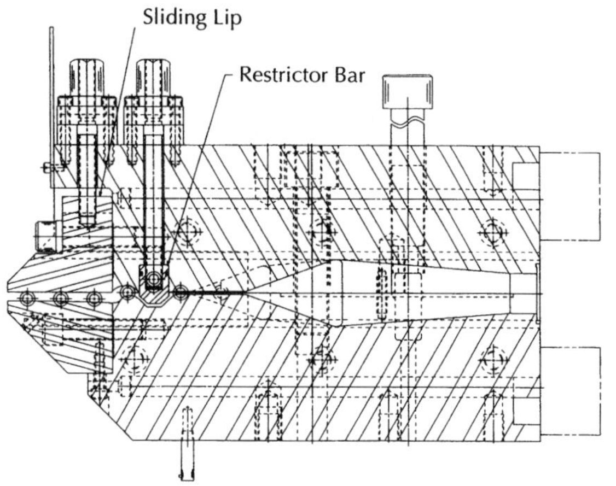

It is common for multiple layers of polymer to be extruded from a single die. This can be done in two ways. A single manifold die can be used downstream of a coextrusion feed block (Fig. 4). Different materials are fed into the feed block, where they are brought together in one melt stream. This system works well for materials with similar viscosities. If the material viscosities vary too much, then a multi-manifold die should be considered (Fig. 5).

Fig. 4. Die and feedback combination

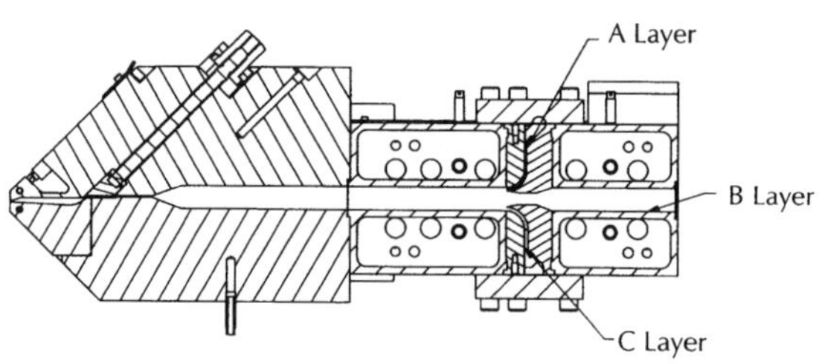

Fig. 5. Multi-manifold die

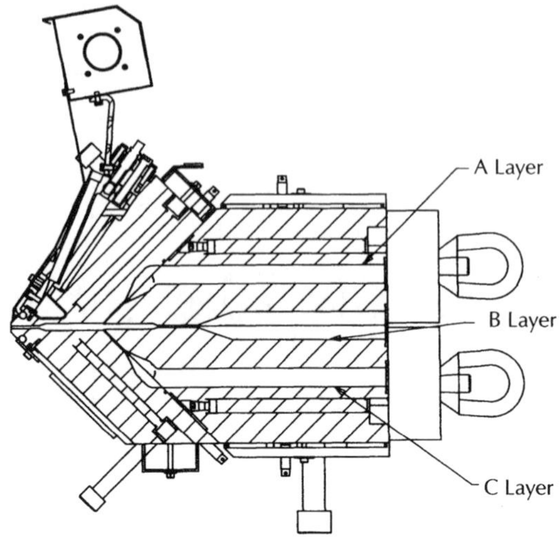

Die body materials

Extrusion dies are made from a wide variety of materials, the most common being P-20, medium carbon tool steel, which is chrome plated after construction. A few different grades are of stainless steel construction and some exotic alloys are also used, particularly in applications where corrosion is a concern. Different platings are also used. The aggressiveness of the material to be processed, coupled with the process itself, are the primary factors in material selection.

Some other factors to consider when selecting die body material are the mechanical properties. Hardness, strength at operating temperature, and the availability of the material in large, high purity billets are all considerations. The ability to work with the material is also a factor.

Flow channels within the die are usually finished from two to four rms, and are mirror-like in appearance. This is done to enhance the appearance of the material being extruded.

Manifold design

The area of flat die design that has seen the most significant improvements in the past five years is the flow channel or manifold. Aided by computational fluid design programs, state-of-the-art hybrid manifolds tuned specifically to polymer and process conditions are designed by solving three-dimensional flow equations.

The basic manifold for an extrusion die is a constant cross section or T-slot design. In all cases, the internal geometry should be streamlined, and care must be taken to avoid dead spots that could lead to polymer stagnation. The T-slot manifold relies on a large manifold area and a lip land long enough to create a large enough pressure drop to force the material to the ends of the die. This simple manifold design is used when low viscosity materials that are not thermally sensitive are being processed.

The Coat Hanger manifold design can be tuned for a window of materials and flow rates. This manifold is designed by establishing the same resistance to flow across the manifold and preland, at various points across the die (Fig. 6). The manifold shape is typically a teardrop that reduces in size from the center to the end of the die. This reduction in volume reduces residence time in the die, which is an important design consideration if thermally degradable materials are being processed. After the pre-land, the polymer flows into the secondary manifold, an area of the flow channel that allows the material to move laterally again, if required, and is also used to control the total pressure drop of the die, in conjunction with the final lip land.

Fig. 6. Coat hanger manifold

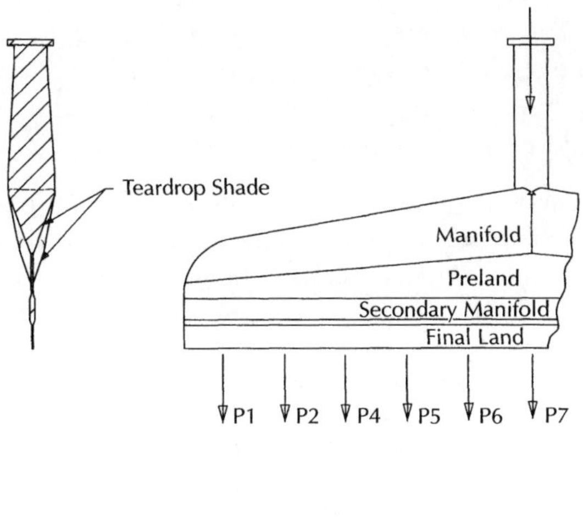

There are a few disadvantages with coat hanger manifolds. The back line of the manifold is further from the exit of the die at the center than it is on the ends. This results in a die that will deflect open more at the center, causing uneven melt distribution. "M&W" patterns are also common with coat hanger manifolds, which are affected by linear flow channel designs within the die.

The newest manifold design is a hybrid and addresses the drawbacks of the earlier mentioned designs. The teardrop shape is elongated, with an aspect ratio that changes from the center to the end of the die. This results in less interface distortion in coextrusion applications. The pre-land is a nonlinear shape that reduces the M&W patterns associated with coat hanger manifolds. These manifolds also have a straight backline, that allows for straight-line body bolt placement. This results in a die that deflects uniformly, further improving the distribution of the die (Fig. 7).

Fig. 7. Hybrid manifold

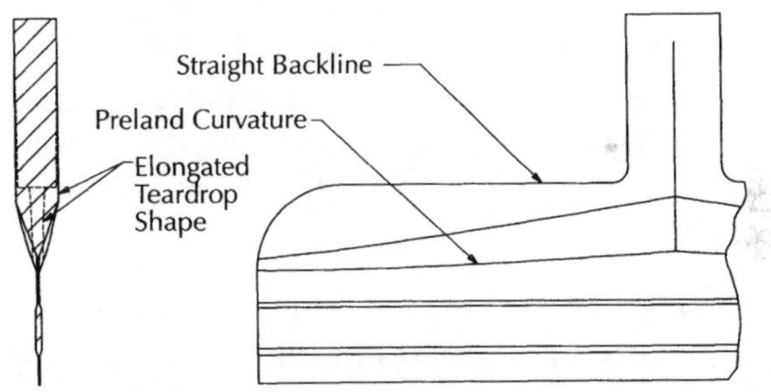

Deckling systems

A deckle has two functions. Primarily, it is used to change the die slot width. A deckle can also be used to control the edge bead on the web. The material to be extruded, and the extrusion process together determine whether or not the die can be deckled. Degradable materials preclude themselves.

The most common deckle is a fixed external type that can be bolted to either half of the die (Fig. 8). External deckles can be designed to adjusted "on-the-fly," typically driven by power screws or rack and pinion gears.

Fig. 8. Fixed external type deckle

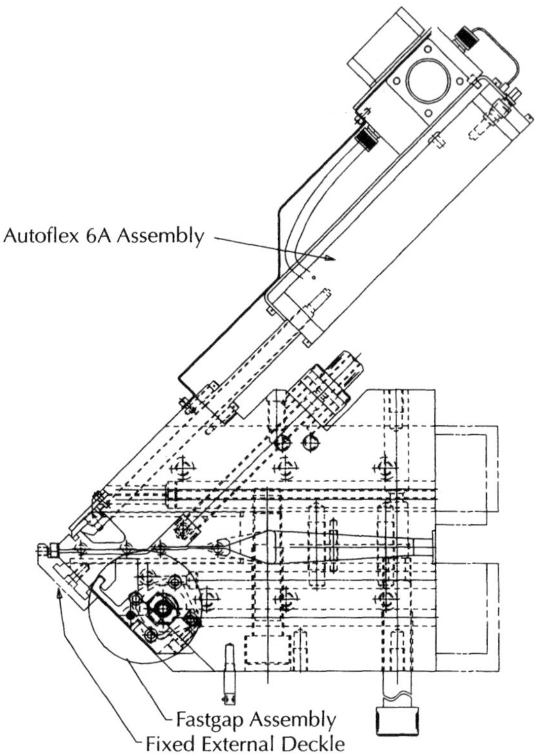

The flow of material from behind an external deckle usually causes a heavy edge. To reduce this tendency, edge bead rods, or internal deckles can be used.

An edge bead rod is typically a shaped piece of rod that is inserted into the die upstream of the final land. By varying the depth of insertion, the heavy end flow can be controlled.

The internal deckle is a more sophisticated version of the edge bead rod, filling some or all the flow channel in the die. The deckle flags can be profiled more accurately to produce the desired result. The internal deckle can be more difficult to set up, but it reduces the clutter around the outside of the die at the lip exit, allowing for a closer approach to a nip or casting roll (Fig. 9).

Fig. 9. Internal deckle

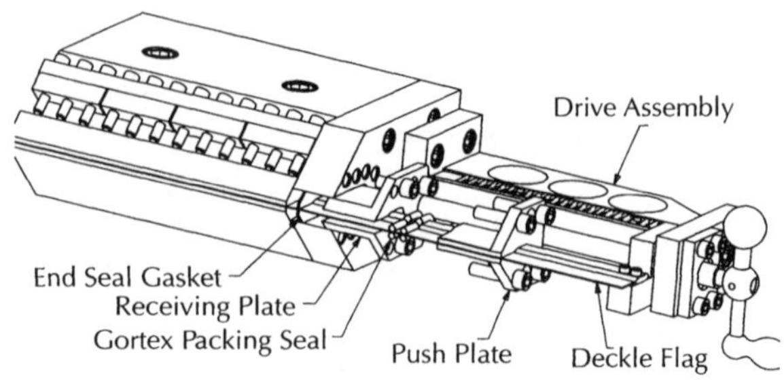

Hybrid deckles, internal and external deckle being combined, can also be employed.

Temperature control

Temperature control of a die is critical for good performance. Using today's control systems, ±1°F is typical and expected. If there is a cold area in the die, the material flow in that area will slow and the result will be thin gauge. Hot areas mean more flow and the potential to burn the material.

Type J thermocouples are most often used to measure the die body temperature. The die body is usually cored and heated with electric cartridge heaters (Fig. 1). External plate heaters, liquids, and vapors are also used to heat the die body. The wattage used to heat a die is determined by the operating temperature and the mass of the die itself. Thermocouple placement and die zone layout are important for proper control of the die and to prevent hot and cold spots. Insulation of the die should be used whenever practical: for safety, to protect against burns, to eliminate cold areas, and to conserve energy.

When required, die body cooling can be accomplished with air or liquid. Heat tubes are also used occasionally to equalize the temperature of the die.

Adjustment systems

Extrusion dies designed for extrusion coating and processing film and sheet typically are adjustable. Dies designed to produce thin gauge material, i.e., lip openings under 0.040 in., and extrusion coating dies are usually provided with a flex lip adjustment.

The die can have adjustments for flow tuning early in the flow stream, after the manifold, or late in the flow stream, using a lip adjustment system. The intent of the adjustment in either case is to control the final shape of the extrudate. The selection of a style of die depends on the end product that is to be produced.

Sheet dies, which are run with larger lip openings, will most often have a restrictor bar. They will also have a sliding lip (Fig. 3). The restrictor bar is located upstream of the flexible lip in the die, and it is an aid in establishing good polymer distribution. Restrictor bars can be externally adjustable or fixed. In either case, the principle of operation is the same. The gap in a specific area of the flow channel is either increased or decreased by profiling the bar. The change in resistance to flow by the change in channel height will encourage or discourage the flow of polymer in that area. Externally adjustable restrictor bars are most common in thick sheet extrusion and in applications where the polymer varies, by type, or by rheological properties.

A new concept in die design is the membrane, or flexible pre-land die, used to establish uniform distribution within the die. This concept will allow the flow to be tuned to a finer degree than currently possible. The idea of a flexible pre-land is not new, but the design approach is.

The simplest lip adjustment system is the "push only" adjustment. Screws are used to push the die lip closed. Dies are usually operated by closing down on the high gauge areas, although the die lip will respond in most cases if the adjusting screw is loosened, which allows low gauge areas to be corrected.

A more positive means of controlling the lip, particularly when trying to eliminate low gauge areas, is the push/pull adjustment. By attaching to the flex lip, movement can be made in both directions. The push/pull system can be provided with a differential screw thread system. This system has finer linear resolution per revolution of the adjustment nut, and this makes it easier for an operator to make small changes to the die lip. Differential adjustments are also available in push-only dies.

Fast-Gap™ is a new device that has been in use for about two years, and it allows the lip of a die to be moved open or closed with a single adjustment screw located on the end of the die (Fig. 8). Typically Fast-Gap™ is used on the lower lip and has been a real time saver in many applications. To change the opening range of a die previously, the die had to be disassembled and have a different lip installed. With Fast-Gap™, the die gap can be changed ±0.100 in. while the line is running, with a single point adjustment. The die lip can also be opened wide to clean or to flush the die if something gets hung-up in the lip land.

Autoflex™ systems (Fig. 1) work in conjunction with a measurement device located downstream of the die. The material thickness is measured on-line, and the results are fed back into a microprocessor that sends an output signal to either open or close the flexible lip of the die. The motion is typically produced by heating or cooling a series of closely spaced adjustment blocks. Heat is most commonly provided by cartridge heaters, and cooling is provided through a steady flow of air. Motors, piezo-electric elements, and liquid heating and cooling have also been used in auto die applications; but the simple reliable cartridge heater, with air cooling, remains the industry standard.

The earliest Autoflex™ dies used a long bolt, which ran through a block of steel that also held a cartridge heater. The range of movement was about ±0.005 in., and the heat required to get this movement was significant.

A refined version came along a few years later. The translating bolt and the heater became one unit. The range of motion increased to ±0.008 in. The heat required still could exceed the die body temperature by about 250°F at the maximum stroke. This system became the industry standard, and thousands of these units are in use today, primarily in film extrusion applications.

The newest Autoflex™ system addresses the heat and range issues that previously limited some Autoflex™ applications (Fig. 8). This system allows auto dies to be used in more applications, primarily where wider lip openings are required, and more temperature-sensitive materials are processed. The heated translator has been moved out of the die body, which allows it to run at a mean temperature that is lower than that of the die body. The stroke available now is ±0.015 in., and the actuators run at temperatures that exceed the die body by about 100°F. This new system does require more room on one side of the die.

PROCESSING INFORMATION

The die is an important part of any extrusion process. It is precisely manufactured and has a mirror finish on the flow surfaces. The slightest scratch can produce a flaw in the extruded product, so great care must be used in its installation and handling.

To start up a new die or one that has just been cleaned, the following procedure should be used. In a place where the body bolts can be accessed, heat the die to its operating temperature and allow it to soak for an hour at this temperature. If it is a new die or one that has been in storage for a long time, it should be brought up to temperature slowly. The heaters are filled with a hygroscopic material, and their life can be extended with a slow heat-up. This can be done on-line, with the die held in place by a support system or die cart, and attached to the melt pipe or adapter; or it can be done off-line with the die securely fastened to a sturdy support. After the temperature has been reached, torque all of the body bolts. This should be done to the die manufacturer's recommendations. The torque sequence should include at least two steps, first 50% of the recommended value, then to the full value. When the die is attached to the upstream equipment, these fasteners must also be torqued, while hot.

If the die is an Autoflex™ model, turn the autoflex heaters on to the neutral power level, typically 40% or 50%, and allow them to settle at this level. If the die has manual adjustments, ignore the last step. Gap the die lip using a soft brass shim of the desired thickness. The lip is usually set even across the width of the die, unless prior experience suggests that a different profile should be used. Knowledge of where the lips are set will help in getting to the correct gauge faster during a startup. If the desired setpoint is unknown, a rule of thumb would be to target the lip opening 20% above the produce thickness, down to about a 0.030-in. die opening. Thinner materials can usually be drawn down from this point. After the lips are set, the end seal gasket assembly can be installed and tightened.

When a die is being cleaned, it must also be at operating temperature. First all of the bolts must be loosened; then the power can be turned off. The die is split by pushing it apart, using jack bolts, and pushing it off of the dowel pins. The molten polymer is then scraped out, using either soft brass or aluminum tools. Then the flow channel is polished. If any residue is left in the channel, it can cause problems with streaks and lines in the material when the die is being run.

The die bolts should be cleaned, inspected for thread damage and signs of stretching, and recoated with an anti-seize compound. Dies should be stored fully assembled. If they are to be stored for long periods, the flow channel should be coated with a corrosion inhibitor. This is the best way to ensure that the flow channel does not get damaged.

When the die is being adjusted, a few rules apply. The low areas should be opened, and the high areas should be closed. The tendency to only close down the high areas results in a die opening that is too tight. The line pressure will then become high, and the die adjustment will begin to feel too sensitive. Each adjustment screw will move a standard flex lip over a wide area (usually 3-4 adjustment positions to either side will be affected) because the flex lip cross section tends to be stiff. The adjustment should be spread out also, with small movements made over larger areas. If the opposite is done, some of the adjustment bolts will begin to float, taking them out of control of the lip.

It is important to allow time for the die to settle after the adjustment, and also to measure the effect of the adjustment by gauge and to display the measurement. The rate of the extruder is not changed by the lip adjustment - the material flow is only being diverted because of the change of resistance to flow.

If the rate of the extruder is changed, one can expect the die to flow differently. If the resin is changed, this will also affect the flow of the die. Stiffer resins in general will exhibit heavy center flow; soft resins will flow more easily to the ends of a die.

The die will be affected if the melt being delivered is not stable. The extruder must be steady, not surging, and the material must be fully melted and well mixed. Temperature variations in the material entering the die will affect the distribution.

SAFETY

There are many hazards that are associated with the handling and operation of a flat die. In many cases, the die manufacturer also has to rely on the upstream and downstream equipment manufacturers to ensure safety because, in most cases, the die is incorporated into a system designed by another party.

The high temperatures required to process thermoplastic create burn hazards. Die body insulation can help reduce this risk, as will protective clothing. Dies are typically run close to rolls, where the material is nipped or cast. This presents the operator with a potential nip hazard. Installing guards to limit access to the nip areas can eliminate this problem. Dies are the outlet of a high-pressure extruder and during startups, or if upstream problems occur, molten polymer can spurt out of the die opening. Train extrusion personnel to be aware that this can happen, and have them stay away from the front of the die. The high torque required to keep the die from leaking during operation can present a hazard to the personnel that have to assemble and disassemble them. The right tools can save time and reduce the potential for injury that can occur if a tool fails. Dies are usually very heavy, and care must be taken when they are being lifted. The lifting lugs provided with the die should always be used. The die is designed so the lifting lugs support it in a balanced position.

It is important to follow the manufacturer's safety recommendations because the die and the associated extrusion line equipment is the first step to long and successful operation.

Autoflex™ and Fast-Gap™ are registered trademarks of Extrusion Dies, Inc.

CHAPTER 8

A Survey of Current Bead Reduction Die Technology

Thomas Bezigian, PLC Technologies Consulting Co.

ABSTRACT

Bead reduction dies have been in use for many years. Early designs provided bead reduction via a single deckle rod in close proximity to the final die land. The next innovation to this technology came from Japan in the way of a rigid-lipped die with no adjusting bolts, very short die land length, and several internal deckles to drastically reduce edge bead thickness, primarily for extrusion coating applications. The benefits of this die design extend far beyond reduced polymer consumption and waste, also providing reduced substrate waste, reduced downtime, and a significant overall cost reduction.

Several new bead reduction die designs have emerged in the current era of technology for both extrusion coating and cast film applications. These dies vary dramatically from each other, and offer the converter a wide variety of choices and options when either purchasing new equipment or upgrading existing equipment. This chapter will review the bead reduction technologies currently available, as well as the pros and cons associated with each.

DISCUSSION

It is interesting to note that the first die used for extrusion coating (ca. 1950) had no adjusting bolts and had very short (3.2 mm) die lands. This is possible because the flow manifold and internal polymer flow geometry was designed specifically for a particular polymer at a particular melt temperature and flow rate. This extruder and die remained in use until the late 1980s and made only one commercial product.

It is not particularly difficult to design a die to uniformly extrude a certain thickness of plastic if all the flow criteria are established and never change. The challenge comes when flow rate, temperature, and material changes are introduced to the manufacturing process. The die, with internal flow channels having a fixed geometry, cannot be increased or decreased to accommodate the non-Newtonian nature of polymer melts. With this in mind, die designers gradually lengthened die lands and introduced adjustment bolts so that uniform coating thickness could be achieved over a wide range of materials, extruder outputs, and melt temperatures.

Egan Machinery then developed a bead reduction technology (in the early 1970s) based on a single deckle rod. If properly designed, this worked quite well. Realizing the benefits to the industry, Egan released the use of this patent to the industry at no charge, thus paving the way for future technology.

This remained the status quo for more than twenty years until a Japanese company modified the original extrusion coating die design (from the H.P. Smith Company, Chicago) and refined it so that adjustments to coating weight could be made via temperature adjustments across the width of the die body. This die, which has become known in the United States as the Jyohoku die after its manufacturer, utilized a rigid sliding lip to eliminate leakage, short die lands to improve processing, and three internal deckles to reduce edge bead, in some cases reportedly eliminating edge bead totally.

Though these dies were very expensive and required special training on the part of the operator to reduce edge bead and adjust coating weight, they gained prominence because of the benefits derived from their use. In fact, to date they are the most used bead reduction die around the world with an installed base of more than 1000 units.

The benefits derived from bead reduction dies far surpass the simple reduction in raw material usage via reduced edge bead. Table 1 sites the quantifiable benefits derived from bead reduction dies for a typical extrusion operation. Note that cost and waste figures are not necessarily representative of actual industry values, but rather serve as an example of the potential savings the edge bead reduction dies offer.

Table 1. Savings analysis using a bead reduction die (BRD) in an extrusion coating operation

Structure Cost Analysis					
Poly/Paper/Poly	Weight (g/m^2)	Cost ($/kg)	Cost ($/lb)	Cost ($/M m^2)	Cost ($/M ft^2)
16 g/m^2 (10 lb/rm) LDPE	16	$1.10	$0.50	$17.65	$1.64
80 g/m^2 (50 lb/rm) Bl. Kr	80	$1.54	$0.70	$123.54	$11.48
16 g/m^2 (10 lb/rm) LDPE	16	$1.10	$0.50	$17.65	$1.64
Totals	**112**			**$158.83**	**$14.76**

Waste Analysis		**Overcoat/trim**	
Width (1,500 mm finished)	Pre-BRD*(1)*	Post-BRD	Difference
1st pass Poly Trim	205	10	195
2nd pass poly trim	205	10	195
Paper trim	51	8	43

Width (1500 mm finished)	Area saved (m^2/h)	Weight saved (kg/h)	Savings ($/h)	Available Hours/yr.*(2)*	Savings $/year
1st pass Poly Trim	597	10	$11	6989	$73,596
2nd pass poly trim	597	10	$11	6989	$73,596
Paper trim	132	11	$16	6989	$113,602
				Total	**$260,794**

Material Usage	**Metric Units**	**US Units**
Line Speed (m/min or ft/min)	305	1000
Width (m or in.)	1.52	60
(m or ft)^2/min	464	5000
(m or ft)^2/day	667,584	7,200,000
Downtime	20%	20%
M (m or ft)^2/year	194,935	2,102,400
Salable Material Cost/year	$30,961,841	$30,961,841
Waste (1% of total usage)	$309,618	**$309,618**

Downtime Analysis		
Machine hour rate ($/h)	$1000	
Annual web breaks due to edge bead	50	(rubber roll wraps, edge trim loss, etc.)
Downtime cost associated to edge bead	**$50,000**	

(1) Extruded width less finished width

(2) 24 h/day, 7 days/week 52 weeks per year, @ 80% uptime

Table 1. (continued) Savings analysis using a bead reduction die (BRD) in an extrusion coating operation

Component	Savings
Reduced plastic usage	$147,192
Reduced substrate usage	$113,602
Reduced waste	$309,618
Reduced downtime	$50,000
Total	**$620,412**

Project	Return
Cost	$250,000
Savings	$620,412
Pay back	5 months

Bead reduction theory

All bead reduction dies utilize the same basic components, namely a series of internal deckles that adjust the width of the extruded film as well as stretch the edges of the polymer melt. Figure 1 is a representation of the internal deckles used in Black Clawson's ROI™ die.

Fig. 1. Key components of an internally deckled bead reduction die (1)

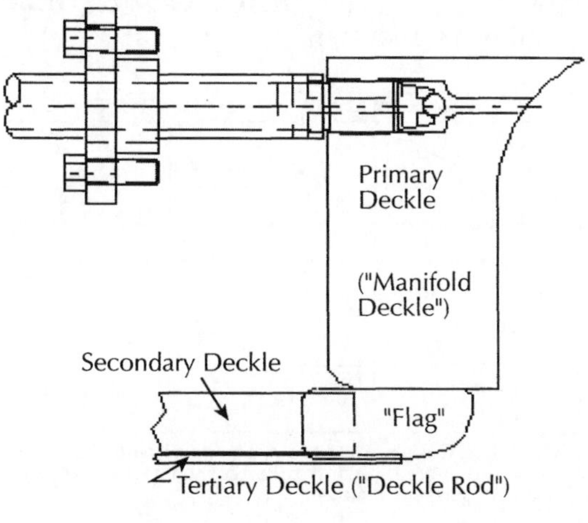

Bead reduction dies can have any number of internal deckles. The original bead reduction dies from about 25 years ago used only one deckle rod with a specially designed tip. Current internal deckle die technology employs three deckles. The relative position of these deckles determines the width and neck-in of the molten web. The greater the distance is between the main (upper) deckle and the final deckle rod, the more the molten polymer is stretched and thus the thinner the edge bead.

It is possible to over-stretch the edge bead, effectively making it too thin. This point is different for each polymer, die, temperature, and output condition; but essentially, if the stretching forces exceed the elastic strength of the polymer melt, then some instability will result. This could manifest itself in the form of edge weave or edge tear. The use of these dies typically requires superior knowledge on the part of the extruder operator and a certain amount of time to acquire those skills.

Figure 2 shows a 3D rendering of a cutaway view of a Cloeren EBR™ III die. The primary, or manifold, deckle resembles a coat hanger manifold, with the exception that it is adjustable to accommodate the width of the current production. The secondary deckle serves to stretch the edge in the pre-land area of the die. One might imagine this as "pre-stretching" so as to relax the polymer chains prior to the final stretching that is affected by the tertiary deckle, shown as a rod in this diagram.

Fig. 2. A 3-D rendering of an internally deckled bead reduction die showing the polymer flow paths (courtesy of The Cloeren Co.)

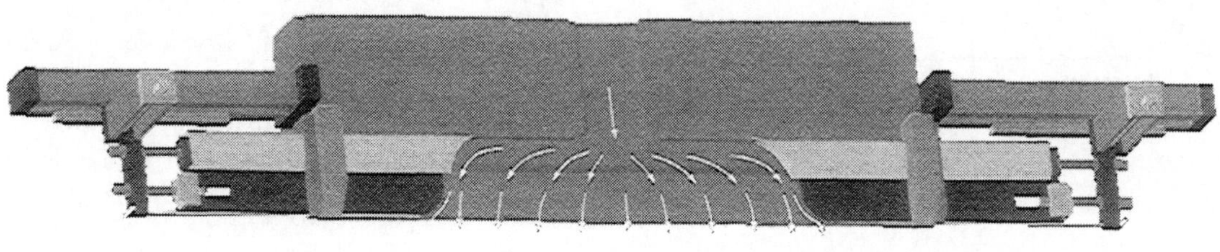

All three deckles must work together to provide the optimal edge bead reduction for each particular set of operating conditions. Care must be taken to balance processability and edge bead reduction. Once the relative position of the deckles is established, that relationship can be maintained over the entire operating width of the die for that particular resin within a relatively broad operating range.

The primary (leading) deckle can be seen to set the coating width. The secondary deckle, or "flag" is the principal tool by which the thickness of the edge bead is decreased. The tertiary deckle, or deckle rod, reduces the width of the edge bead.

Figure 3 shows various deckle positions that might be encountered in the course of typical operation of an internally deckled die. The top part of Fig. 3 would be considered the starting position, for example, after cleaning and assembly, and represents no edge bead reduction advantage. Figure 3(B) shows the bottom two deckles adjusted outward as would be the adjustment first step in edge bead reduction. Figure 3(C) shows the final deckle rod adjusted to its final position obtaining maximum edge bead reduction benefit.

Fig. 3. Various deckle positions *(1)*

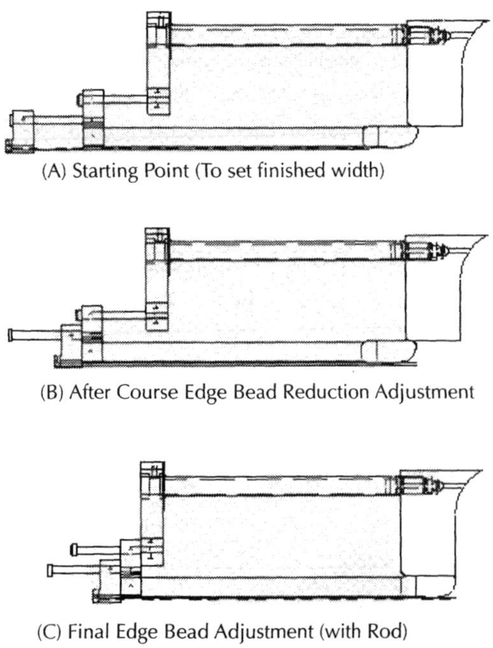

(A) Starting Point (To set finished width)

(B) After Course Edge Bead Reduction Adjustment

(C) Final Edge Bead Adjustment (with Rod)

New generation edge bead reduction dies

There was a time when all there was to ordering flat extrusion dies was to specify its width. As time evolved, more and more different polymer materials were being processed by the converter, and die design became more complicated. Coat hanger dies were introduced to help with profile control and flow patterns at the end of the die. The computer revolution came to extrusion coating in the early 1980s, at which time mass marketed automatic profile control was introduced. This in turn brought on increased usage of flex lip dies for coating and laminating due to their quick response time. Coextrusion, a number of resin "revolutions," and a return to short die land lengths gave further permutations on the already large number of die options. The evolution in bead reduction die technology continues today.

Fig 4. Basic patterns for edge bead deckle adjustment (courtesy Black Clawson, from the Jyohoku die manual)

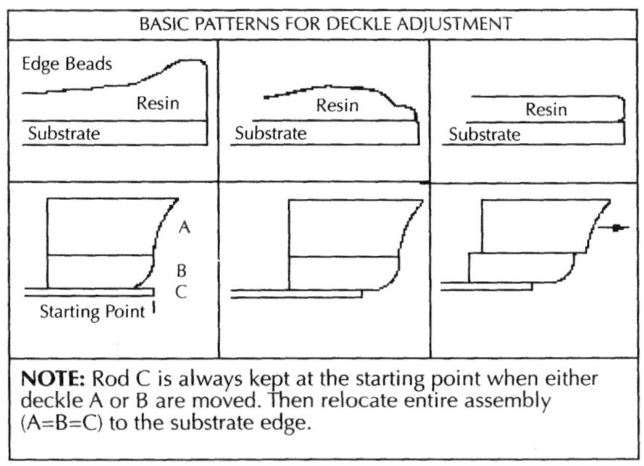

There are many differences in bead reduction die designs available today, as is shown in Table 2 *(2)*. Manufacturers do not actually claim any of these dies to be beadless, but rather only bead reduction dies. These are all a variation on a theme of the Jyohoku die, which itself is a fallback on basic thirty-five year-old die technology.

All current bead reduction dies for extrusion coating utilize adjusting internal deckles for width and bead reduction control. All but one current offering utilizes very short final die lands (~ 3 mm, or 0.125 in.). Short die lands have shown to reduce

Table 2. Comparison of important features of bead reduction dies (BRD)

Manufacturer	Black Clawson	Cloeren	Egan-Davis Standard	ER-WE-PA-Davis Standard	Extrusion Dies, Inc.	Jyohoku
Trade Name	ROI	EBR	Series 36-ID	Model 510A	CL-EPC	Jyohoku
Lip Type	Rigid, sliding	Flexible	Flexible	Flexible	Flexible	Rigid, sliding
Deckle Plug	Partial	Full	Full	Full	Full	Partial
Purge Ports	Yes	No	No	No	No	Yes
Deckle rod stabilization	Yes	Yes	Yes	N/A	No	No
External Deckles	No	No	No	Optional	No	No
Motorized deckles	Optional	Optional	Optional	Optional	Optional	Optional
Auto die ready	Yes	Yes	Yes	Yes	Yes	Yes
Die body metal type						
Standard	Stainless	Tool steel	Stainless	Tool steel	Tool steel	Tool steel
Optional metals	-	Stainless	-	Stainless	Stainless	Stainless
End gasket	No	Yes	Yes	Yes	Yes	No
Plating						
Standard plating	-	Nickel	-	Chrome	Chrome	Chrome
Optional plating	-	Chrome	-	Nickel	Nickel	No
Special die lip treatment	Yes	Yes	No	Yes	Yes	No
Heater type	Surface mounted	Cartridge	Surface mounted	Cartridge	Cartridge	Surface mounted
Retrofit to BRD capability	No	No	Yes	No	Yes	No
Removable Die Jaws	No	No	Yes	No	Yes	No
Final die land length	2.5 mm 0.100 in.	3.2 mm 0.125 in.	3.2 mm 0.125 in.	10 mm 0.394 in.	2.8 mm 0.110 in.	3.2 mm 0.125 in.

die streaks as well as pressure requirements of the die, which in turn reduces leakage.

The differences that today's bead reduction technology dies center around are:

1. Lip adjustment design (rigid sliding or flex lip)
2. Deckle design (full or partial plug)
3. How the issue of leakage is addressed.

Table 2 enumerates these design differences.

Black Clawson has marketed the Jyohoku beadless die for years in the United States. It has since begun making its own die, dubbed the ROITM die, which is essentially similar to the Jyohoku die with a number of reported improvements. Black Clawson has maintained the rigid lip and partial internal plug design, both rather unique in the industry, and has added a new direct acting thermal bolt system, drainage ports on the ends of the die, a rigid rod system to prevent deflection of the final deckle, and all die construction of stainless steel, which obviates the need for plating. Black Clawson's ROI™ (see Fig. 5) die joins the Jyohoku die as being the standard of the industry in terms of the shear number of installations worldwide.

Fig. 5. Plan view of end of EDI CL-EPC™ showing key components of its edge bead reduction die (courtesy of Extrusion Dies, Inc.)

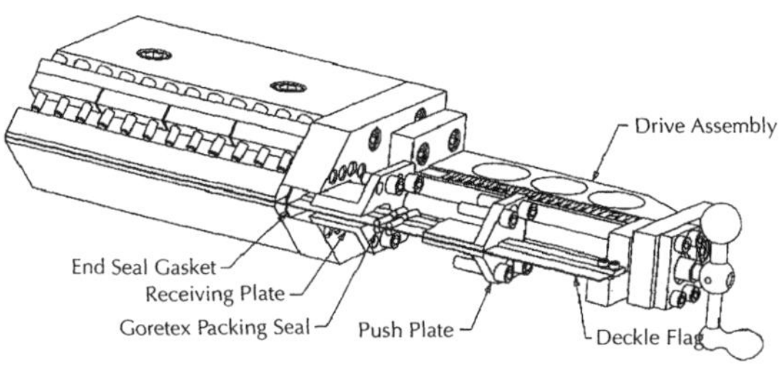

The Cloeren Company offers the EBR™ die, which features full deckle plugs and a flex lip design. As can be seen in Fig. 6, the EBR™ die offers a full length deckle plug, as well as a flex lip die design, cartridge heaters, and a deckle rod stabilization support. The use of these supports to maintain trueness of the deckle rod to the die land seal area is becoming understood and more widespread. Cloeren's dies are made of tool steel, and then chrome plated. Alternative die surface treatments are available depending on the particular application.

Fig. 6. Black Clawson ROI™ Die, featuring partial plug internal deckles, rigid die lip adjustment system, purge ports, and automatic die adjustment bolts (courtesy of Black Clawson Co.)

Egan-Davis Standard Series 36ID dies feature a stainless steel die using a flex lip design and full deckle plugs. Egan's secondary deckle is actually a two-piece tapered wedge mechanism that has a groove cut into its bottom half to receive and support the deckle rod to minimize deflection (Fig. 7). Similar to its non-bead reduction dies, it uses surface mounted heaters. Egan also has the ability to retrofit any of its earlier Series 36 dies to the Series 36ID technology.

Fig. 7. Cloeren EBR™ III Die, featuring full deckle plugs, flexible adjusting die lips, deckle rod supports, and automatic die bolts (courtesy The Cloeren Co.)

Fig. 8. CL-EPC™ die, featuring motorized internal deckle assemble with full deckle plugs, automatic die bolts, flexible adjusting die lips, and a custom fume exhaust hood assembly (courtesy of Extrusion Dies, Inc.)

ER-WE-PA-Davis Standard, with its Model 510A die, is the latest entry into the bead reduction dies arena (Fig. 8). This die is sufficiently different from the others on the market in that its die lands are about three times longer than the industry average and that its tertiary deckle is a "flag" instead of a "rod." Additionally, it offers an external deckle for applications that require it. Otherwise, it is a flex lip die made of chrome-plated tool steel utilizing full deckle plugs. The ER-WE-PA system reportedly eliminates edge bead on difficult-to-process resins such as LLDPE and metallocenes. Figure 9 shows a cross-section of the Model 500 die.

Extrusion Dies, Inc. offers the CL-EPC™ edge bead reduction dies. These dies are unique in that they offer the ALTTM (advanced lip technology) option, which involves removable inserts that can be made out of corrosion resistant, hard, or high lubricity materials and which can be easily and inexpensively changed if damaged. Figure 10 is a plan view of the CL-EPC™ die showing key components. Figure 11 is a photograph of the same die. EDI claims it can retrofit any flat extrusion die to its EPC (edge profile control) technology at a fraction of the time and cost of a new die. Figure 12 shows an extrusion coating die before and after retrofitting by EDI.

Fig. 9. Cross-section of an Egan-Davis Standard Series 36 ID Die, showing the hinged, flex lip system with a full deckle system, and a tapered wedge flag *(3)*

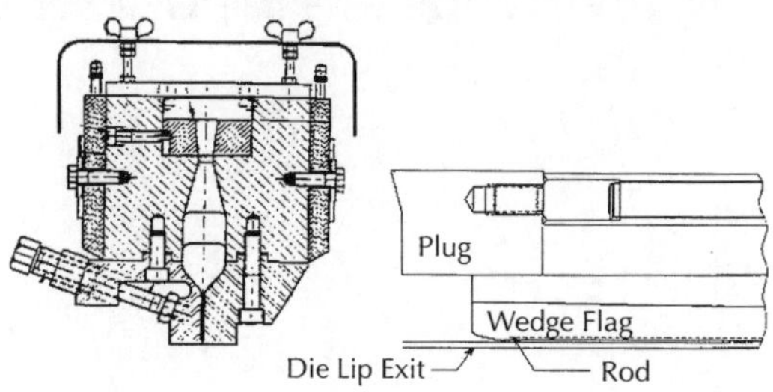

Fig. 10. ER-WE-PA Davis Standard Model 500A deckle assembly *(2)*

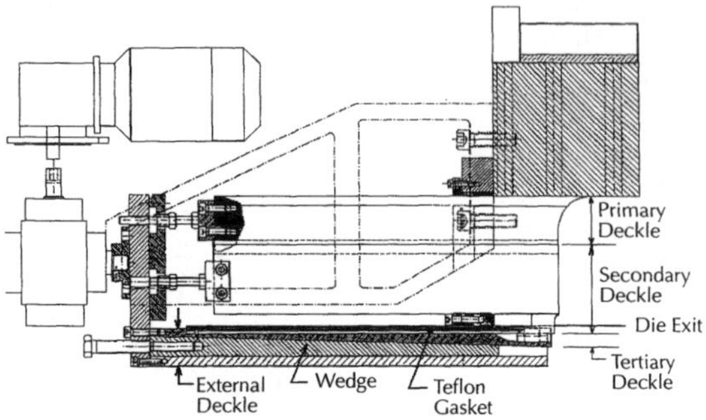

Fig. 11. ER-WE-PA Davis Standard Model 510A bead reduction die (courtesy ER-WE-PA Davis Standard GmbH)

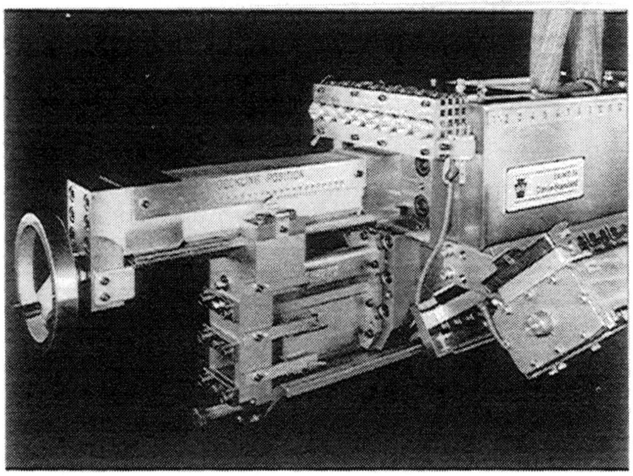

Fig. 12. EDI Retrofit Technology. The top photo shows a standard, non-edge bead reduction die before conversion, and the bottom photo shows the same die after conversion (courtesy Extrusion Dies, Inc.)

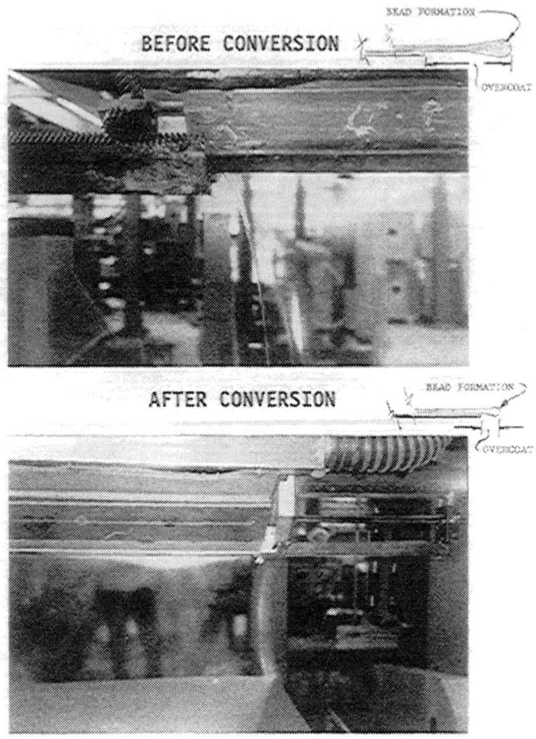

Although Jyohoku dies are still available in the United States, for all practical purposes, they are no longer sold here. Black Clawson, Jyohoku's U.S. distributor now manufacturers a similar die, but at a fraction of the cost. There are many other options available, all at similar savings, giving the converter many choices.

Nonetheless, the Jyohoku die remains the standard of the bead reduction die industry. The more than 1700 installations worldwide prove the effectiveness of its rigid sliding lip and partial plug technology.

The main concern shared by all these dies is leakage around the deckle assembly, both in the primary polymer flow direction and perpendicular to it. Deflection of the die lip relative to the deckle rod, or the deckle rod relative to the die lip, will cause leakage in the primary polymer flow direction.

CONCLUSIONS

Bead reduction die technology has come of age and is now an established technology within the realm of extrusion coating technology. The benefits of use for such dies is enormous, both in terms of dollar savings to the user and to the environment in terms of reduced waste generation.

The primary concerns with these dies are leakage around the deckle, in both the flow direction and perpendicular to the flow direction, as well as wear and tear of the die lip area. Leakage of polymer around the secondary and tertiary deckles and the subsequent degradation can damage the die lip area when the deckles are adjusted to a new slot opening. Over time, this wear can gall and chip chrome plating, which in turn will further promote damage to the die lip area.

The flex lip and rigid sliding lip bead reduction die technologies offer very good bead reduction control. Both can also be operated successfully in a 24-hour a day, around the clock operation, if adjusted and maintained properly. The rigid lip technology, however, has a natural resistance to leakage in the flow direction around the deckle. Flex lip designs can accomplish the same end result, but with suitable die modifications.

Serious consideration must be given to full plug versus partial plug deckle technology. Leakage around the deckle perpendicular to the primary polymer flow direction is inevitable due to the nature of the die design. This leakage flow must be taken into consideration in the design, operation, and maintenance of bead reduction dies. The partial plug deckles have the ability to scrape away polymer before it is carbonized while full deckles do not. Purge ports are a good option to eliminate polymer that has accumulated in the non-flow areas of the die.

Finally, retrofitting non-bead reduction dies to bead reduction status is a fast, cost-efficient, method of upgrading die technology to partially reflect the state-of-the-art technology. Similar benefits can be derived from a simple modification of a single deckle rod.

However, to fully enjoy the benefits of today's bead reduction die technology, an investment in a new die will be necessary.

ACKNOWLEDGEMENTS

The author extends his appreciation to the following individuals and companies without whose artwork and assistance it would not have been possible to write this paper.

- John Montalbano, Black Clawson Co., Fulton, New York
- Hart Benoit, The Cloeren Co., Orange, Texas
- Harry Lippert, Extrusion Dies, Inc., Chippewa Falls, Wisconsin
- Frank Orsini, Egan Davis Standard Corp., Somerville, New Jersey
- Dragan Djordjevic, ER-WE-PA Davis Standard, Erkrath, Germany.

REFERENCES

1. "Operation Manual for the Black Clawson ROI™ Die." Black Clawson, 1995.
2. Bezigian, T., Converting Mag. 14(7):36 (1996).
3. "Operating & Maintenance Instructions Egan Series 36 Coating Die," Egan Davis Standard Publication, June 1995, p. 31.

CHAPTER 9

Corona Treatment Enhanced Adhesion for Extrusion Coating

David A. Markgraf, Enercon Industries Corporation

ABSTRACT

Maintaining high-quality extrusion coating on paper, film, metalized film, and aluminum foil substrates at production level line speeds requires a method of enhancing substrate surface energy. Corona treatment can promote bond sites and increase surface energy without sacrificing the positive properties of the substrate or the extrudate. In extrusion coating applications, treater system design has undergone significant changes to accommodate lighter webs, higher line speed operation, and advanced substrates. The range of developments has included the requirement for handling conductive substrates, nips at the treater roll, using the treater roll as a "pull roll" and modifying the treater station design to reduce wrinkling and "backside" treatment. Several different techniques were attempted to accommodate conductive substrates and to reduce wrinkling and backside treatment; however, dual dielectric/coated roll technology has become by far the primary method used to overcome these problems. This chapter discusses the history of various technologies currently used and explains the state of the art in equipment, control parameters, and applications.

INTRODUCTION: HISTORY OF APPLICATION DEVELOPMENT

When polymer substrates were first introduced, the converting industry faced surface energy and adhesion problems caused by the inability of currently available coating materials to "bite-into," soak into, or stick to the chemically inert, nonporous surface that plastics presented. *(1, 2, 3)* Over time the resin producers, film manufacturers, coating material suppliers, and converting machinery builders modified their products to establish a workable system that allowed plastic substrates to be coated. A key part of this system was the ability of film producers to increase the surface energy of their products through corona treating. *(4, 5)* The process by which advancing technology in one area requires and, in some way, forces new developments in other areas has been the engine powering developments throughout the packaging and converting industries.

Prior to the early 1980s, treater station design for extrusion coating applications remained essentially unchanged since corona treating was first applied to coating lines. The station was enclosed, the roll was covered with a dielectric, and the electrode was a metal rod in some form. Differences between stations boiled down to a choice of three or four dielectric roll coverings and some relatively minor variations in metal electrodes. This was and is the basic, conventional covered-roll station. *(6)*

Metal Electrode Variations

The basic metal electrode was simply a straight metal bar (Fig. 1).

Fig. 1. Bar electrode

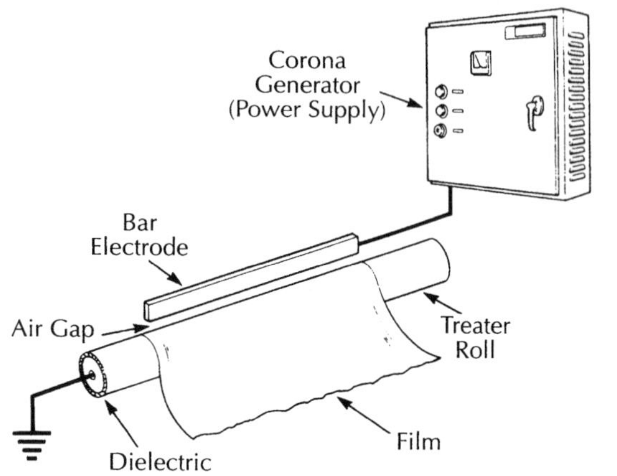

From the early 1960s, when converters would often build their own stations, the bar electrode with simple variations such as the use of "L-iron" or "T-shaped" iron or aluminum stock was the standard. In the mid- to late 1960s, companies began to build corona-treating systems for commercial sale and provide some measure of electrical and mechanical sophistication to the design. Specially designed electrodes were developed such as "shoe" electrodes, electrodes with patterned treatment surfaces, electrodes whose treatment surface was contoured to the treater roll circumference (Fig. 2), and segmented electrodes (Fig. 3) that allowed deckling of the electrode to vary the treat width or provide stripe treatment. Treater electrodes of these types are still manufactured and sold today.

Fig. 2. Concentric cut electrodes-side view

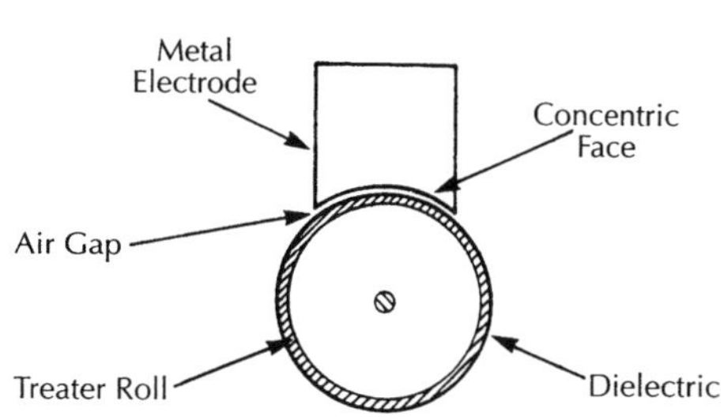

Fig. 3. Segmented metal electrode

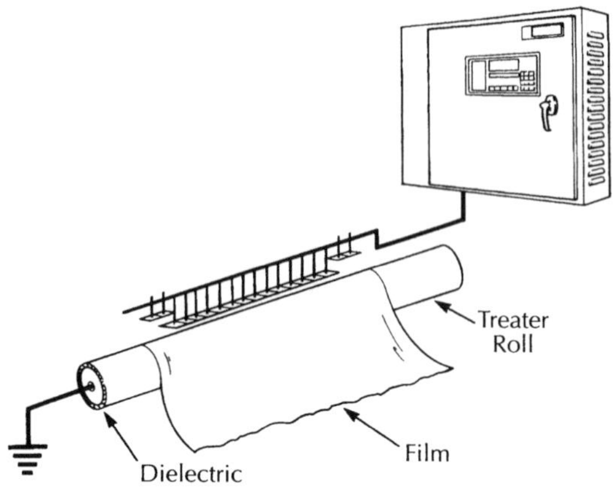

Many stations in use or sold today are still the basic, conventional covered-roll station with an enclosure built around the treating mechanism—the dielectric covered roll and metal electrode—and where the ozone exhaust intake is located at the perimeter of the station enclosure (Fig. 4).

Fig. 4. Air flow with covered-roll station

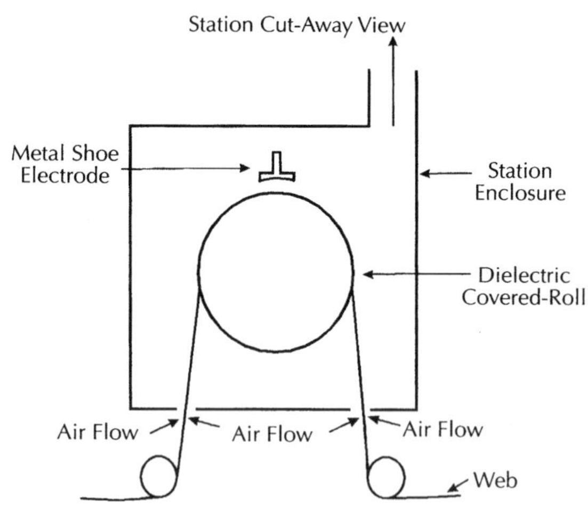

Recent Station and Electrode Advancements

In the early 1980s, a station was developed that revolutionized the structure of both the electrode and the station. *(7)* The first element of this revolutionary design was the first successful movement of the dielectric covering from the treater roll to the treater electrode (Fig. 5).

Fig. 5. Ceramic electrode/bare roll

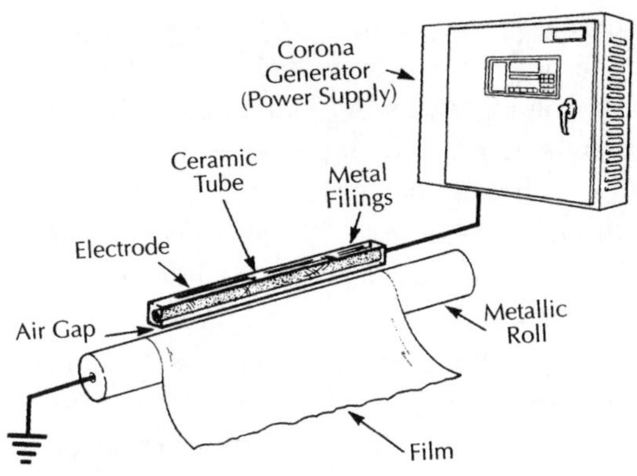

This change allowed the first use of ceramic as the dielectric medium since, at that time, covering a roll with ceramic was not technically feasible. It allowed the station to be of an "open" design and still provide operators with safety from electrical shock. And, it allowed the station to treat both conductive and nonconductive substrates. These benefits were well publicized and recognized at the time. Less recognized was the simultaneous development of an electrode assembly (Fig. 6) that moved the locus of the ozone exhaust from the perimeter of an enclosed station to the end of the electrode assembly.

Fig. 6. Ceramic electrode assembly-cross section

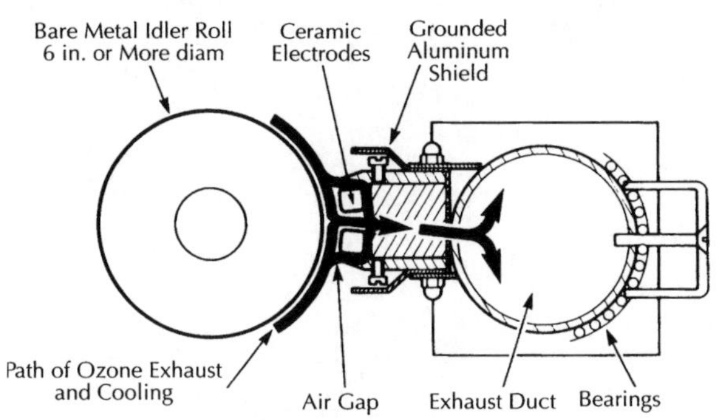

In the same way that the bare-roll provided the electrical freedom to eliminate the station enclosure, the change in the air flow pattern provided the mechanical freedom to eliminate the enclosure and enabled the electrode assembly to rotate to allow the passage of web splices and knots. The enclosure was no longer required to capture and contain the ozone for removal. Instead the ozone was captured and removed immediately in the air gap—at the point where it is produced. This change was not given its proper recognition at the time, yet it provided significant benefits since it greatly reduced the opportunity for ozone to seep into the work area and, ultimately, permitted the construction of covered-roll stations in an "open" design. Initially, these design advancements were only used in the now well known and accepted bare-roll station.

The bare-roll station found immediate application on extrusion coating lines because it offered:

1. No roll covering to fail; this was a source of frequent problems with covered-roll stations. In recent years, new ceramic coatings have mitigated this problem. However, ceramic coated rolls are very expensive.
2. Edge-to-edge treatment which the covered-roll could not achieve without possible roll damage.
3. Consistent treatment across the web; this is important as long as stripe (lane) treatment is not required.
4. A higher level of operator safety from electrical shock while providing easier threading because of the open design.
5. Ozone removal and electrode cooling that maintained the electrode at a stable temperature that eliminated warping and provided more consistent treatment while it ensured an ozone safe work area.
6. A smaller diameter treater roll for a smaller, more compact station.
7. A greater range of station mounting options than most covered-roll stations can provide.

However, the bare-roll station did have its limitations:

- It required slightly more power (kW) to treat difficult-to-treat substrates to a given dyne level.
- It did not easily allow stripe treatment. However, stripe (lane) treatment is seldom required on extrusion coating lines.

The universal-roll station is a recent development that overcomes the first limitation of the bare-roll station mentioned above. The universal-roll station (Fig. 7), which uses a special ceramic coating on the roll as well as the ceramic electrode, matches or betters the treating efficiency and effectiveness of any other system.

Fig. 7. Universal-roll system: ceramic electrode/coated roll

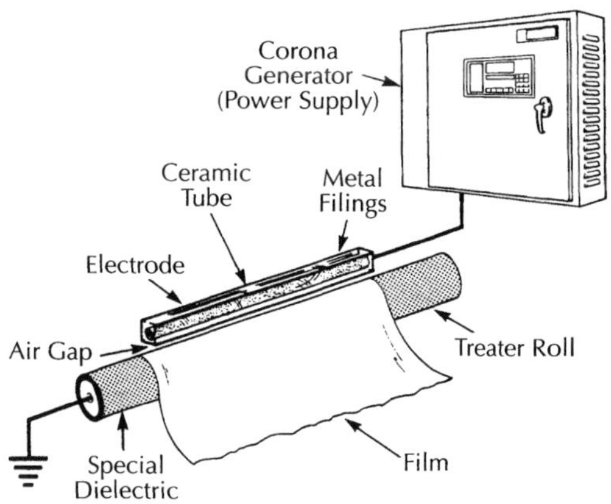

With one exception, the universal-roll station provides all the benefits of the original bare-roll station. That exception is the fact that the special coating can, but seldom does, require repair or replacement. However, unlike the covered-roll station the universal-roll station does not become unusable when the dielectric coating is pitted, cracked, or pin-holed. Since the electrode is also ceramic covered, the universal-roll station can continue to treat. The dual ceramic provides an additional benefit in that heat stress buildup during operation is shared; therefore, neither electrode nor roll coating is subject to the same level of heat stress as that found in covered-roll stations during normal operation. This, of course increases the long-term, operational reliability of the coated roll.

The universal-roll station provides an additional, highly significant benefit over the bare-roll and covered-roll stations. The universal-roll station greatly reduces wrinkling even of ultra light substrates and reduces undesirable "back-side" treatment, i.e., treatment on the side opposite that being treated.

The covered-roll station, with the arrival of consistently producible ceramic covered rolls in the late 1980s and early '90s, has seen a significant increase in long-term operational reliability. These stations, which in the past could only claim a higher level of treating efficiency and the ability to stripe (lane) treat when compared to bare-roll stations, have recently undergone improvements that allow features once only available on the bare-roll station. For example, metal electrodes are now housed in assemblies that permit cooling air to pass around the electrode and remove ozone from the air gap-the point at which it is generated. This structure and the extension of both the outer conductive shroud and the inner nonconductive shroud around the electrode assembly have allowed covered-roll, bare-metal electrodes to be housed in an open station while ensuring a high level of personnel safety from shock (Figs. 8 and 9).

Fig. 8. Stainless steel tube electrodes

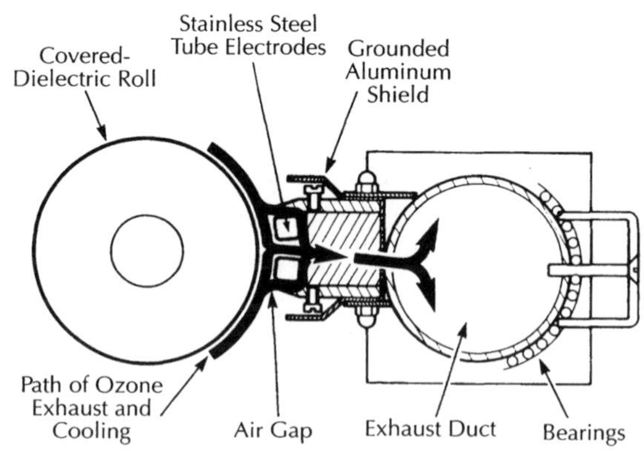

Fig. 9. Stainless steel segmented electrode

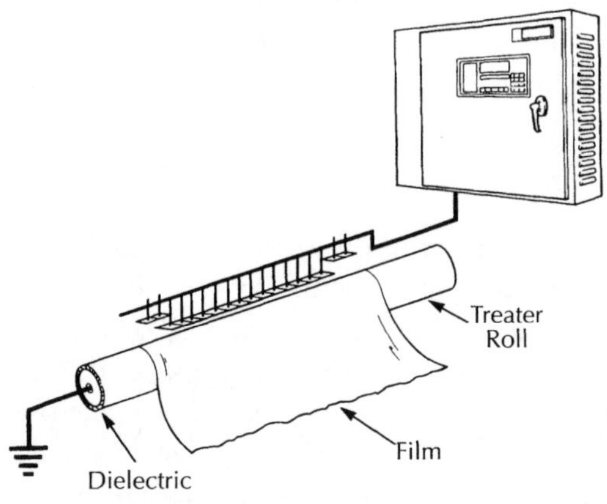

This electrode assembly configuration also provides an additional operational benefit in that it can be built to easily rotate out of the way to allow web splices and knots to pass without damage to the electrode or assembly. After the splice or knot has passed, the electrode assembly is counterweight balanced such that the assembly automatically rotates back into the treat position. Limit switches on each electrode assembly ensure that the power supply cannot be energized unless all electrode assemblies are in the treat position.

Driving and Nipping the Treater Roll

The requirement to drive or nip the treater roll is more often dictated by machine or substrate characteristics than by technical treatment demands. For example, treatment of substrates, especially ultra-lightweight substrates running under light tension, has been seriously complicated by the tendency for these substrates to wrinkle under the corona discharge. The wrinkling not only causes a problem with the quality of the wound roll, but also produces backside treatment that can be detrimental to the final product.

Wrinkling and the resultant back-side treatment can be reduced by increasing the amount of wrap and tension of the substrate on the treater roll. *(8)* For minor wrinkling problems, this might be sufficient. For more serious wrinkling, the addition of a nip roll on the treater roll where the film enters the station can eliminate or mitigate the problem. The addition of a nip on the treater roll will require that the treater roll be driven, which will of course require a drive shaft on the roll. This can require a complete rebuild or replacement of the station.

If there is any question that wrinkling might be a problem on your application, it is best to plan ahead and provide space for the inclusion of a nip roll, even if you are not having the nip roll installed immediately. Either of these steps would force the inclusion of a drive shaft on the treater roll.

As an aside, most (that is over 50%) of corona treatment stations do not require the treater roll to be driven; therefore, most stations are not provided with a drive shaft. Even light substrates will provide sufficient friction and tension to rotate the roll at line speed. Even if the roll does not rotate when the substrate is moving and treater power is off, the roll will rotate when treater power is applied. This phenomenon is referred to as "electrical pinning" and is used by companies manufacturing what are called electrostatic pinning bars. These devices are sometimes touted as electrical nips, but they are not nearly as effective as a mechanical nip roll. Their real benefits are that they are small and can be added easily after the machine has been installed, but they are not replacements for nip rolls.

Since the corona discharge provides electrical pinning of the film to the treater roll, one might ask: "Why doesn't it prevent wrinkling?" The answer is that, even though the corona discharge will provide pinning power, in many cases, that pinning is accompanied by static electrical forces that cause the film to wrinkle. This effect is enhanced by the fact that the roll is not rotating at a rate exactly matching the line speed. Driving the roll, therefore, will reduce the possibility of wrinkling.

Driving the treater roll and, in some cases, adding a nip roll also can be required when working with substrates that are highly extensible. These substrates require highly uniform and tightly controlled tensions throughout the process. Using the substrate to drive the treater roll can cause an unbalanced tension from one side of the station to the other and, at times, this imbalance is sufficient to stretch the substrate.

In some cases, the machine builder will request that the treater station be constructed as a "pull-roll." This requires that the treater roll be nipped and driven and provides the machine with a tension control point.

DESIGN CRITERIA

The most significant design criteria involved in corona treatment equipment selection is sizing the system to meet the specific application requirements. Power supply size in kW is determined by the treatment power required by the most difficult to treat substrate to be coated. Treatment power is measured in watt density (watts/ft^2/min) which takes into consideration not only power level but also the length of time the power is applied. *(9)* If a given material is treated at a given watt density, its surface energy will be increased—that is the basic goal of surface treatment. However, both the ultimate surface energy achieved and the amount of increase are dependent on the material's starting surface energy. For example, applying a watt density of 1.0 watts/ft^2/min to PET at 41 dynes can raise it to 46 dynes, but applying that same watt density of 1.0 to PET at 44 dynes can raise it to only 48 dynes. Although the final dyne level is higher in the second instance, the incremental increase is less because of the starting point dyne level. Further, the variation of barefoot material response to corona treating is compounded by the impact of additive loading.

As you might expect, different materials react differently to corona treating. Some materials, such as most polyesters, accept treatment readily and will exhibit a rapid increase in surface energy under relatively low watt density levels, say 0.5-1.0. Other materials, such as polyethylene with low levels of slip additives, accept treatment less readily but will exhibit a significant increase in surface energy under moderate watt densities, say 1.4-1.8. *(10)* Finally, some materials, such as polypropylene with relatively high levels of slip additives, are difficult to treat and might exhibit only moderate increases in surface energy under relatively high watt densities, say 2.0-2.5. *(11)*

Treatment levels, i.e., watt density levels, vary considerably from application to application. The materials and watt densities selected in the prior paragraph were chosen because they are somewhat typical of extrusion coating applications. However, treatment levels can also vary significantly based on additive load and other substrate characteristics.

Treatment of substrates for extrusion coating also is affected by the type of treater station used. Although covered-roll stations can provide a higher level of treatment than bare-roll stations, covered-roll stations are seldom used on extrusion coating applications because *(1)* they cannot treat conductive substrates and conductive substrates are sometimes run on extrusion coating lines, and *(2)* the extrusion coating process requires a high degree of consistency of treatment both across the web on each run and from run-to-run. Covered-roll stations with segmented metal electrodes do not deliver that level of consistency. As a result, bare-roll or universal-roll stations are most often used on extrusion coating applications.

PROCESSING INFORMATION

An extrusion coating line is frequently provided with a secondary web input so that the process can both extrusion coat the primary web or laminate the primary web to the secondary web. See Fig. 10.

Fig. 10. Extrusion coating/laminating line

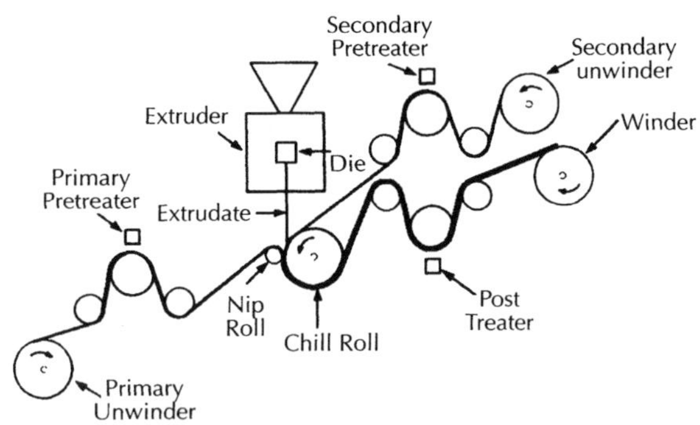

Primary Web Pretreatment

In Fig. 10, the primary pretreater—that is, the unit that treats the primary web just prior to the application of the coating—is shown as a corona treater. However, on a significant percentage of extrusion coating lines, especially those dedicated to paperboard coating, the primary pretreater can be a flame treater. (See Chapter 13 on flame and plasma surface treatment.)

On lighter weight, more flexible paper, film, or aluminum foil webs, electrical pretreatment using the corona process is the treatment of choice on the great majority of extrusion coating lines.

In actual applications, the corona treater is placed as close to the coater/chill roll as possible and the treated surface passes over as few intervening rolls as possible. This assures the highest possible surface energy available for high coating to substrate bonding. The primary pretreater will be set to apply a watt density of 2.0-3.0 watts/ft^2/min depending upon type of substrate, type of extrudate, and additives involved.

Extrudate Treatment

More and more extrusion coating lines are including an ozone web treating system to increase the extrudate's bonding characteristics. *(12)* By introducing ozone into the nip between the extrudate and the substrate (see Figs. 11 and 12), a substantial increase in bonding strength is achieved. Figure 13 shows the results of testing adhesion strength while using TAPPI Test Method T 539 and assigned values from 3 through 5 (good to excellent).

Fig. 11. Simple application schematic

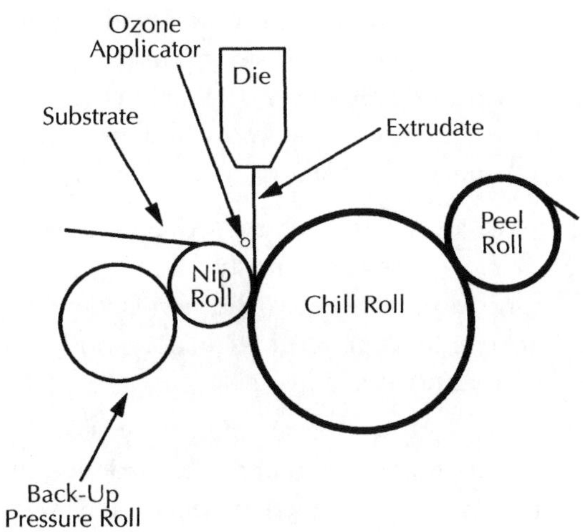

Fig. 12. Ozone applicator

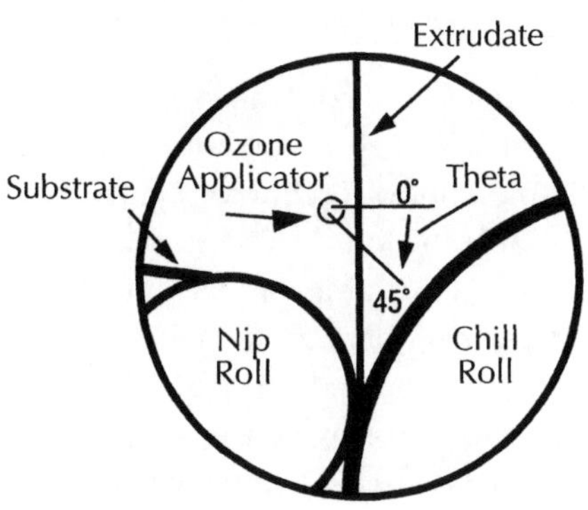

Fig. 13. Effect of corona treatment on substrate and ozone treatment of extrudate

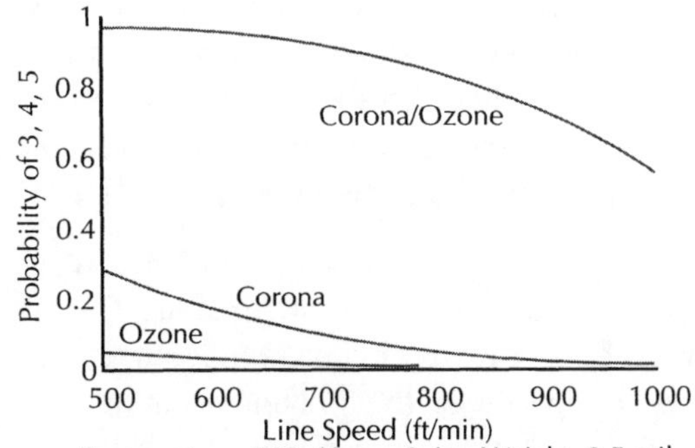

Temperature, 540; Airgap 6, in.; Weight, 0.5 mil

The increase in bonding strength provided by ozone treatment can frequently allow a reduction of extrudate temperature which decreases unpleasant taste and odors in the product, or the increase in bonding strength might permit higher line speeds that result in a significant improvement in productivity. *(13)*

Secondary Web Pretreatment

When the line is set up to perform the laminating process, the secondary web—which might be anything from a polymer film to a aluminum foil or metalized film—in most cases will also require surface treatment to enhance the laminate strength. In actual applications, the secondary treater is placed as close to the coater/chill roll as possible and the treated surface passes over as few intervening rolls as possible. This assures the highest possible surface energy available for high coating to substrate bonding to assure that delamination will not occur. The secondary pretreater will be set to apply a watt density of 1.5-3.0 watts/ft^2/min depending upon the type of substrate, type of extrudate, and additives involved.

Post Treater

When the line is set up for the coating process, the polymer coating is frequently treated to allow for later converting processes such as printing. In order to accept inks, the polymer surface energy must be raised to a level that counteracts the surface tension of the inks to be used. This means that a higher treatment level will be required if water-based inks are to be used, while a lower level of treatment will suffice if solvent-based inks are to be used.

Post treatment is also used on the exposed surface of the laminated products produced by the coating/laminating line. This treatment is also necessitated by surface energy requirements of later converting processes such as printing.

Post treatment watt densities generally fall in the range from 1.2 to 2.0 watts/ft^2/min. This range covers both the treatment requirements of the exposed surface of the polymer coating and the exposed surface of the laminate.

SAFETY

Being an electrically powered system, the corona treating system requires the usual Warnings and Cautions associated with high voltage, high frequency equipment. In addition, the OSHA lock-out/tag-out safety procedures are required when working on the system.

The corona treating station as a piece of "rotating machinery" also requires the usual safeguards outlined by OSHA and further defined by the Converting Equipment Manufacturers Association (CEMA).

Rather than repeat standard safety information that is readily available and understood in the industry, here we will only cover that safety data peculiar to corona treating and ozone treating systems. Since the corona treatment process produces ozone as a by-product and the ozone treating system produces ozone as a process enhancer, both OSHA and EPA regulations must be considered for the work area safety of employees and the meeting of "clean air" emissions requirements.

OSHA Requirements

OSHA limits worker exposure to ozone to a permissible exposure level (PEL) of 0.1 parts-per-million (ppm). Corona treating stations are constructed to contain and/or remove ozone before it contaminates the work area. Properly installed and maintained corona treatment systems provide ample protection for the operating crew from ozone.

While the ozone created by corona treating is not consumed in the process, ozone created by ozone treating systems for extrudate surface treatment is fully consumed in the coating/laminating process. For that reason, no special exhaust or containment devices are required for work area safety.

EPA Emissions Requirements

While ozone is not a hazardous emission as defined by the EPA, some state and local "clean air regulatory bodies" do require the reporting of ozone emissions under certain circumstances. Since 1990, the Federal Government through the EPA has placed the burden of "clean air" control and regulation on the states. Each state has submitted detailed plans defining how they will the meet the "clean air" regulation which includes ozone and particulate matter limits among other requirements. Further complicating the issue, certain areas which cross state lines are designated extreme, severe, or serious Ozone Nonattainment Areas. In these areas, both volatile organic compounds (VOCs), the precursors of ozone, and ozone emissions are more tightly monitored and, in some cases, regulated. Because ozone emission monitoring and regulation varies from state to state and even from area to area within states, no simple, definitive regulatory guidelines can be provided here. However, the following steps can be taken to ensure you know and are conforming to the specific regulations in your area:

1. Since most areas of the country do not limit ozone emissions directly, assume that ozone can be exhausted to the atmosphere in your area until your local regulatory body informs you otherwise.
2. Your state or local regulatory body might require that your company provide information for an environmental baseline, a New Source Review (NSR), or a Title V Permit. The forms required by these filings will alert you to any limitations on ozone emissions to the atmosphere in your area.
3. If you find that you must reduce your ozone emissions to zero because of Reasonably Available Control Technology (RACT), Maximum Achievable Control Technology (MACT), or Lowest Achievable Emission Rate (LAER) requirements, then contact your corona treating system supplier and request information on an ozone eliminator, which is a catalytic converter that converts ozone to molecular oxygen. *(14)*

CONCLUSION

The extrusion coating and laminating process almost always requires on-line pretreatment of the surface of substrates being coated or laminated and frequently requires post-treatment of the product to accommodate later converting processes. These treatment requirements are most often accomplished by corona treating systems that are available in several configurations to meet the widely varied technical and cost requirements found in extrusion coating and laminating applications. Despite increased line speeds, the move to ever lighter webs and lower web tensions and the introduction of co-extruded substrates and extrudates, corona treating technology has advanced to offer both technically effective and cost efficient solutions to surface treatment requirements for the substrate coating and laminating industry.

BIBLIOGRAPHY

Literature Cited

1. Spell, H.L., *et al.,* "Surface Analysis of Corona Treated Polyethylene: Bonding Printability Problems," TAPPI PRESS, Atlanta, 1978.
2. Blythe, A.R., *et al.,* POLYMER, 19(11): (1978).
3. Markgraf, M.P., "Corona Treatment: An Adhesion Promoter for UV/EB Converting," Rad Tech Report, Vol. 7, Number 5, (Sept/Oct 1993).
4. McLaughlin, T.F. Jr., The Surface Treatment of Polyolefins for Bonding to Inks and Adhesives, E.I. Du Pont de Nemours, 1962.
5. Kadash, M.M., et al., Plastics Engineering, 41(12): 45(1985).
6. Markgraf, D.A., "Corona Treatment: An Overview, 1986," 1986 Co-Extrusion Seminar Notes, TAPPI PRESS, Atlanta, p. 85.
7. Markgraf, D.A., "Evolution of Corona Treating Electrodes," 1983 Paper Synthetics Conference Proceedings, TAPPI PRESS, Atlanta, p. 255.
8. Markgraf, D.A., "Understanding Causes Can Deter Backside Treatment," Paper, Film & Foil Converter, Vol. 66, No. 9, p. 145,(Sept. 1992).
9. Gilbertson, T.J., "Mixing Water With Electrical Energy: Successful Printing With Water-Based Inks," 1991 Polymers, Laminations & Coatings Conference Proceedings, TAPPI PRESS, Atlanta, p. 321.
10. Spell, H.L., *et al.,* "Surface Analysis of Corona Treated PE," 1978 Conference Proceedings, TAPPI PRESS, Atlanta, p. 283.
11. Marra, J.V., "Surface Modification of Polypropylene Film," 1985 Polymers, Laminations and Coatings Conference Proceedings, TAPPI PRESS, Atlanta, p. 103.
12. Cheney, G., *et al.,* "Statistical Analysis of the Effects of Ozone on Adhesion in the Extrusion Coating Process," 1997 Polymers, Laminations and Coatings Conference Proceedings, TAPPI PRESS, Atlanta, p. 649.
13. Tietje, A., "Fifteen Years of Ozone Treatment in Extrusion Coating," 1987 Polymers, Laminations & Coatings Conference Proceedings, TAPPI PRESS, Atlanta, p. 221.
14. Markgraf, D.A., "Ozone Decomposition in Corona Treatment," 1985 Polymers, Laminations and Coatings Conference Proceedings, TAPPI PRESS, Atlanta, p. 227.

Resource Literature

Markgraf, D.A., "Sizing: the Critical Element for Effective Corona Treatment," 1989 Co-Extrusion Conference Proceedings, TAPPI PRESS, Atlanta, p. 17.

Thompson, K., "Surface Treatments for Co-Extruded Polymer Films/Coatings," 1989 Co-Extrusion Conference Proceedings, TAPPI PRESS, Atlanta, p. 11.

Markgraf, D.A., "Corona Treatment: An Overview," 1986 Co-Extrusion Conference Proceedings, TAPPI PRESS, Atlanta, p. 85.

Pochan, M., *et al.,* "XPS and Contact Angle Investigation of Corona Treatment," 1985 Polymers, Laminations and Coatings Conference Proceedings, TAPPI PRESS, Atlanta, p. 109.

Markgraf, D.A., "Physical and Surface Chemistry of Corona Discharge," 1985 Polymers, Laminations & Coatings Conference Proceedings, TAPPI PRESS, Atlanta, p. 107.

Schwab, F.C., *et al.*, "Effect of Resin Additives on Corona Treatment," 1985 Polymers, Laminations & Coatings Conference Proceedings, TAPPI PRESS, Atlanta, p. 95.

Markgraf, D.A., "Practical Aspects of Determining Level of Corona Treatment," 1985 Film Extrusion Conference Proceedings, TAPPI PRESS, Atlanta, p. 65.

Whiteside, D.L., "Corona Treating of Substrates," 1985 Polymers, Laminations & Coatings Conference Proceedings, TAPPI PRESS, Atlanta, p. 89.

Sherman, P.B., "Quartz, Ceramic or Rubber Dielectric in Corona Treatment," 1985 Polymers, Laminations & Coatings Conference Proceedings, TAPPI PRESS, Atlanta, p. 341.

Sherman, P.B., "Corona Treat-mechanical not Electrical Problem," 1985 Film Extrusion Conference Proceedings, TAPPI PRESS, Atlanta, p. 45.

Markgraf, D.A., "Practical Aspects of Determining Level of Corona Treatment," 1984 Polymers, Laminations and Coatings Conference Proceedings, TAPPI PRESS, Atlanta, p. 507.

Schuelke, G.W., "Modern Trends in Corona Treating," 1984 Polymers, Laminations & Coatings Conference Proceedings, TAPPI PRESS, Atlanta, p. 249.

Collins, W.M., "Classical Review of Corona Treatment," 1983 Co-Extrusion Conference Proceedings, TAPPI PRESS, Atlanta, p. 47.

Cramm, R.H., *et al.*, "Theory and Practice of Corona Treatment," 1981 Paper Synthetics Conference Proceedings, TAPPI PRESS, Atlanta, p. 1.

Collins, W.M., "Recent Technological Advances in Corona Treating," 1981 Paper Synthetics Conference Proceedings, TAPPI PRESS, Atlanta, p. 129.

Zito, W.F., "Electrostatic Treatment of Films Watts the Answer?" 1980 Paper Synthetics Conference Proceedings, TAPPI PRESS, Atlanta, p. 181.

Tietje, A., "Corona Treating Systems for Coater-Laminators," 1978 Conference Proceedings, TAPPI PRESS, Atlanta, p. 173.

Ristey, W.J., *et al.*, "Degradation and Surface Oxidation of PE," 1978 Conference Proceedings, TAPPI PRESS, Atlanta, p. 267.

Maynard, P.W., "Electrostatic Treating to Promote Adhesion," 1976 Paper Synthetics Conference Proceedings, TAPPI PRESS, Atlanta, p. 59.

CHAPTER 10

Ozonation of Polymer Melt for Improved Adhesion

Philip B. Sherman, Sherman Treaters Ltd.

The flexible packaging industry and, in particular, the formidable presence of the extrusion coating process in the manufacture of many sophisticated packages, has a vested interest in optimum adhesion being attained between laminates of various substrates. This ozonation update will pay particular attention to the improvements obtainable between extrudates and carrier webs on extrusion coating lines, by the use of ozone.

HISTORY

Ozone as a means of adhesion promotion has been successfully utilized in conjunction with extrusion coating for at least thirty years.

In the first decade of utilization of ozone in the context under review, it was confined to the manufacturing facilities of a worldwide manufacturer of liquid packaging cartons. That company pioneered the process and remains the largest user of ozonation equipment.

The dramatic increase in oil prices which took place in the mid- to late-1970s and the urgent need to reduce polymer coating weights concentrated minds to seek additional aids for adhesion and raised interest in ozonation industry-wide.

Subsequently, the use of ozone has spread to every continent, wherever extrusion laminators are used to produce flexible packaging through the use of innovative polymers, requiring enhanced levels of adhesion.

ADHESION PROBLEMS

Some materials presented to an extrudate, because of their porous nature, will readily accept polymer coatings, but most plastic films and many smooth or glossy grades of paper do not possess the necessary high surface tension characteristics to ensure satisfactory adhesion.

This requirement is critical in pursuit of profitable extrusion coating and converting operations and is as much a necessity as the correct melt temperature, die height, and nip pressures in obtaining optimum inter-ply bonding to meet customer specifications.

The achievement of adequate interface fusion can be particularly difficult where, in addition to adhesion, effective product resistance to aromatic liquids, fruit juices, and other penetrative products is required. Line speeds and productivity sometimes have to be compromised in order to eliminate unwanted features such as poor sealing, odor, or off-taste from the finished container.

Since ozone can alleviate some process problems, it is necessary to quantify the advantages gained by adding ozone to the extrusion coating line and then to look at evidence showing that the benefits can be substantiated.

The advantages include:

1. The opportunity to reduce coating weight
2. The opportunity to reduce melt temperature
3. Enhanced adhesion

4. Achievement of higher line speeds
5. The optimization of melt stability and neck-in
6. Improved heat seal characteristics
7. Reduced odor and off-taste.

EFFECTS OF REDUCED COATING WEIGHT

Polymer coating weights that today would be deemed high were always effective as a means to enhance adhesion because the polymer weight enabled the melt to be nipped into close contact with the fibrous surface, *(1)* as shown in Fig. 1.

Fig. 1. High coating weights have always been effective in enhancing adhesion because the polymer weight enables the weight to be nipped in close contact with the fibrous surface.

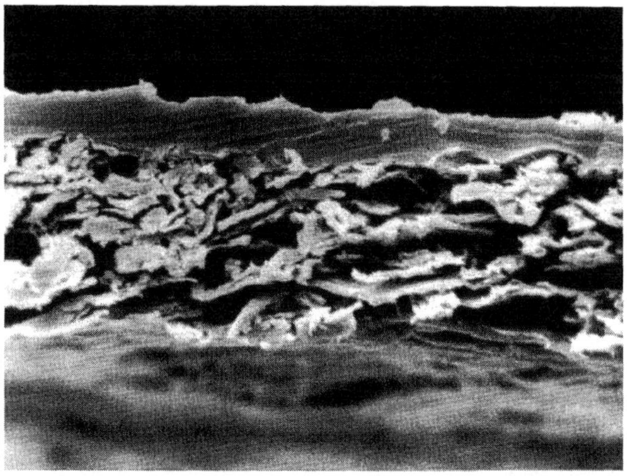

As the coating weights became the focus of attention, following dramatic rises in the price of oil, a desire to lower those coating weights developed in all sectors of the extrusion coating industry.

When coating weight is reduced, the actual area of contact between polymer and fibers is minimal because there is insufficient polymer to be nipped into the voids between projecting fibers (Fig. 2).*(1)*

Fig. 2. When coating weight is reduced, the actual area of contact between polymer and fibers is minimal.

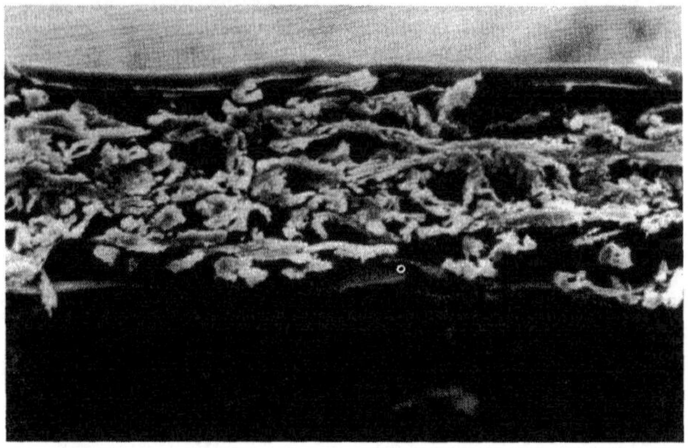

The bird's eye view (Fig. 3) emphasizes the reduced contact when only the gloss surface has actually contacted the nip.

The patches are unsupported polymer sagging over voids in the paper structure. *(1)*

Fig. 3. A bird's eye view emphasizes the reduced contact when only the gloss surface has actually contacted the nip.

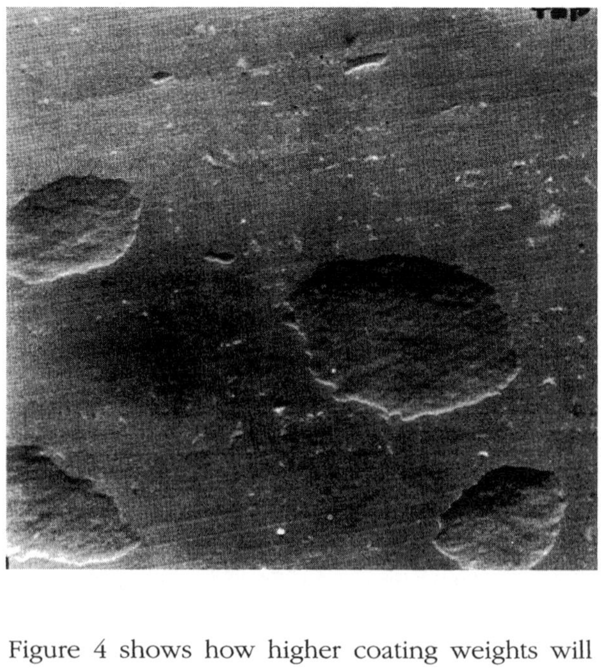

Figure 4 shows how higher coating weights will certainly increase adhesion values. *(1)*

Fig. 4. Adhesion of LDPE on aluminum foil (using corona at 150 m/min—deleting primer results)

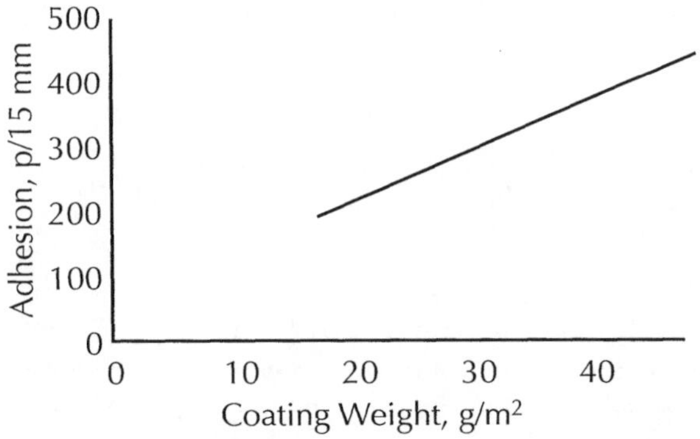

REDUCING MELT TEMPERATURE

It is a well known fact that, with a prepared "smooth" surface, a high temperature melt adheres better than a low temperature melt. With light coating weights, this has nothing to do with keying but relates to the level of surface oxidation of the melt, which is accelerated by heat. The oxidation yields highly polar surface groupings that form strong bonds with the substrate.

Figure 5 shows the contacting surface of an LDPE coating when peeled from the substrate. *(1)* The coating was applied at a high temperature to achieve a good surface bond to the prepared substrate, indicated by the delamination of the paper structure rather than clean peeling of the coating. This is due to high melt oxidation.

Fig. 5. High-temperature extrudate produces a good surface-to-surface bond.

Figure 6 is the same view of a peeled low temperature coating. Note the complete absence of retained paper fibers despite the obvious close contact of polymer and substrate shown by the fiber imprints. *(1)* The poor bond is due to lack of melt oxidation.

Fig. 6. Low-temperature coating process produces a poor bond.

Coatings such as LDPE are non-polar, but the melt will oxidize in the air gap, between the die and nip, to form oxidized, polar functional groups on the melt surface. This induced polarity leads to improved adhesion, particularly on nonporous substrates, so enhanced melt oxidization of the interface surface is essential for good adhesion.

OXIDATION OF LOW-TEMPERATURE MELT

Oxidation is a time- and temperature-dependent chemical reaction. *(1)* The time is set by the air gap and line speed, and the maximum temperature is limited by the product quality requirements such as odor, off-taste, and blocking. Figure 7 summarizes the effect of specific process conditions on polymer melt oxidation, together with the resultant limitations on the product quality and process optimization.

Fig. 7. Oxidation versus process parameters and limitations

Polymer Melt Oxidation	Process Parameter	Quality Limitations
High	Low line speed	Low production rate, high neck-in cooling of melt, odor/off-taste, poor sealing
High	High air gap	High neck-in, cooling of melt, odor/off-taste, melt instability
High	High melt temperature	Odor/off-taste, high neck-in, chill roll sticking, polymer degradation through chain scission and crosslinking.

The use of an ozonizer to achieve a rapid preferential oxidation of a low temperature melt enables a good bond to be formed without the problems associated with high-temperature oxidation. Ozone is produced under strictly controlled conditions and introduced on to the melt deep in the nip area and only on to the contacting surface, hence the term preferential oxidation.

Ozone is a strong oxidizing agent because the ozone molecule (O_3) readily decomposes into an oxygen molecule (O_2) and nascent oxygen (O). The nascent oxygen has a much shorter reaction initiation time and will accelerate melt oxidation. This effect is depicted in Fig. 8 where oxygen uptake of molten LDPE is plotted versus time.

Fig. 8. Oxygen uptake versus time

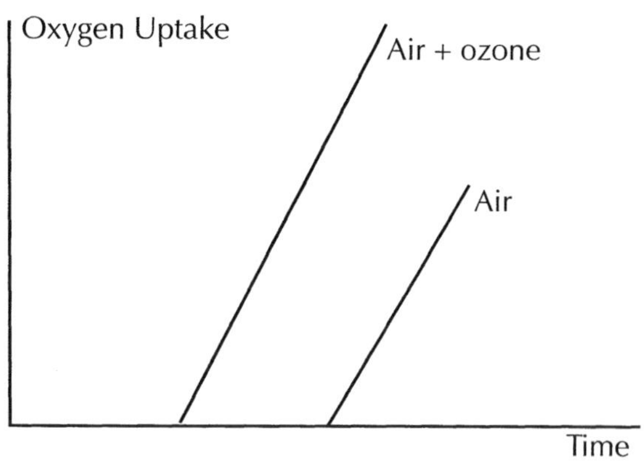

EFFECT OF MELT TEMPERATURE

Melt temperature determines both initiation time and reaction rate. At constant line speed, adhesion is not linearly proportional to temperature, but drops dramatically below a certain critical temperature. *(1)* This temperature is dependent on the dwell time of the melt in the air gap. Figure 9 shows that the adhesion is very poor without the ozonizer, below 310°C melt temperature. But, by applying ozone, the adhesion value reduces much less sharply.

Fig. 9. Adhesion as a function of melt temperature

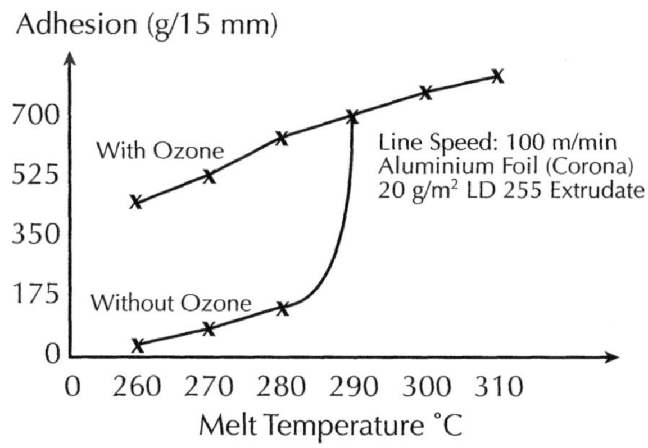

ADHESION AS A FUNCTION OF MELT TEMPERATURE

It can be seen that benefits are available in a reduced dwell time and, since the oxidation is preferential, this means that the noncontacting surface will have suffered less natural oxidation. This benefit is greatly emphasized by the reduction in temperature. *(1)*

Thus, heat sealability will be subject to less compromising conditions.

The same parameters, reduced extrusion temperatures, and shorter dwell times guarantee benefits related to reduced odor and reduced off-taste readings by minimizing contact with hot polymer fumes.

EFFECT OF LINE SPEED

The residence time in the air gap is directly proportional to line speed. With increasing line speed, the oxidation will decrease to a point where the residence time is shorter than the oxidation initiation time. At that point, oxidation is minimal and adhesion levels will drop. *(2)* With ozone, the initiation is accelerated, allowing a higher line speed before adhesion deteriorates.

IMPROVED ADHESION AT SPECIFIC PROCESSING CONDITIONS

After collating the results of trials run during the last ten years, we see benefits quantified. *(1, 2, 3)*

What happens to adhesion on matte alufoil when adhesion promoters are monitored individually and in unison?

Fig. 10. Comparison of adhesion (PE/Matte Alufoil at150 m/min)

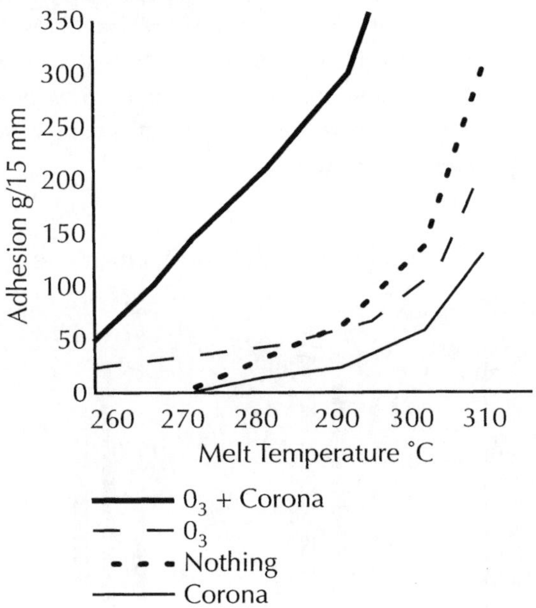

Figure 10 shows adhesion is achievable even without promoters, but only with the assistance of melt temperatures above 300°C.

By adding corona or ozone, adhesion levels are improved even as melt temperatures decrease, but a combination of both corona onto the substrate and ozone to the extrudate results in very good adhesion even at melt temperatures below 280°C.

WHAT HAPPENS TO LINE SPEED WITH ADDED OZONE?

From Fig. 11, it is evident that acceptable adhesion values with no ozone applied to the extrudate are obtainable only at speeds under 150 m/min.

Fig.11. Adhesion on aluminum foil as a function of line speed

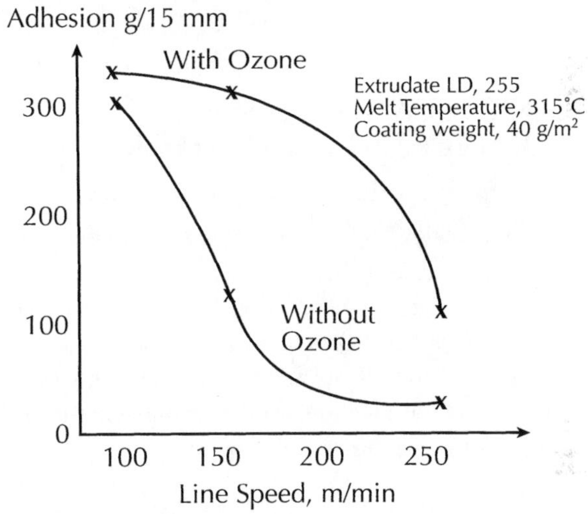

With ozone, adhesion is more than adequate at speeds 100 m/min higher. *(1)*

Figure 12 illustrates the adhesion obtainable on kraft paper at respectable line speeds only when ozone is utilized. *(2)*

At 350 m/min using ozone, adhesion is stated to be 100% and only a marginal decrease in this quality is exhibited at 400 m/min.

Without ozone, perfect adhesion declines rapidly as soon as line speed exceeds 210 m/min.

Fig. 12. Adhesion on kraft paper as function of line speed

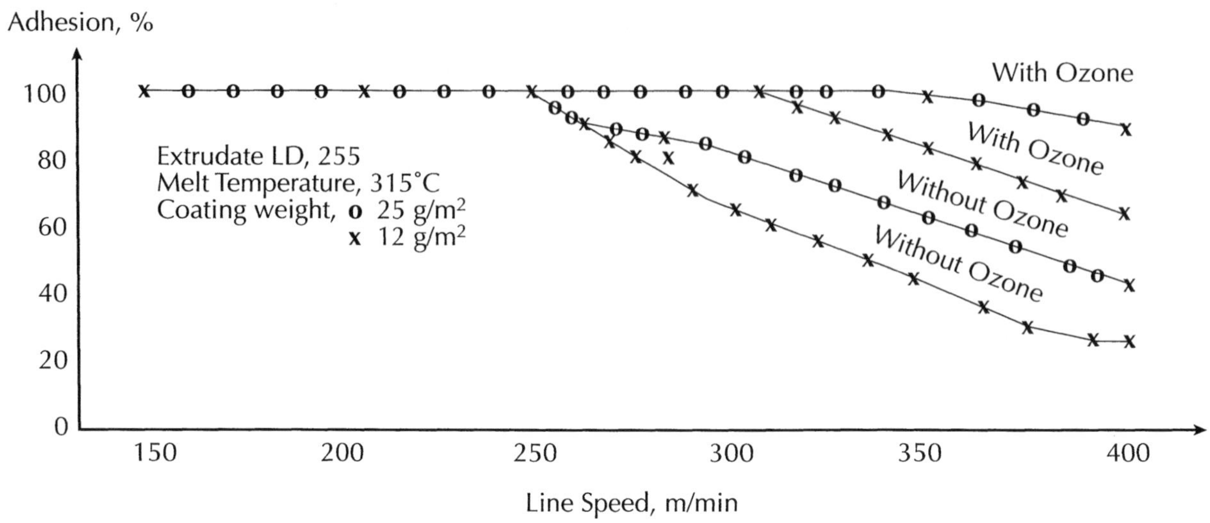

CORONA AND OZONE COMPLEMENT EACH OTHER

In every test conducted on pilot lines *(1, 2)* equipped with both corona and ozone generators, it is always established that the use of corona and ozone results in better bonds than is the case when only one form of promotion is utilized. These facts are illustrated in Fig. 13 using extrudate of LDPE at 320°C and coating weight at 25 g/m^2 and substrate of solid board.

Fig. 13. Good adhesion only when combining corona and ozone

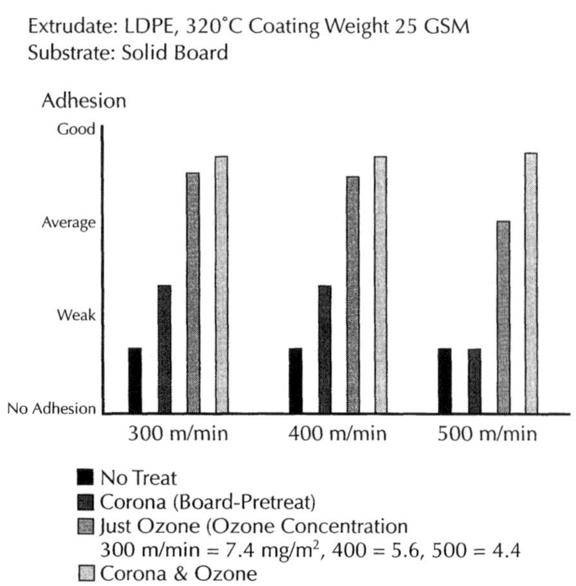

ADHESION WITH DIFFERENT SUBSTRATES

Continuing tests involving corona and or ozone on different substrates, such as polyester and nylon, show that optimum adhesion always results from the use of both promoters rather than one. On polyester, less ozone is required to achieve results equal to those obtainable on nylon, when more ozone is required (Fig. 14).

Fig. 14. Adhesion with different substrates

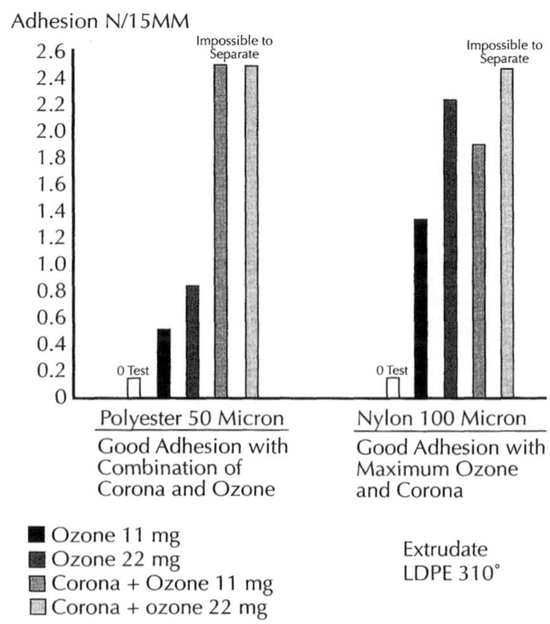

ADHESION WITH PRIMERS IN LINE

Some further laboratory and commercially run tests *(3)* reveal the benefit of ozone when aqueous primers are used on the substrate.

Figure 15 shows that the typical extrusion coating temperature for LDPE was reduced as shown.

Fig. 15. Reduction in the typical extrusion coating temperature for LDPE (courtesy of Mica Corp. USA)

Substrate—12μ Polyester/24 g/m² LDPE Extrudate

Temp.	Actions	Results
275°C	Corona, Primer Corona, Primer, Ozone	Easy peel Destructive bond
260°C	Ozone Corona, Ozone Corona, Primer, Ozone	Easy peel Peels with some effort Destructive bond

Figure 16 illustrates work based on a commercial process which typically is run at 91 mpm with a 305° - 315°/C melt. Ozone demonstrated the capability of reducing the melt temperature to 288°C. Addition of the primer to the paper while using ozone, then demonstrated the capability of reducing temperature further to 274°C and doubling line speed.

Fig. 16. Illustration of work based on a commercial process which typically runs at 91 mpm with a 305-315°C melt. (courtesy of Mica Corp. USA)

Substrate—Solid Bleached Sulfate Board/12g/m² LDPE—Extrudate

Speeds	Actions	Results
91 mpm	288°C, Corona, Ozone 274°C, Corona, Primer Ozone	Fiber Tear Fiber Tear
182 mpm	274°C, Corona, Primer 274°C, Corona, Primer Ozone	Peels Fiber Tear

THE EFFECT OF EXTRUSION TEMPERATURE ON ODOR AND OFF-TASTE

To measure these effects in the laboratory, *(1)* aluminum foil was coated with 50g/m² LDPE over a melt temperature range from 245°C to 322°C.

At a line speed of 50 m/min and an air gap of 15 cm, odor and off-taste were determined.

Odor

It should be noted that, in Fig. 17, the average score is calculated with 100 being equivalent to zero odor level.

Fig. 17. Odor vs. extrusion temperature of Escordene LD255

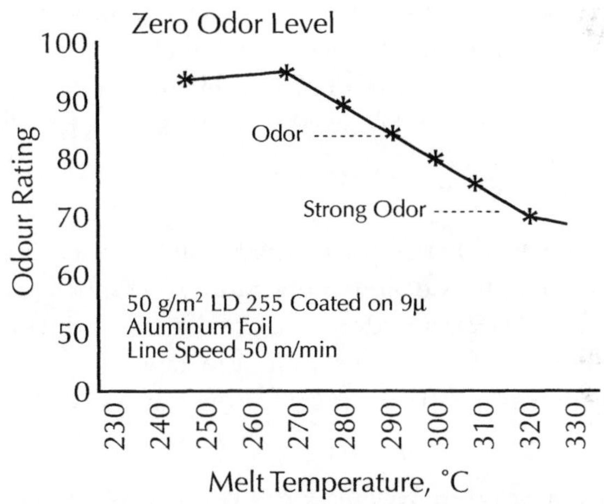

Figure 17 demonstrates the deterioration in odor with an increase in temperature. The critical temperature is in the region of 285-295°C where the odor level is beginning to move into the "strong" region. Odor level remains constant below 280°C.

Off-taste

The taint or off-taste imparted to water follows the same trends as illustrated in Fig. 18. *(1)*

Fig. 18. Off-taste vs. extrusion temperature of Escorene LD255

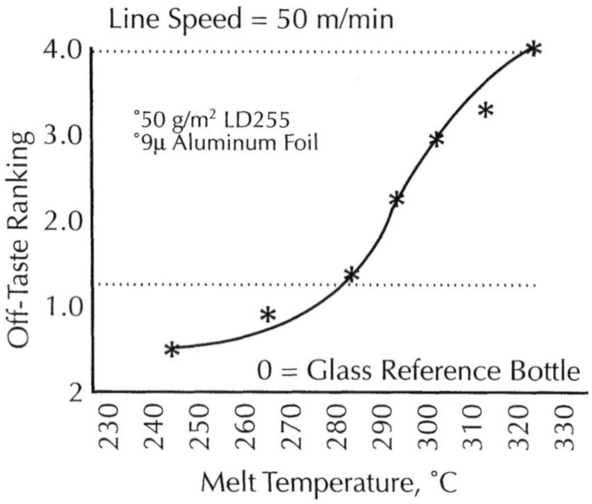

Fig. 19. Off-taste as a function of line speed

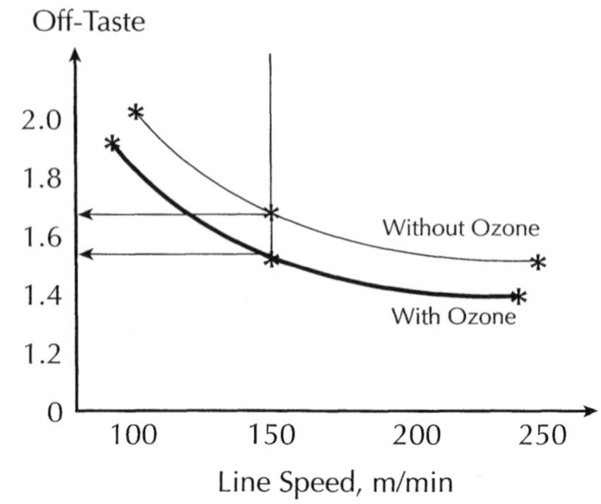

At melt temperatures below 280°C, off-taste rating begins to be indistinguishable from the glass reference. Water is extremely sensitive to tainting and, for this reason, deliberately selected for reference.

Conclusions

The interdependence of LDPE tainting of drinking water with extrusion temperature shows that a critical temperature zone is reached at around 280°C, above which odor and off-taste rapidly increases as oxidized by-products from the LDPE are formed.

OFF-TASTE REDUCED WHEN OZONE IS APPLIED *(1)*

Clearly, excessive melt oxidation will lead to the degradation of the polymer, resulting in odor or off-taste within the laminate. This feature is very critical when the finished product is to be used for food packaging. Figure 19 shows that equal levels of adhesion at low temperatures are achievable regardless of line speed, but only with the introduction of ozone to achieve preferential oxidation will off-taste and odor be significantly reduced.

EFFECT OF OZONE CONCENTRATION

Tests *(2)* at four different levels of ozone concentration and at three different extrudate temperatures for LDPE onto aluminum foil show clearly that more ozone significantly improves adhesion. A 13% reduction in extrudate temperature to 280°C results in only a 10% reduction in adhesion when 11 mg/m² of ozone is applied. A doubling of ozone to 22 mg/m² does not dramatically improve adhesion and confirms that 8-10 mg of ozone/m² is a satisfactory amount to work with (Fig. 20).

Fig. 20. Adhesion against ozone concentration

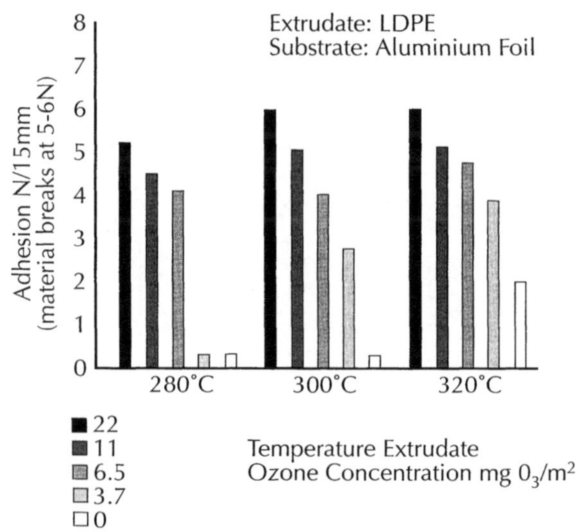

Ever-increasing concentrations do not bring commensurate benefits because the molten polymer takes time to absorb the ozone and the higher the concentration is, the longer will be the time required for absorption.

ON-LINE MEASUREMENT OF OZONE CONCENTRATION

Due to the high cost of an ozone analyzer to measure ozone concentration on-line, these units have not willingly been utilized on many extrusion coating lines, unless located in R&D centers at polymer manufacturers' laboratories.

Today, however, with the much greater emphasis placed on customer service and the importance attached to total quality control and repeatability of product specifications, the ozone analyzer becomes essential as the only reliable tool available to measure ozone concentration applied to the polymer melt.

The analyzer is located so as to measure the ozone gas stream as it enters the ozone applicator. The operating principle is by UV absorption with levels of ozone concentration in g/m³, or % weight or % volume. These values will be displayed on a high resolution flowmeter display.

Programmable alarm settings allow signals to be sent back to the ozone generator to trigger, as required, an increase or decrease in the amount of ozone produced in order to maintain the concentration at the level required for repeatability of product.

EFFECT OF OZONE APPLICATOR POSITION

Tests *(1)* were conducted to evaluate the effect of ozone applicator position, in relation to the melt. The applicator tested consisted of a tube, with 1-mm-diameter holes drilled 10 mm apart. With a 20-cm air gap and all machine parameters constant, the ozone applicator was used in different positions as shown schematically in Fig. 21.

In order to obtain significant differences, the following critical conditions were chosen:

Melt temperature: 295°C

Coating weight: 20 g/m²

Line speed: 150 m/min.

The distance between applicator and melt was 4 cm. Adhesion was expressed as a percentage of the highest value:

Fig. 21. Applicator positioning

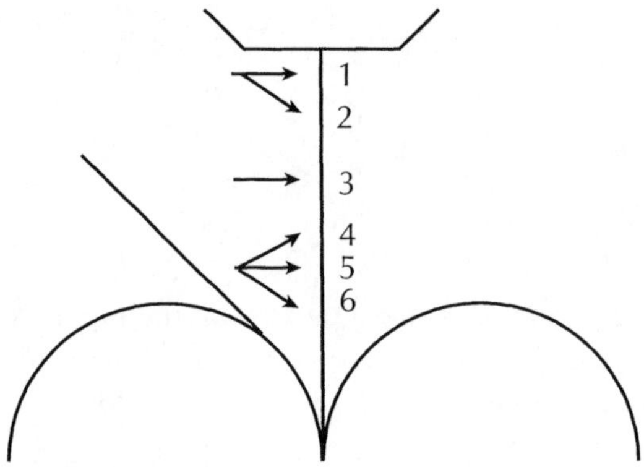

Position 1: Adhesion 66%
Position 2: Adhesion 62%
Position 3: Adhesion 61%
Position 4: Adhesion 88%
Position 5: Adhesion 100%
Position 6: Adhesion 99%

In Position 5, where the highest adhesion level is obtained, the test was repeated with applicator distances from the melt of 2, 4, and 6 cm respectively.

- Position 5: 2 cm, adhesion 97.5%
- Position 5: 4 cm, adhesion 100%
- Position 5: 6 cm, adhesion 90%.

The results show that ozone must be applied relatively low in the air gap. Since the drawdown ratio can be 100:1, the melt surface will increase in this ratio as it travels through the air gap. Therefore, the ozone applied lower in the nip will react over a bigger surface.

The low position of the applicator into the nip area and the deckling of the applicator width to suit the extrudate width ensures that ozone in the surrounding work area will never exceed the accepted maximum level of 0.1 ppm.

Fig. 22. An applicator fitted with deckle sleeves to ensure that ozone is retained within extrudate width

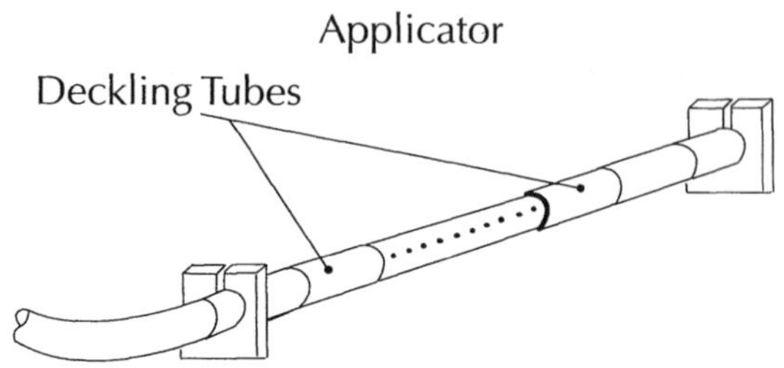

The applicator is designed to suit the nip area and is manufactured from a corrosion-resistant material and should meet the following criteria:

- Provide full width ozone application to the extrudate
- Permit deckling to a minimum extrudate width
- Ensure constant velocity of applied ozone
- Ensure no disturbance of extrudate polymer melt.

A typical layout of the equipment required to achieve a successful use of ozonation equipment, allied with corona treatment is shown in Fig. 23.

Fig. 23. A typical schematic of a setup for using ozone to improve adhesion on an extrusion coating line

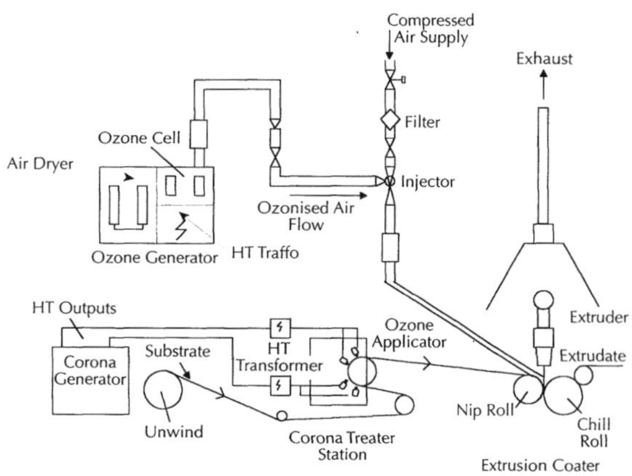

THE OZONIZER-OPERATION AND HARDWARE

Clean, oil-free, ambient air at 21°C is fed into the system at a pressure of 620 kPa and at the design flow rate. The gas passes through a prefilter, a coalescing filter, and enters the desiccant dryer where its dewpoint is reduced to at least -51°C. The dried gas moves to a postfilter and then through an ozone shell and a pressure regulator where its pressure is reduced to approximately 55 kPa.

A gas flowmeter measures the rate of clean dry gas going to the ozone shells. A pressure relief valve is located upstream of the gas flowmeter and is set at 83 kPa, the maximum working pressure.

The gas enters the ozone shell(s) where it is exposed to an electric discharge corona that produces ozone from the oxygen present in the gas. The ozonated gas exits the ozone shell(s), flows through a manually operated ozone outlet valve, and leaves the ozonizer via the ozone outlet port to the ozone applicator.

Fig. 24. Typical ozonizer flow diagram

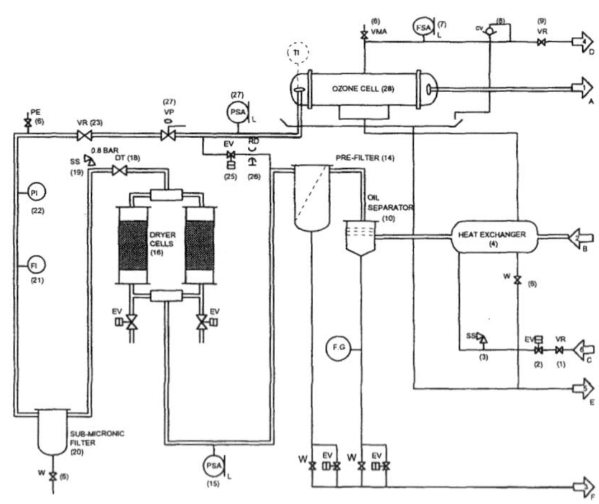

CONTROLS AND INDICATORS

The schematic diagram in Fig. 25 is representative of the controls and indicators which are available on all sizes of ozonizer from 10 g/h to 1000 g/h. Figure 26 shows a typical 750-g/h unit.

Fig. 25. A 90-g/h ozonizer schematic

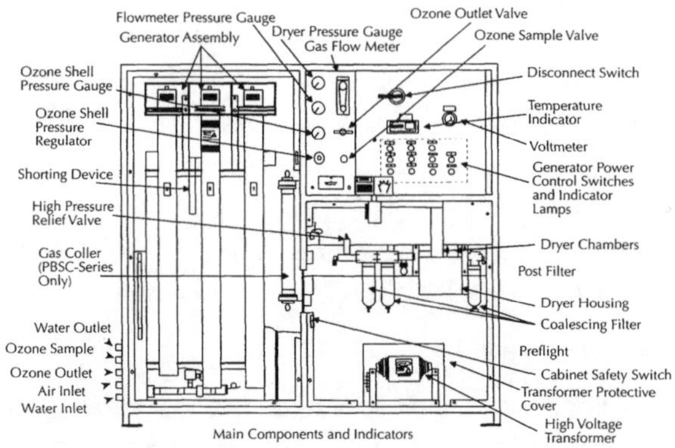

Ozone output can be automatically controlled from a personal computer either directly or via a remote control panel on the operator's side of the extrusion line.

Fig. 26. A 750-g/h ozonizer

TYPICAL AIR DRYING SYSTEM FOR AN OZONE GENERATOR

At any one time, one cylinder is drying the atmospheric air before it is passed to the ozone production chamber, while the second cylinder is being regenerated. Warm air is forced through the dryer, with the direction being opposite to the production flow to ensure that moisture in the alumina is not transferred to the ozonizer elements.

The timing mechanism for the heaters and solenoids (which route the air in the correct direction for either drying or regeneration) are controlled by a series of cam-operated micro switches. Thus, if the ozonizer is shut down at any time, the timing cycle will continue in the correct sequence, when production recommences. Although, if there has been a prolonged shutdown, one regeneration cycle must be completed before commencing ozone production.

Fig. 27. Typical air drying system for an ozone generator

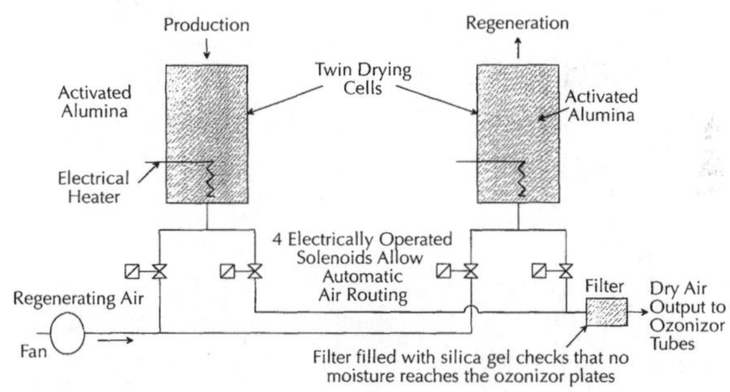

EFFECT OF HEAT ON OZONE OUTPUT

Figure 28 illustrates the mechanical setup in the ozone creation zone of the ozonizer. The result of the high voltage discharge in an enclosed chamber structure is the creation of heat, which will be absorbed into the cooling water, in the water jacket surrounding the electrodes.

Fig. 28. Corona discharge gap

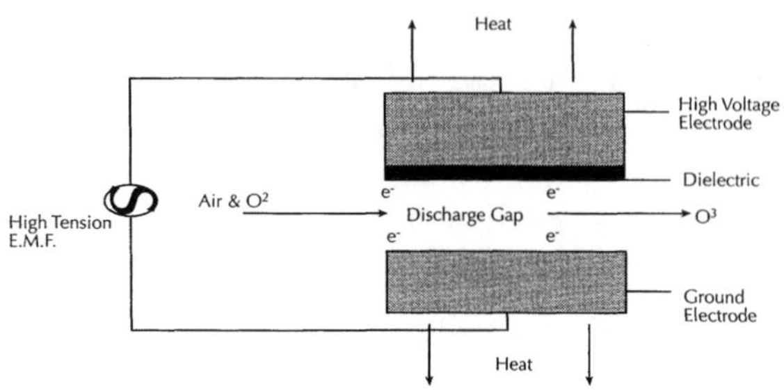

Figure 29 shows the dramatic effect on ozone output in % terms as cooling water increases in temperature. It is therefore imperative to ensure that the water supply to the ozonizer approaches the ozone creation zone at a temperature below 25°C, so that the effect of absorbing heat from the electrodes is nullified.

Fig. 29. Effect of cooling water inlet temperature on ozone output

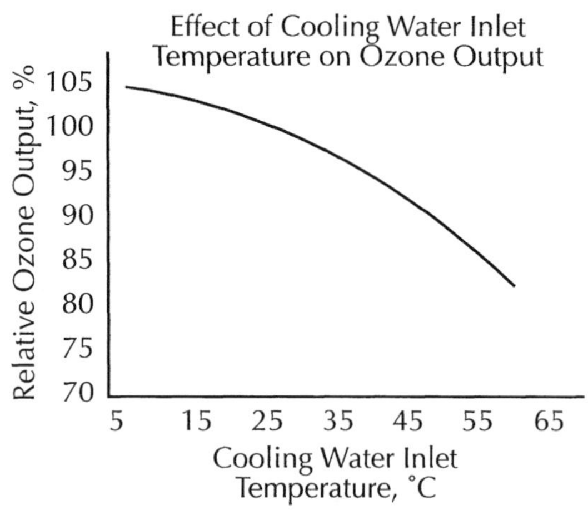

SAFETY FEATURES AND CONSIDERATIONS

1. Ozone destruction devices

During line stoppages and also at other unexpected times, it is necessary to be in a position to convert the ozone produced from the generator into a harmless medium.

Because ozone can be easily and cheaply converted to oxygen by passing it through a suitable catalyst, this requirement is easily met.

A compact ozone destruction unit can therefore be sited at an appropriate location in close proximity to the line so that, as and when required, manufactured ozone is immediately diverted to the destruct unit and rendered harmless.

2. Ozone generators

The ozone generator is a piece of high voltage equipment and is therefore fitted with:

- Door switches

  There are key-operated power supply door switches and electromagnetic interlocks to all doors and panels. It is supplied with custom designed HT grounding tools.

- Indicators

  Individual indicators show power on, ozone on, air fault, water fault, and electrical fault emergency stop so that all services and conditions are positively monitored.

- Sequential controls

  Water supply, dryer chambers, and ozone production shells each have dedicated flowmeters and regulators interlocked to offer cut-out protection in the event of an overload, underload, or pressure loss.

  Ozone production is within a sealed circuit and can only emerge at the applicator, with ozone flow being diverted from the applicator to the atmosphere or suitable ozone destruct unit during emergency stops and web breaks.

3. Three-way diversion valves

The safety of personnel demands that ozonation equipment will always fail safely. To this end, the introduction of a solenoid actuated three-way diversion valve into the ozone output pipework ensures that during any unpredicted situation; or for example, an emergency stop or a web break, as well as startup-the ozone flow can automatically, and independent of human assistance, be directed from the applicator-operator area to the atmosphere or to the aforementioned ozone destruction device.

4. Startup precautions

These include the use of dewpoint testing equipment and the obligatory purging sequence, both contributing to safe and trouble-free operation.

OZONE EFFECTIVENESS *(1)*

When melt temperatures need to be low to obtain benefit, line speeds will be enhanced and adhesion improved by ozone treatment (Fig. 30).

Fig. 30. Ozone effectiveness as a function of line speed and melt temperatures

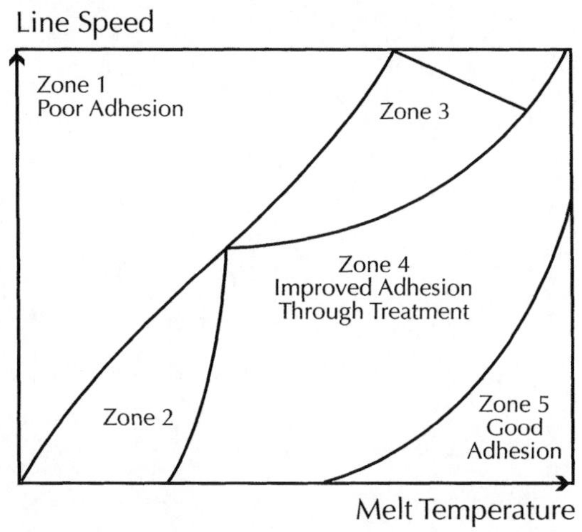

Zone 1 Poor adhesion with or without ozone treatment
Zone 2 Adhesion acceptable provided ozone treatment is applied
Zone 3 Adhesion acceptable provided ozone treatment is applied
Zone 4 Adhesion improved through ozone treatment
Zone 5 Good adhesion with or without ozone treatment

To summarize the points made hitherto, it becomes clear that even with existing adhesion promoters, if melt temperatures are too low or line speeds are too high, unacceptable adhesion results. If melt temperatures are very high, good adhesion is obtainable at most line speeds without ozone, but negative effects manifest themselves. Reductions in coating weights and melt temperatures result in reduced oxidation of the extrudate in the air gap. Only by adding ozone will economic advantages be exploited, allowing wider tolerances in melt temperatures, air gap, line speeds, melt stability, and neck-in.

Improved heat seal characteristics clearly are a benefit to be exploited, along with the resultant reduction in odor and off-taste.

Barrier polymers, such as EVAs, which must be extruded at lower temperatures represent polymers whose processing benefits from the addition of ozone.

REFERENCES

1. Michiels, D., Adhesion Improvement by Ozone Treatment, Essochem, International Extrusion Coating Symposium Antwerp 1982.
2. Johansson, C.J., Some Attempt to Improve Adhesion in Extrusion Coating by Ozone Treatment, Neste OY, Petrochemical Division, Finland.
3. Siegel, J. , & Singer, F., Chemical Primers for Extrusion Coating and Extrusion Lamination, Mica Corporation, USA, TAPPI Proceedings, International European Extrusion Coating, 1993.

CHAPTER II

Flame, Corona, Ozone–Do We Need all Pretreatments in Extrusion Coating?

Erkki Laiho and Tomi Ylänen, Borealis Polymers Oy

INTRODUCTION

To accomplish adhesion in extrusion coating, various pretreatments are needed. Interlocking as a result of the molten plastic penetrating pores in the surface of the substrate is not enough; a true adhesive bond is also needed. It is known that pretreating the substrate considerably improves adhesion. We likewise know that, for example, nonpolar LDPE oxidizes at high temperatures (300–330°C) in the extrusion-coating process. The most common variables that influence adhesion of LDPE are outlined in Fig. 1.

Fig. 1. Factors affecting the adhesion

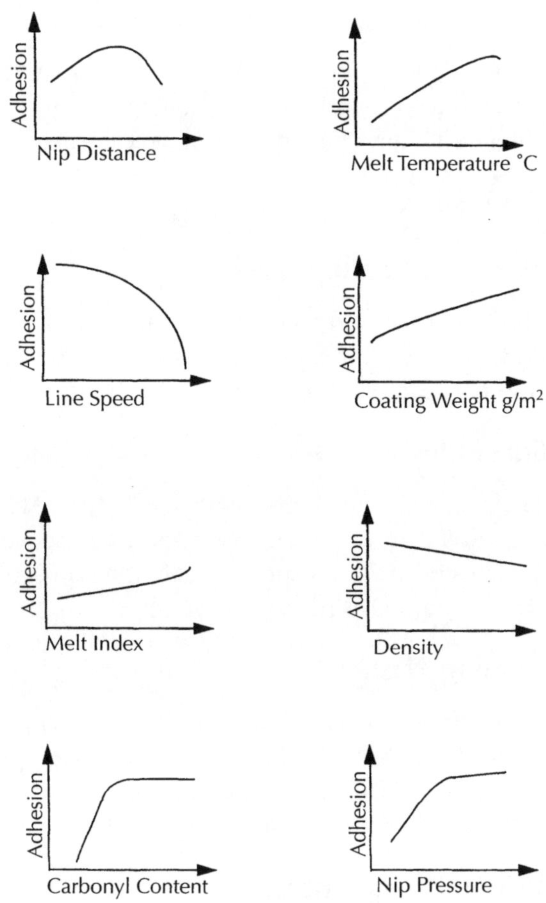

For molten plastic to oxidize, two things are needed: high temperature and contact with air. In extrusion coating, the latter is provided by means of the air gap, which is the nip distance, i.e., the distance between the die and the nip.

Excessive oxidization causes a risk with respect to organoleptic properties. The slower coating is done and the higher the temperature used, the greater is the danger for odor and taste problems to arise (Fig. 2). For that reason, extrusion coaters have always been fascinated by the idea of being able to use lower temperatures without special adhesion plastics or chemically priming the substrate.

One well-known way to accomplish this is to blow ozone (O_3) directly onto the molten plastic, thereby oxidizing it without a high temperature being necessary. Although the O_3 shower has been around as an idea for a fairly long time, it has been the subject of little research. The influence of ozone on different plastics together with various pretreatments (flame, corona) is one example of an area about which the literature contains very scant information.

Fig. 2. Odor and taste vs. extrusion temperature

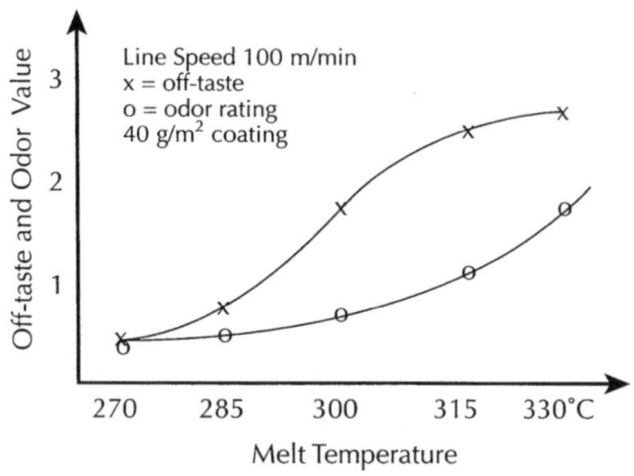

Nor have tests been carried out to determine odor and taste effects in long-term storage. As the length of time for which products are stored increases, oxidation products caused by O_3 could creep through a "coarse" LDPE layer.

In addition to the factors outlined above, our research also focused on the combined effects of different treatment methods (flame + corona).

TEST RUNS

Extrusion Coating Machine

The test runs were conducted at the Borealis Polymers Oy pilot extrusion coating line, the main configuration of which is described in Fig. 3.

Substrates

The substrates used are shown in Table 1. When the results are examined, the differences between the substrates with respect to their adhesion properties should also be kept in mind.

Coating Plastics

Four different plastics were used for coating. As set forth in Table 2. Unless otherwise stated, the results relate to LDPE, which is still overwhelmingly the main extrusion-coating plastic.

Corona Treatment Unit

The unit used was a Vetaphone-E-treater, type-designation ET5-12kW and output power 12,000 W. The electrode was a shoe-type.

Flame Treatment

An Aerogen flame-treatment device was used in the trial runs. The distance between the flame and the substrate was about 6 cm. The treatment efficiency changed automatically as a function of line speed (100% = 500 m/min). The maximum gas output of the flame-treatment device is 8.3 m^3/h. The gas used in the trials was propane.

Ozone Equipment

Table 3 shows the technical values of the Sherman ozone equipment used. To ensure that the equipment would achieve the best-possible treatment efficiency, a trial run was conducted to determine the effect that the distance of the ozone nozzle and the blowing angle would have on LDPE adhesion.

Figure 4 shows the position of the ozone die in the nip.

During the preliminary tests, the distance of the die was adjusted between 4 and 8 cm and the angle between straight and down 45°.

The adhesion and carbonyl-index measurements that we made indicated that the best distance is between 6 and 8 cm and that the most effective angle for blowing is at 45° to the nip. When the 0_3 is blown in this way, a greater surface area is affected and the reactive ozone manages to exert its influence on the molten plastic in a shorter time.

Ozone oxidizes the surface of the LDPE. A concentration meter is linked to the oxidizer. How concentration influences the carbonyl index (C=O -groups) was clearly seen during the preliminary tests, as Table 4 shows.

An IR examination is used to determine the carbonyl index, which is calculated from the IR spectra as the ratio between the reference peak (1984 cm^{-1}) and the oxidation peak (1720 cm^{-1}).

ANALYSIS AND DISCUSSIONS

In our efforts to determine the effects of various pretreatment systems, the following combinations of runs and symbols were employed (Table 5).

Fig. 3. Extrusion Coating and laminating pilot line

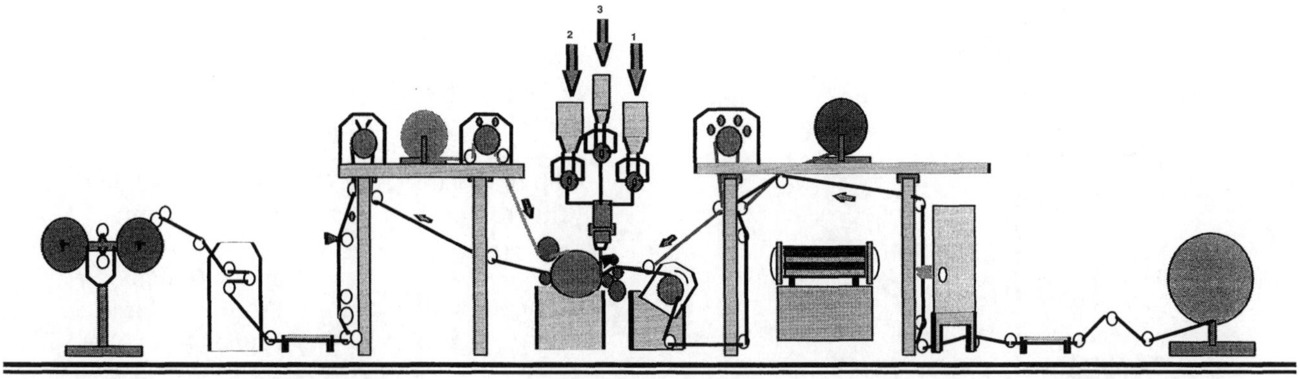

GENERAL
Substrates:
Paper and board, (30 - 400 g/m²)
Films ≥ 12 µm
Al-foil ≥ 9 µm,
LAM ≥ 7 µm
Slitters:
Station with a trim removal system

WINDERS
Main unwind:
max reel dia 1.3
core sizes 3" and 6"
Second unwind:
for Al-foil, max dia 0.8 m core 3"
Laminating unwind
max reel dia 0.8 m care 3"
Turret rewind, max reel dia 1.0 m, core 3" and 6"

LAMINATOR
Working width 600-830 mm Max. speed 1000 m/min.
Chill rolls:
1) matte, 2) glossy
3) "mirror pocket"
4) "chrome teflon" with carrier internal cooling system

EXTRUDERS
Extrudes 1 and 2:
4, 5"
L/D 24, compression 4.0
Output 450 kg/h (PE-LD)
Extruder 3:
2,5"
L/D 30
Output 170 kg/h (PE-LD)
Screw: "speciality polymer"
Die: EBR (Peter Cloeren die)
Feedblock: 5-layer possibility

TREATMENTS
Flame treatment for substrates (Airogen)
Corona pre- and post-treatments (Vetaphone)
Ozone treatment for melt (Sherman)

Table 1. Substrates in the test runs

Material	Grammage (g/m²)	Other Properties
UG-kraft	70	unbleached, light creped unglazed
MG-kraft	60	bleached, machine glazed
Simcastor	80	other side: starch coated
Al-laminate	7µm Al- foil + paper	A-grade Al-foil

Table 2. Properties of used coating resins

Resin	Manufacturer/type name	MFR	Density (kg/m³)
PE-LD	Borealis/LE7518	7.5 g/10 min (190°C, 2.16 kg)	918
PP (Copolymer)	Borealis/WG341C	23 g/10 min (230°C, 2.16 kg)	910
PP (Homopolymer)	Borealis/VC2031D	19 g/10 min (230°C, 2.16 kg)	903
E/MA MA-content 20%	Borealis/OE5620	6g/10 min (190°C, 2.16 kg)	946

Effect of Line Speed on Adhesion

To obtain a clearer picture of the differences between treatment systems, a lower-than-normal melt temperature (setting value 300°C) was selected for the coating process. In such a case, for example at 200 m/min, the molten plastic does not oxidize sufficiently and the significance of the O_3 shower is accentuated. Table 6 presents he parameters of this "basic run."

Fig. 4. Ozone shower die

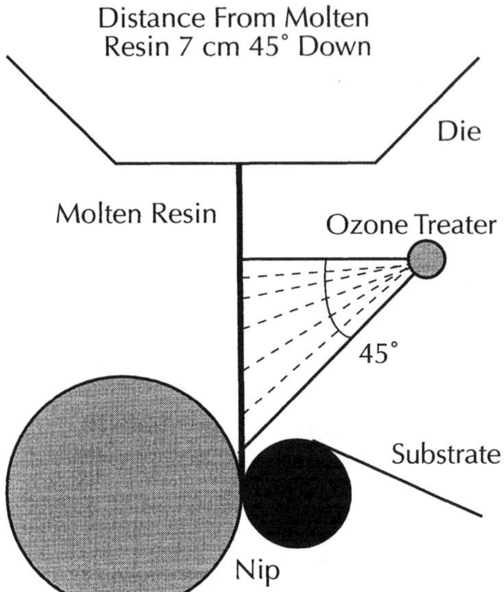

Table 3. Technical data of ozone treater

Ozone temperature and concentration during test runs	20–22 g/m^3
Generator	Output power 2.3 kW
Applicator's distance and angle from molten resin	7 cm, 45 down
Cooling water temperature and pressure	10°C, 0.4 bar

Table 4.Corbonyl index values for different concentrations of ozone

Power of treatment %	0	50	75	100
O_3-concentration g/m^3	-	~12	~15	~22
Carbonyl index	0.1	0.28	0.3	0.6

LE7518, 40 g/m^2, 300°C, 200 m/min

Table 5. Used pretreatment combinations and their symbols

Treatment	Meaning
1. NO	No treatment
2. Oz	Ozone treatment
3. Oz + C	Ozone + Corona treatment
4. Oz + F	Ozone + Flame treatment
5. Oz + F + C	Ozone + Flame + Corona treatment
6. F + C	Flame + Corona treatment
7. F	Flame treatment
8. C	Corona treatment

Table 6. The effect of line speed (test run program)

Resin and melt	PE-LD (LE7518)
Setting temp.	300°C
Substrate	MG-kraft 60 g/m^2
Coating weight	20 g/m^2
Nip distance	200 mm/+10 mm
Chill roll	Matte/+10 mm
Line speeds	100,200,300,400 and 500 m/min
Treatments	Precorona 11kW, Flame (Auto, O_2 +1.77), Ozone 2.3 kW, 20 g/m^3)

The adhesion values were measured using the well-known manual measurement method, with a scale of 0–5. An adhesion value of ≥3.5, at which more than 50% of the total surface of the paper substrate is adhering to the coating plastic, is considered sufficient in this test.

Figure 5 shows the results of the speed tests.

At this temperature, the (molten ~295°C) LDPE does not oxidize sufficiently, nor are enough chemical bonds created without pretreatment.

Ozone on its own improves the situation, but accomplishing a good result requires both oxidization of the molten plastic and pretreatment of the paper substrate. It is also worth noting that "over-treating" the paper substrate, i.e., flame + corona, does not improve adhesion compared with when only one form of treatment is used.

Thus when ozone is used, the result is improved by also pretreating the substrate with flame or corona. Double treatment of the substrate (F + C) does not improve the result.

Treatments with Different Melt Temperatures

This series of test runs was conducted in the same way as those described in the preceding paragraphs, but this time melt temperature was the variable. It must be taken into consideration that the temperatures in the results are setting values. The true melt temperature is about 5°C lower. The trials were made on UG-kraft paper at a line speed of 200 m/min.

Fig. 5. Effect of the pretreatment on adhesion with different line speeds

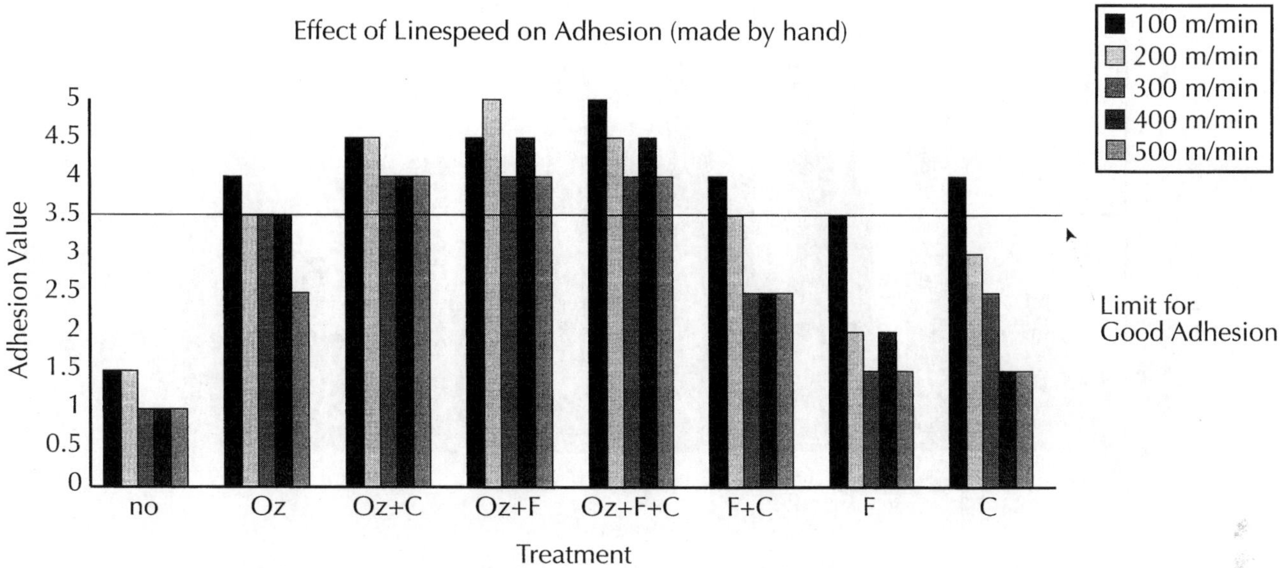

Substrate: MG-paper/Resin: PE-LD (300°C, 20g/m²/Pretreatments:Oz = ozone (20 g/m³)
C = corona (11kW) and F = flame (5m3/h)

Fig. 6. Carbonyl index of PE-LD as function of melt temperature

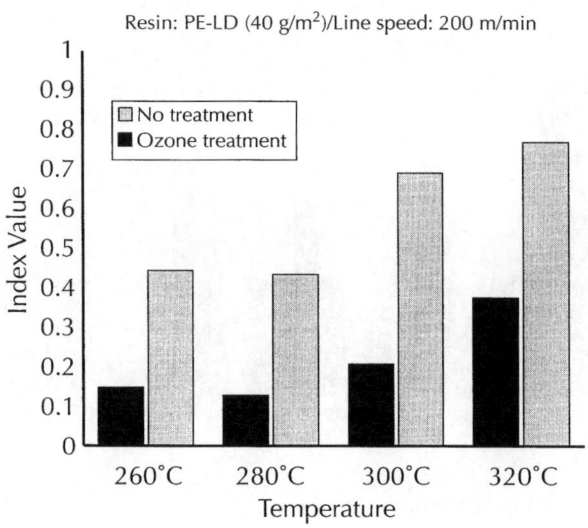

It was interesting to note how ozone affects the carbonyl indices most strongly at a higher temperature (Fig. 6). At a setting of 320°C and using only O_3, adequate adhesion is achieved.

At lower temperatures, by contrast, treatment of the substrate is needed. Also now, double treatment of the substrate (F + C) does not lead to a better result than either flame or corona on its own (Fig. 7).

Pretreatment with Different Plastics

In this test series, the substrate also was different. It was fine paper (80 g/m²) with one side mineral-coated. The plastics are listed in Tables 2 and 7. The optimum selected running temperatures are shown in Table 7 and in Fig. 8. The coating weight was 40 g/m² and the line speed 200 m/min.

As can be seen from Fig. 8, ozone + pretreatment of the substrate guarantee adequate adhesion with all types of plastics. The fairly good adhesion values achieved with a 20% polar copolymer, EMA, are also worth noting.

Treatment of Al foil

To facilitate handling, pretreated aluminum laminate was used in the trial runs. The aluminum side was coated with the LDPE grade LE7518 at a temperature of 300°C and a line speed of 200 m/min, although so-called A-grade aluminum foil was available. Flame treatment clearly altered the surface of the aluminum, and the change is reflected in the contact angle measurements (Fig. 9). Adhesion when only flame treatment is used also results in improvement (Fig. 10).

Fig. 7. Effect of pretreatments on adhesion with different melt temperatures of PE-LD

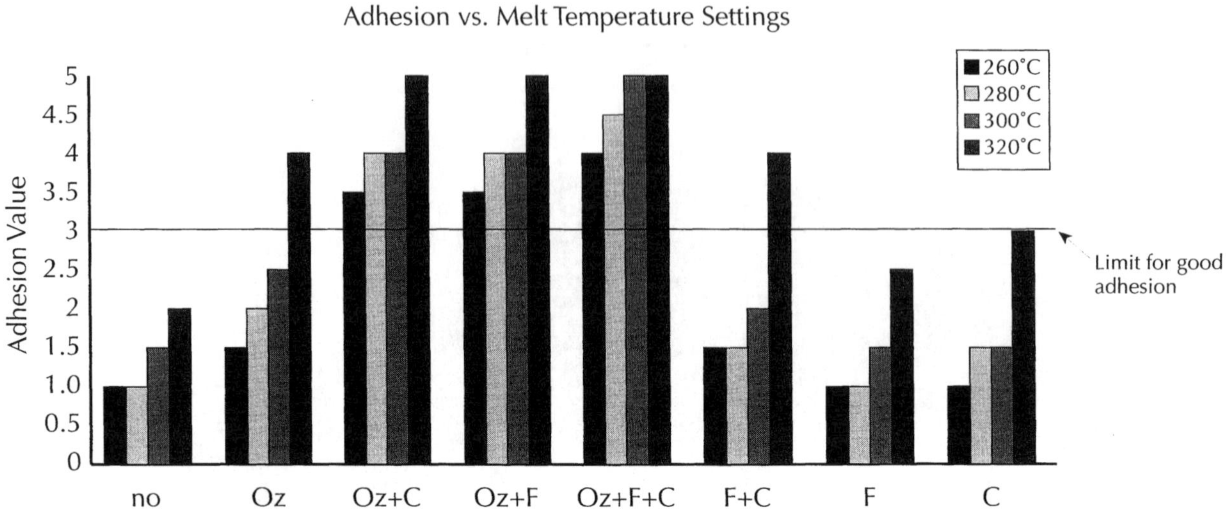

Substrate: UG-kraft/ Resin: PE-LD (40 g/m^2)/ Line speed: 200 m/min/ Pretreatments: Oz = ozone C = Corona/ F = Flame

Fig. 8. Effect of the pretreatments on adhesion with different resins

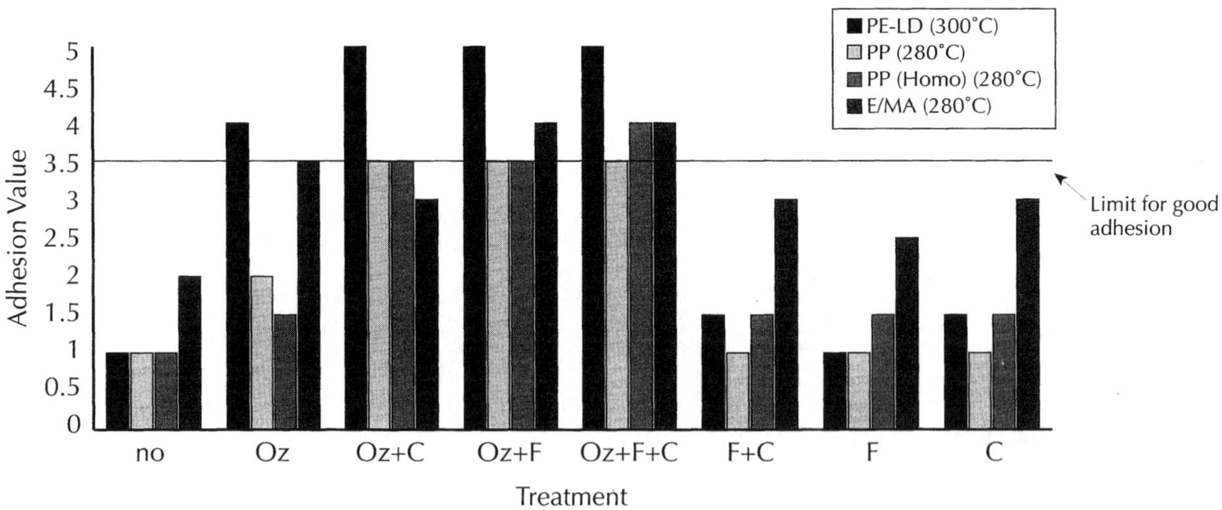

Substrate: Simcaster/ Line speed: 200 m/min/ Coating weights: 40 g/m^2/ Pretreatments: Oz = Ozone C = Corona and F = Flame

Ozone treatment of the molten plastic on its own also increases the polarity of the LDPE and gives a better result. The best adhesion is obtained when the use of ozone is supplemented by flame-treating the surface of the aluminum (Fig. 10).

Earlier studies have indicated that a combination of corona + ozone achieves results equivalent to those we obtained.

Odor and Taste

The organoleptic problems that high melt temperature causes have been known for a long time. When LDPE oxidizes, a variety of compounds (aldehydes, ketones, acids), some of which give packaged products foreign tastes or odors, are formed on the surface of the plastic. On the other

Fig. 9. Change of the contact angle between Al-laminate and water drop after flame treatment

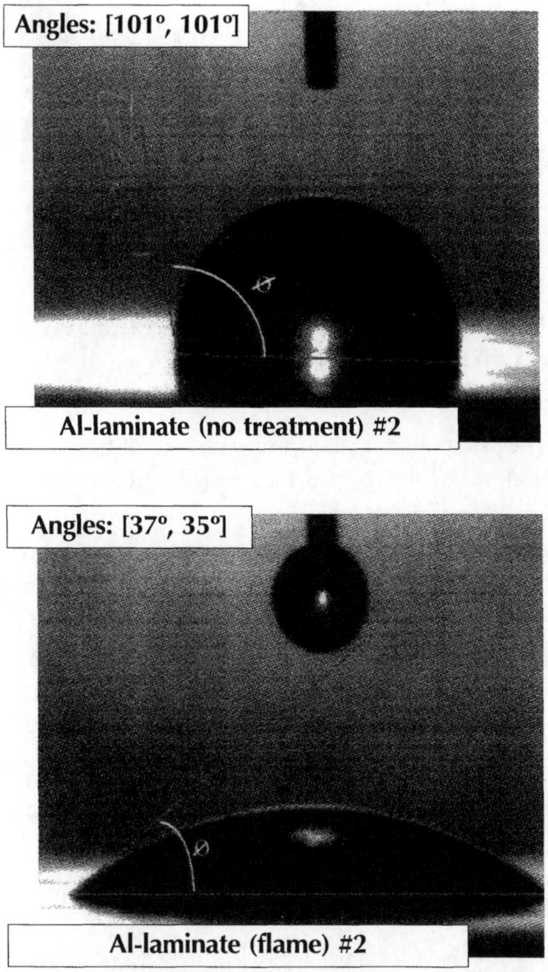

Fig. 10. Effect of the pretreatments on adhesion between Al-laminate and PE-LD

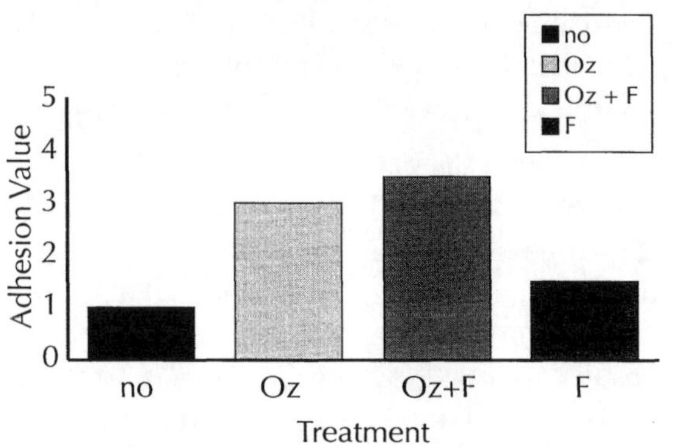

Substrate: Al-laminate/ Resin: PE-LD (40 g/m^2, 300°C)/ Line speed: 200 m/min/ Treatments: Oz = ozone, F = Flame

Table 7. Used polymers and setting temperatures

Type	Name	Setting °C/Notes
PE-LD	LE7518	300°C
PP	WG341C	280°C, modified PP
PP	VC2031D	280°C, homo PP
E/MA	OE5620	280°C, 20% MA

Table 8. Al-laminate, A-grade Al, 200 m/min, 40 g/m^2, LE7518

Temperature Settings	280°C	310°C
O-sample	No	No
Flame treatment	F	F
Flame + Ozone	F + Oz	F + Oz

hand, a certain amount of polar carbonyl groups is needed to ensure adhesion. The guiding principle in 0_3 blowing has been to oxidize only the part of the surface that is necessary for good adhesion and to use a low melt temperature.

It has also often been suspected that even if the surface (the one that will be touching the product) were not oxidized in the extrusion-coating process, the harmful compounds created would in the course of time migrate through the LDPE layer.

To eliminate the influence of the substrate, the odor and taste tests were made on aluminum laminate of the type described earlier in this paper, i.e., a 40 g/m^2 layer of LDPE was applied to the surface of the aluminum. To find out the influence of coating temperature, two different setting values were used (Table 8).

Heat-sealed bags were made from the coated samples and filled with water, after which they were placed in a warm cabinet (40°C). Some of the samples were taken for organoleptic tests. In this Borealis taste analysis, water from one of the bags was compared with water that had been kept in a glass bottle and eight panelists evaluated the taste difference. On the scale used, 0 means "no difference" and 3 is "very marked foreign taste" (Fig. 11). If the intended purpose is, for example, to package water, the product is deemed acceptable if it gets a figure below 2.

Fig. 11. Taste values

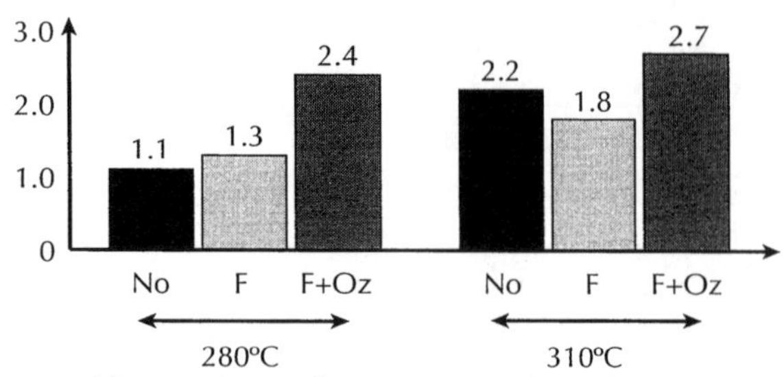

Al-laminate/ 40g/m² PE-LD/ Water after 4 days (40°C)

Fig. 12. Odor values

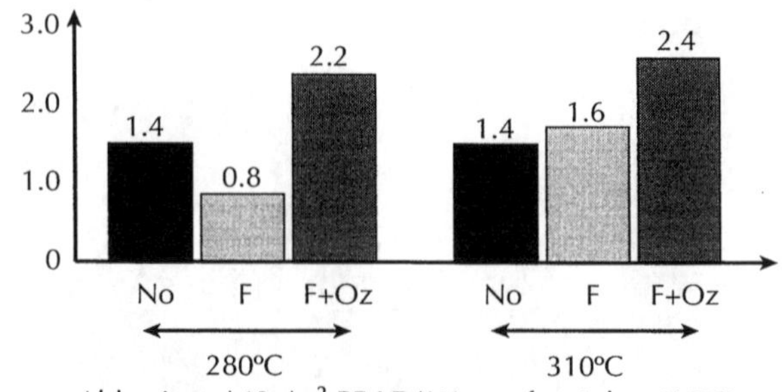

Al-laminate/ 40g/m² PE-LD/ Water after 4 days (40°C)

Fig. 13. Content of Carbonyl groups as function of time

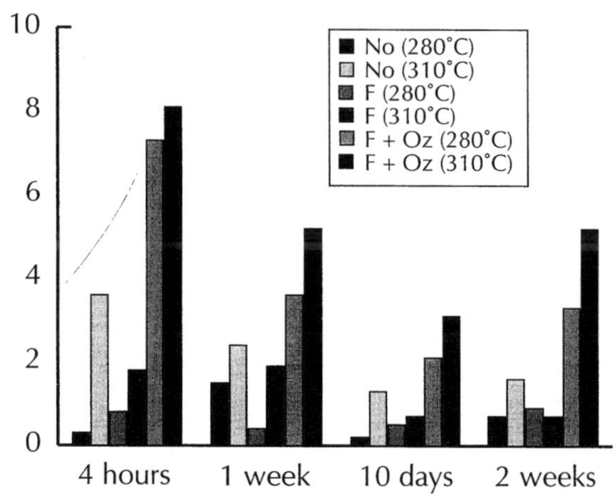

Content, µg/l

For the odor analysis, the LDPE film is ripped loose from the substrate and likewise placed in a warm cabinet (40°C). The odor analysis is done after 2 hours in the heat. The results obtained are shown in Fig. 12.

These organoleptic tests (Figs. 11 and 12), which were carried out 4 days after the runs, again clearly indicated the significance of temperature in examining foreign odors and tastes. Ozone treatment also appears to increase odor and taste values. It would appear that "oxidation products" creep through the LDPE layer in warm water as time goes by.

The components that had been transferred into the water were analyzed by means of a gas chromatograph/mass spectrometer). The 10-ml samples were analyzed using Purge & Trap equipment (Tekmar 2000) combined with a Mass Spectrometer (VG Autospec.).

The water samples were removed from the cabinet after 10 days, but the water continued to be analyzed for two weeks.

The first water sample was examined after 4 hours. Carbonyl compounds (aldehydes, ketones) had been transferred to the water. As the storage period lengthened, their amount did not increase, but in fact even decreased. Ozone treatment appears to increase the amount of carbonyl groups, which would probably explain the organoleptic results. Table 9 shows the carbonyl compounds measured after 10 days. It should be noted that the concentrations are very low (µg/l).

Figure 13 outlines the entire follow-up during 2 weeks of storage.

CONCLUSION

To ensure adequate adhesion in extrusion coating, both pretreatment of the substrate and oxidization of the molten plastic (LDPE) or suitable polar groups (copolymers) are needed. There are no essential differences between flame and corona in pretreatment. Their combined effect is no better than that of either on its own. In other words, there is no point in using both at the same time.

Pretreatment also clearly reduces the surface tension of aluminum or improves adhesion even in the cases of A-grade aluminums. However, pretreating the substrate is not enough; the molten plastic must also be "treated." Suitable chemical groups are obtained either by means of O_3 blowing or by keeping the molten plastic at a high temperature.

Odor and smell problems are always a risk when LDPE oxidizes. The risk is not eliminated even

Table 9.Identified components (ug/L) after 10 days

COMPONENT	No 280°C	No 310°C	Flame 280°C	Flame 310°C	Flame+ Ozone 280°C	Flame+ Ozone 310°C
ALDEHYDES						
Butanal	0.06	0.28	0.05	0.06	0.06	0.18
Pentanal	0.02	0.09	0.02	0.06	0.06	0.14
Hexanal	0.03	0.07	0.06	0.06	0.10	0.28
Heptanal	0.02	0.00	0.03	0.03	0.06	0.25
2-et-hexanal	0.00	0.01	0.02	0.02	0.07	0.06
Octanal	0.00	0.04	0.03	0.04	0.15	0.30
TOTAL	**0.13**	**0.48**	**0.21**	**0.27**	**0.50**	**1.20**
KETONS						
3-methyl-2-butanone	0.00	0.00	0.00	0.00	0.03	0.00
2-pentanone	0.00	0.01	0.05	0.02	0.06	0.06
4,4-dimet-2-pentanon	0.00	0.02	0.01	0.11	0.20	0.23
3-hexanone	0.00	0.06	0.01	0.03	0.05	0.08
3-heptanone	0.01	0.30	0.10	0.20	0.29	0.47
2-heptanone	0.01	0.08	0.07	0.05	0.10	0.21
4-octen-3-one	0.00	0.05	0.00	0.00	0.07	0.00
4-octanone	0.00	0.23	0.01	0.13	0.57	0.45
3-octanone	0.04	0.06	0.03	0.04	0.14	0.31
4-methylene-3-hepta	0.00	0.00	0.01	0.00	0.07	0.08
3- nonanone	0.00	0.00	0.00	0.00	0.05	0.05
6-me-3heptanone	0.00	0.00	0.00	0.00	0.00	0.00
TOTAL	**10.06**	**0.81**	**0.28**	**0.44**	**1.60**	**1.94**

when low temperature is used and only one side of the molten plastic is treated with ozone. The noxious components "migrate" through the LDPE layer, when aluminum is a barrier layer.

ACKNOWLEDGMENT

The author would like to thank Tomi Ylänen, whose Master of Science thesis has been referenced for this paper. The thesis research work was financed by UPM-Kymmene Oy, Finland, and Borealis Polymers Oy, Finland. The head of the Steering Group was Development Director Hannu Karhuketo of UPM-Kymmene, Walki Wisa, to whom the author extends many thanks for advice.

CHAPTER 12

Chemical Primers to Enhance Adhesion and Other Properties

David J. Bentley Jr., RBS Technologies, Inc., and Fred M. Singer, Mica Corporation

INTRODUCTION

The technique of applying molten thermoplastic resins such as polyethylene or various copolymers onto paper, foil, or plastic film provides a method to combine two substrates or to coat one substrate with a functional layer. The manufacturers of flexible composites use these processes of extrusion laminating or extrusion coating to replace laminating or coating by other means so they can utilize existing extrusion equipment to achieve faster line speeds or realize economic advantages in lengthy production runs. One study of the extrusion process cited low resin cost, minimal storage requirements, flexibility in coating changes, and improved clarity as advantages *(1)*.

Figure 1 is a diagram of an extrusion coating operation with the use of an in-line extrusion primer. Figure 2 shows an extrusion laminating process again using an in-line extrusion primer. Figure 3 is a schematic for a high-speed, single station extrusion coating or laminating operation for preparing various flexible web combinations.

Fig. 1. Extrusion coating

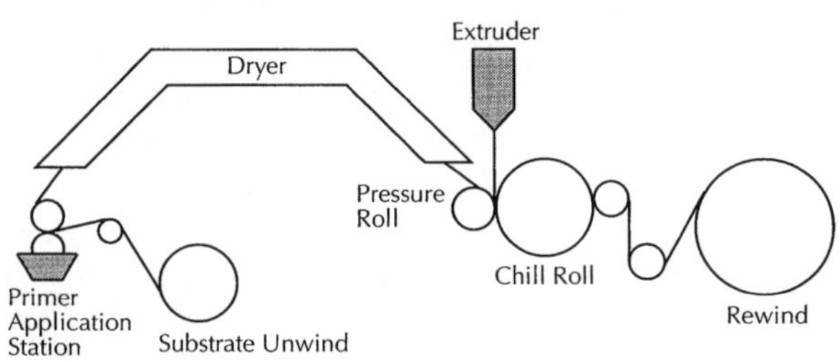

Fig. 2. Extrusion lamination

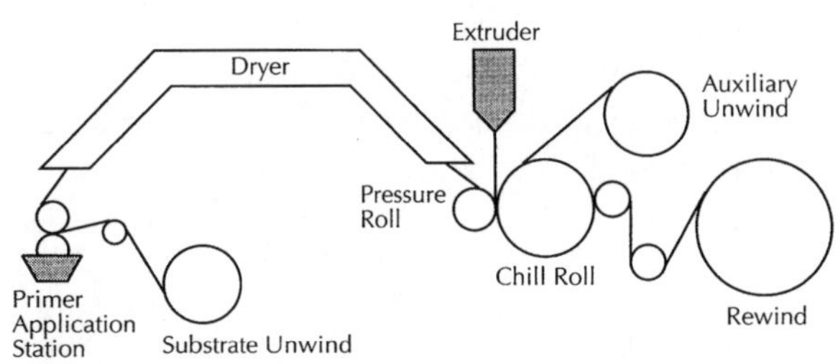

Fig. 3. High-speed extrusion process

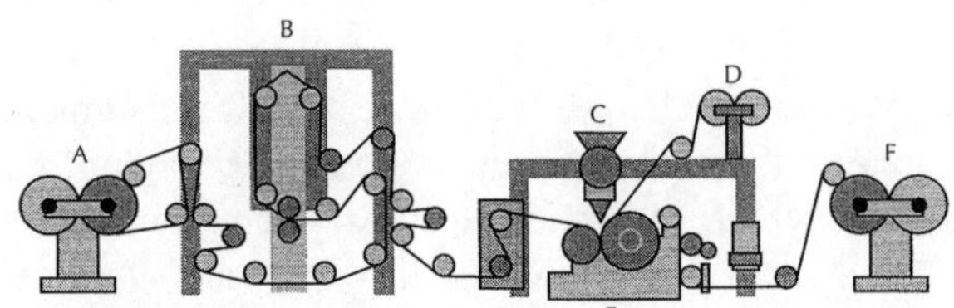

A. Main Unwind
B. Primer Coating Station and Oven
C. Extruder
D. Unwind Station for Laminating Web
E. Extrusion Coating Laminator
R. Rewinder

For both extrusion coating and extrusion laminating, the final product is only as good as the adhesion of the extrudate to the substrate. This adhesion may relate to mechanical factors including the flow characteristics of the extrudate and the porosity of the substrate *(2)*. For example, a polyethylene resin with a high melt index will flow into the interstices of a paper or paperboard substrate and coat any protruding fibers, thereby providing physical anchorage.

Investigators have conducted numerous studies regarding the bond of polyethylene to a substrate *(3-5)*. They examined the variables both in the extrusion process and the polyethylene resin and generally found that bond values increased with higher extrusion temperatures, higher coating weights, and higher melt index resins. Processing conditions that exposed the polyethylene to air for a longer time promoted the formation of carbonyl and hydroxyl groups by surface oxidation. These functional groups gave increased adhesion *(6)*. Copolymers of ethylene and acrylic acid and some other extrusion resins already contain reactive sites and might not need oxidation to promote adhesion. The high odor levels and increased heat-sealing temperature requirements that can accompany the higher bond values obtained by apparently excessive oxidation are often unacceptable when preparing the final package. Therefore, extrusion users are interested in the higher bonds which chemical primers can offer at more acceptable operating conditions.

Since mechanical adhesion will not usually occur on smooth or nonporous surfaces such as polyester film or aluminum foil and because excessive polyethylene resin oxidation can cause odor and heat sealing problems, it is often necessary to use other techniques to improve adhesion. These techniques involve some type of surface modification such as flame ionization treatment and electrostatic (corona discharge) treatment, which presumably function through the same chemistry involving oxidation which can render a surface receptive to ink *(7)*. Another method is the use of chemical primers. This chapter will address such chemical primers both to enhance adhesion and, in many cases, to contribute other desirable end-use properties. Frequently, electrostatic treatment accompanies the use of chemical primers.

It is important to discuss theory of adhesion prior to any mention of the chemical primers themselves because this background will contribute to a better understanding of the performance and to the proper application of the primers. The theory of chemical adhesion for extrusion priming is not necessarily simple and perhaps not altogether thoroughly understood. In simplified terms, however, the extrudate either contains, or is processed so that it will contain, chemically reactive sites. In conjunction with this, the substrate must be readily wettable or reactive or both.

A primer appears to serve two main functions as the extrudate contacts the substrate *(8)*. First, it helps to provide an uncontaminated surface so that the extrudate will rapidly spread and uniformly wet out. Second, the primer joins to both the substrate and the extrudate by a covalent bond, hydrogen bond, van der Waals' force, dipole interaction, or some mixture of these effects. The ability of the primer to increase the number of reactive sites on the substrate and maximize intermolecular attractions is especially important with nonpolar surfaces.

Contaminated surfaces that can interfere with proper adhesion include those carrying dirt particles, residues of aluminum foil rolling lubricant, migrating additives such as waxes and oils, and low-molecular-weight polymer fractions. These surface impurities are in themselves or cause the development of adhesively weak layers at the surface of the substrate. Despite the probable existence of adequate surface energy on the pure substrate, these contaminating factors can restrict the number of actual contact points between extrudate and substrate as well as limit wetting out by the molten resin. This also leads to formation of tiny air pockets which act as stress points, causing weakened bond strength. By avoiding these potential problems, maximum wetting provides the most intimacy between extrudate and substrate. This in turn encourages the highest possible intermolecular attractions to develop at the interface *(9)*.

Physical performance requirements of primers in the flexible laminating process depend upon the machine configuration and the substrates. Adherence to the primary web is one consideration. Compatibility with the applicator system-which might consist of rotogravure, smooth roll with or without metering bars, or air knife influences primer selection. Considering the relatively high speeds of extrusion coating operations and the relatively low applied weights of primers, the manufacturer must be able to maintain the applicator station and drying equipment to produce a smooth and continuous layer of primer.

In virtually all packaging procedures involving flexible webs, heat plays an integral part in forming and closing the package. Conditions of time, temperature, and pressure required for proper sealing as well as for adequate productivity often expose the composite packaging substrates to stresses which would cause instability to or degradation of some of the component layers, if unsupported. Thus the extrusion primers must maintain interfacial bonds through stresses applied later such as heat sealing, hot filling, thermoforming, freezing, and other operations. The finished package will not function unless the combined layers provide both structural strength and product resistance. This latter property generally depends to the greatest extent upon the first few layers nearest the product, i.e., heat seal layer, primer, and supporting or supplemental barrier layers.

PRIMER TYPES

Shellac is a natural, resinous product secreted by insects. It has many years of history as a wash coating for aluminum foil. It is used as a primer on aluminum foil. Almost all of the world's supply originates on the Indian subcontinent. Users are therefore subject to the vagaries of climate, labor, world demand, and similar restrictions.

Shellac is available either in dried flake form or dissolved in alcohol. Normal application is to the exposed side of an aluminum foil surface in the course of wet bonding the aluminum foil to paper. Coating deposition is approximately 0.08-0.16 dry gram per square meter. The coated product dries easily and requires no curing cycle.

Polyethylene extrudate will adhere to the shellac, although there is a severe limit to package performance. Based on its relatively poor heat, moisture, and chemical resistance, shellac as a primer is only suitable for the containment of dry, bland products.

Alkyl titanates such as tetraisopropyl titanate, titanium acetylacetonate, and others were among the earliest extrusion primers *(10)*. These materials function by hydrolyzing on the coated web after drying to form a film of amorphous titanium dioxide. Application is from a very dilute solvent solution at about 5% solids to provide very low coating weights.

The hydrolyzed titanates provide good adhesion for an oxidized polyethylene extrudate and give a product with a fair level of heat, chemical, and moisture resistance. In actual use, the alkyl titanates often prove unsatisfactory because of their extreme moisture sensitivity. Since hydrolysis to the titanium dioxide stage occurs with traces of water encountered in the atmosphere, there is a tendency within the liquid for the formation of particles that do not adhere to either the substrate or extrudate. The variabilities in adhesion and performance of the alkyl titanates have resulted in virtual abandonment of such systems.

One of the largest classes of extrusion primers in use today is the curing type that is based on polyurethane chemistry. This wide range of products includes single-component polyurethane prepolymers which cure by reaction with moisture, two-component systems where an isocyanate terminated prepolymer reacts with a hydroxyl compound, two-component systems where an isocyanate compound reacts with a polyester, and amine terminated polyurethane prepolymers which react with epoxy resins. Organic solvents such as ethyl acetate, methyl ethyl ketone, and toluene generally are the solvents for these primers except for the amine terminated system which uses alcohol.

The polyurethanes are often laminating adhesives used at somewhat lower coating weights for extrusion applications, although there are some having extrusion priming as their primary use. Typical coating weights range from 0.8 to 1.6 dry grams per square meter. The solutions of primers, although reactive, often have as much as 12-24 hours pot life after mixing for use. Bond strengths immediately after extrusion are sufficiently high to prevent adhesion problems. A large percentage of the cure occurs in 24 hours. Total cure is complete in five to seven days.

A primer with over 30 years commercial use is polyethyleneimine. The earliest forms were simple solutions of the polymer in alcohol and further diluted with alcohol or alcohol and water mixes. Later, aqueous forms became available. Unformulated polyethyleneimine solution primers contribute basic adhesion enhancement but provide little chemical or moisture resistance. Formulated versions of polyethyleneimine provide chemical and moisture resistance partly because of improved cohesive strength.

The coating weight for polyethyleneimine is about 0.03 dry gram per square meter. These primers wet out effectively on paper and corona discharge treated films. After wetting and spreading, the polyethyleneimine adheres to the oxidized groups on the base substrate. Adhesion also can be achieved to other types of reactive groups, for example, chlorides. Then other reactive primer sites react with the oxidized sites on the extrudate. Theoretically, a monomolecular layer of primer should function on the fairly smooth and uniform surface of a plastic film. While this is mechanically impossible, the primer deposit is indeed kept to a minimum as this type of molecule has limited cohesive strength. Thus an excessive layer of polyethyleneimine-based primer can negate the entire effort.

Another reason for exercising care to limit applied weight is a tendency of polyvinylidene chloride coated films to take on a faint yellow discoloration in the presence of polyethyleneimine primers. This condition, which can occur with a number of primer types, becomes more apparent as primer weight increases.

Drying is important, although when coating papers, a portion of the diluent will migrate into this porous substrate. Despite the presence of large amounts of water in many of the formulae rather than the more volatile organic solvents, drying is easy due to the relatively low total weight applied *(11)*.

With adequate extrudate temperature (e.g., 315-325°C for low or medium density polyethylene) on equipment properly programmed relative to air gap vs. line speed, polyethyleneimine primers should provide off-machine destructive bonds on the typical polyethylene to paper or polyethylene to film constructions. Heavier constructions for certain industrial applications should impart tensile peel readings of 500 grams or more per 25 mm of width. Polyethyleneimine-based primers require no curing cycle. Marginally low extrudate temperature can compromise initial bond, and a subsequent increase might develop over the next few days. Attention to proper processing conditions, of course, will negate this aberrant behavior.

Aluminum foils exhibit a chemical surface entirely different from plastic films. Acid functional groups bond to metal surfaces. Accordingly, ethylene acrylic acid copolymers are the base for a group of aqueous primers for aluminum foil and other metallic substrates as well as various polar materials.

Coating weight for ethylene acrylic acid copolymer primers varies between 0.16 and 1.2 dry grams per square meter depending on the particular formulation and the need for product resistance in the final package. This primer system can provide bonds with good resistance to water, grease, oil, and many chemicals. Ethylene acrylic acid copolymers further possess the intrinsic property of flexible toughness due to lack of crystallinity.

Because some aluminum foils which are fully annealed (dead soft)—and supposedly completely wettable—occasionally have oily residues at random locations in a roll, many processors will use small amounts of alcohol as a co-diluent with water in ethylene acrylic acid primers to ensure consistent wetting.

Emulsion copolymers of polyvinylidene chloride offer high barrier properties because of their low permeability to water vapor and oxygen. Combined with their resistance to fats and oils, this makes such copolymers a suitable choice for an extrusion coating or laminating primer. Various materials are available, some of which even provide barrier properties as low as 9 cc oxygen per square meter per 24 hours under proper coating conditions *(12)*.

The PVDC primers will promote adhesion at coating weights of 2.0 dry grams per square meter, although very high barrier properties can require a coating weight of 3.2 to as much as 4.8 dry grams. Optimum barrier is only possible with a continuous film. Applying two passes of primer is one method to ensure this. Although polyvinylidene chloride copolymers adhere to a variety of substrates, best adhesion occurs on polyvinylidene chloride copolymer films. The primer will provide a modest initial bond value that will age in one to two days to a high percentage of its ultimate bond. Full strength only occurs after one to two weeks, during which time the polyvinylidene chloride copolymer undergoes cold flow, internal cross-linking, and crystallization.

Table 1. General guide to extrusion primer performance

	Adhesion Characteristics			Performance Characteristic		
Type of Primer	**Paper**	**Metal**	**Plastic Film**	**Heat Resistance**	**Moisture Resistance**	**Chemical Resistance**
Shellac	Poor	Excellent	Poor	Poor	Poor	Poor
Organic Titanate	Good	Good	Good	Fair	Fair	Fair
Polyurethane	Very Good	Excellent	Excellent	Excellent	Excellent	Excellent
Polyethyleneimine	Very Good	Good	Excellent	Excellent	Poor	Poor
Ethylene Acrylic Acid	Excellent	Excellent	Fair	Fair	Excellent	Good
Polyvinylidene Chloride	Excellent	Fair	Excellent	Good	Very Good	Fair

Handling of polyvinylidene chloride copolymer emulsions is critical because a pH range of 1.5 to 2.0 makes them extremely corrosive. Therefore, the coating head, which deposits the primer as well as all of the materials for the primer handling system, must be a suitably corrosion-resistant material such as 316 stainless steel.

Acrylic copolymers, ethylene vinyl acetate copolymers, propylene copolymers, melamine/form aldehyde resins, and other chemical types contribute to rounding out the collection of available primers. These groups comprise a smaller amount of the commercial volume but do provide specific performance, economic, or other requirements for certain uses. Due to a host of governmental and economic pressures including environmental and workplace legislation, many of these miscellaneous primers are aqueous systems.

PERFORMANCE PROPERTIES

Much information undoubtedly exists on the performance properties of various primers on different substrates. Unfortunately, the primer users themselves probably have compiled the most useful data, and they keep it private for competitive reasons. Occasionally there are reports in the literature from extrusion primer manufacturers or extrusion resin suppliers *(13, 14)*.

Table 1 offers a generalized description of the adhesion which various primers will provide on paper, metal surfaces, and plastic films as well as the heat, moisture, and chemical resistance which might result. The table is only a guide. Specific end-use requirements for adhesion or environmental resistance will result only from evaluation of a given primer with the extrudate of choice in the actual application.

Some general comments regarding extrusion primers are obvious from this tabulation. The curing polyurethane systems generally offer the best resistance properties and the widest range of adhesion. The polyethyleneimine, ethylene acrylic acid, and polyvinylidene chloride types of polymers provide good adhesion and selected resistance characteristics. Shellac and the organic titanates do not offer high performance characteristics and only limited adhesion properties. This obviously accounts for their rather small use today. The miscellaneous primers absent from the table can frequently offer excellent adhesion and resistance because their actual development often was for very specific end-use applications.

Despite the specialty performance aspects of each primer type, none will provide panacean performance. Consistent results depend upon consistent conditions. Proper standards for each type of run including extrusion temperature, air gap, and web speed are necessary.

PROPER PRIMER USE

Application by gravure cylinders generally provides the heavier laydowns necessary for polyvinylidene chloride, polyurethane, and most ethylene acrylic acid primers. This technique also suffices for shellac priming. For example, a 200 quadrangular roll will deposit approximately 0.8 grams per square meter using a 25% solids primer solution. The typical low application weights for polyethyleneimine primers encourage use of a smooth roll with very dilute solutions or the use of a very fine gravure cylinder. Thus the primer user can apply a 1-3% solution simply by flooding with a smooth roll or with a 200 or 300 quadrangular roll. The coating weight range of ethylene acrylic acid primers makes gravure cylinders or wire wound rods a suitable choice for higher depositions and smooth roll transfer the choice for the lower depositions. The smooth roll transfer is only suitable for the lower range of applied weight if the high pressure for metering at the applicator nip does not have adverse effects on the end bearings of the applicator system. Also notable is the fact that an increase of the hydrokinetic pressure of a primer at faster operating speeds could tend to force the metering rolls apart and thereby defeat the efforts to limit coating weight. This would result in heavier laydown toward the center of the web while the end bearings wear excessively as they resist the same force.

With the exception of some paper substrates, extrusion laminating operations involve combining relatively impervious layers, although some coated papers fall into this category as well. Drying the applied primer is essential to effective bond performance. It is therefore necessary to establish adequate drying conditions through control of temperature, air volume, and air velocity. With solvent borne primers, avoidance of recycled air in the final zone will help to minimize the possibility of odor retention.

The typical machinery configuration consists of the primer application station and drying tunnel located immediately preceding and in-line with the extrusion coater. From the aspect of chemical primer function, this is most effective and most reliable. Also, there is little possibility of abusing or upsetting a primed surface before extrusion coating. When using primers with residual tackiness after drying, however, minimize or eliminate face contact with idler rolls.

Many manufacturing operations involve out-of-line priming for a number of reasons. One configuration is the use of the last station of a press during the reverse printing process. In such out-of-line operations, thorough drying and low-tension rewind are necessary to minimize blocking or picking of even "dry" primers. Limited holding time before the next unwind will minimize contamination, if any potentially migratory components are present.

Aluminum foil can react with ingredients in paper to develop a weak boundary layer on its surface by the formation of aluminum salts. This phenomenon is known as poisoning. Coating the aluminum surface of a aluminum foil to paper composite with shellac or a vinyl wash coat will prevent poisoning. In some instances where high performance requirements are not necessary, these coatings provide the dual function of preventing contamination while offering adherability. In any event, processing procedures and machinery configuration can influence the use of different primers to obtain optimum chemical resistance in a composite using aluminum foil *(15)*.

There are a number of difficulties that can arise during an extrusion coating or laminating operation to hinder trouble-free operations. Many of these relate directly to the extrudate or the extrusion equipment. Other publications address them *(16-18)*. There are, however, problems that can sometimes arise from the primer itself or its application.

The user must know all of the instructions applicable to the extrusion primer and follow all of the recommendations regarding storage, mixing, and the need to avoid contamination with certain materials. Aqueous products, for example, are prone to freezing and might not be freeze-thaw stable. Certain two-component primers require given mixing orders and specific mixing ratios. Alcohol will contaminate isocyanate-cured systems and will inhibit their cure. Water will decrease their pot life. Such systems therefore require the use of solvents designated as "polyurethane grade."

Dilute the extrusion primer with a material that will not attack inked or otherwise coated substrates. Methyl ethyl ketone or ethyl acetate readily attack polyvinylidene chloride coated films with resultant loss of some anchorage of the coating and possible odor problems in the packaged item. Surfaces such as nylon and cellophane, which are hydrophilic, require special care.

Use the primer at the correct coating weight deposition. Apply polyethyleneimine-based primers at extremely low coating weights, since they might not function at higher weights. Conversely, apply most of the other primers at higher weights, since they might not perform as well at lower weights. Uniform deposition is important for all primers, but especially for polyvinylidene chloride copolymers which will not provide barrier properties if there are pinholes or other irregularities due to foaming, poor wetting, or cracking of high coating thicknesses during drying. Polyvinylidene chloride copolymer weights, of course, directly influence barrier and product preservation. Product containment performance also relates to the amount of primer deposition for polyurethane and ethylene acrylic acid primers.

When coating a primer onto a porous substrate such as paper, exercise care so that all of the primer is not absorbed into the material. Enough primer must remain on the surface to contact the extrudate.

Using a smoothing bar to wipe the surface uniform, corona discharge treating the substrate or using specific diluents in the primer will often remedy poor wetting of a primer. In the case of water borne primers, alcohol is usually an excellent diluent to improve the wetting. Adding more diluent to reduce viscosity should help to solve the problem of a primer that does not flow on the substrate. Pattern marks from the application of the primer and which are evident in the final product can be the result of too rapid drying before the primer can flow sufficiently. Increasing the time prior to drying or adding a diluent that evaporates more slowly will remedy this situation.

Two-component primers which cure on reaction sometimes do not exhibit their normally high bond values, if pre-curing occurs in the drying operation before the extrusion process. Lower oven temperatures or faster line speeds are solutions. Drying is a very critical step in the extrusion priming process. No volatiles should remain in the primer because this can lead to blistering upon application of hot extrudate. Minute amounts of solvent can also cause odor problems in the final package.

Finally, when there is still a poor bond even after investigation of all of these items, some detective work can usually lead to a ready explanation. The mode of failure of the bond is the basic piece of information that will give the clue to the problem. Adhesive failure of the primer from the substrate could mean contamination of the substrate by foreign matter, improper coating weight, or simply the use of the wrong primer. Adhesive failure from the extrudate could also mean insufficient coating thickness, the wrong primer, or improper processing of the extrudate itself. Cohesive failure of the primer is indicative of excessive coating thickness, insufficient time to cure, the need for a curing rather than a thermoplastic primer, or the presence of a contaminant which retards or inhibits the cure. Failure of an ink or a coating is often not related to the primer at all but is a function of the ink or coating component itself.

PRIMERS FOR VARIOUS EXTRUDATES

Extrusion coating and laminating have existed for almost a half century. At the time of inception, polyethylene was the prime resin choice for the extrudate. Through the years, this material has continued to be one of the most popular resins for extrusion. Today, however, many other resins also find use. A given primer might not work for all of the available resins.

Depending on the particular properties and performance that are necessary, one of the wide array of primers mentioned above will function as an extrusion primer for polyethylene. Thus one might use a primer based on polyethyleneimine for a construction where he extrudes polyethylene onto paper for a dry food packaging application, a primer based on a two-component curing polyurethane for a construction where he extrudes polyethylene onto polyester for making inexpensive liquid containing pouches, or a primer based on a polyvinylidene chloride emulsion for a construction where he extrudes polyethylene onto paper for a snack food package requiring moisture and oxygen barrier properties.

The ionomer resins generally function with any of the primers that are suitable for polyethylene. Ionomers generally adhere well to clean aluminum foil without the assistance of primers, although ethylene acrylic acid primer will provide security, especially with contaminated aluminum foil. Polypropylene resins present some problems with adhesion because they are more difficult to oxidize than polyethylene. If there are satisfactory conditions to achieve sufficient oxidation, then the two-component curing polyurethane primers will function to enhance adhesion with a polypropylene extrudate.

The maximum melt temperature for extrusion of ethylene vinyl acetate copolymers is about 232°C. Because of this relatively low temperature, many of the primers discussed above will not function with these extrudates *(19)*. One workable approach involves modified ethylene vinyl acetate copolymers dissolved in organic solvent and cross-linked with an isocyanate-containing curing agent *(20)*. Specially formulated ethylene acrylic acid-based primers also might fulfill these needs.

Another method of accomplishing low-temperature extrusion coating is utilization of generated ozone to contact the molten extrudate as it passes from the extruder die lip to the combining rolls. Delivery of this highly oxidative gas to the intended side of the melt contributes to adequate oxidation that would otherwise not occur in the air gap at the given temperature. In addition to enabling extrusion of low-temperature resins, ozone generation also permits temperature reduction when processing conventional polyethylene. A manufacturer can perform these operations with the regularly used primers *(21)*. However, the resulting laminate should be evaluated for maintenance of the specific properties desired. Polymer interfaces do not necessarily cross-link or otherwise combine in exactly the same manner at two different temperatures or sets of operating conditions.

Other extrudates enter into extrusion coating to a lesser extent. Ethylene acrylic acid resin itself provides effective adhesion on aluminum foil without primer assistance. Ethylene ethyl acrylate resin extruded onto plastic film requires a chemical primer as well as special handling techniques. The same applies to extrusion coating with nylon resin.

The producer of flexible composites should select primers for any other extrusion coating and laminating resins based on consultation with the resin and primer manufacturers. This should be followed by performance trials.

PRIMERS FOR VARIOUS SUBSTRATES

Flexible laminate manufacturers handle paper in diverse ways due to its porous nature and polar functionality. For some relatively low-speed, low-demand applications on some uncoated papers, there are no primers necessary for the extrusion coating operation. There is likewise no primer used for parts of some extrusion laminating operations involving a combination of paper to polyethylene to aluminum foil to polyethylene for pouch stocks for dry product packaging, although this usually is not a low-speed converting operation. The aluminum foil barrier prevents contact by the contents with the outer paper that, in addition to being the decorative surface, provides strength by minimizing the possibility of aluminum foil cracking or tearing. Many converters will extrusion laminate the paper to polyethylene to aluminum foil portion with no primer assist.

In some situations involving carton stock for liquid packaging and other paper or board applications, flame or corona treatment of the substrate will increase the number of oxidized sites, minimizing the need for any chemical primer, as it relates to substrate adhesion. On the other hand, the use of primers achieves the ultimate speed and efficiency in extrusion coating and laminating. An example is a lightweight paper to polyethylene combination for pouch stock where reduction in quantity of extrudate occurs with concurrent increase in speed. Polyethyleneimine-based products are the major primers for paper use, while melamine/formaldehyde resins and ethylene acrylic acid systems are secondary.

There are also a number of different ways to handle aluminum foil. A limited amount of laminate comprising aluminum foil to polyethylene for totally dry, bland end uses requires only corona treatment of the aluminum foil enroute to the extruder lip. This removes any oily residues and ensures wet-out of the extrudate. Some extruders use ozone to oxidize further the molten extrudate before application to clean aluminum foil.

Solvent-based polyurethanes satisfy the highest-demand applications, although environmental pressures have prompted the development of acrylic and other aqueous systems to supplant some of this consumption.

A large volume primer class in aluminum foil service consists of the ethylene acrylic acid copolymers. Ease of handling, economic yield, and broad demand performance enable users to make lengthy runs by scheduling successive jobs capable of using this type of primer.

The use of polyethyleneimine primers in some low-demand aluminum foil applications might provide only marginal containment performance, especially where some moisture resistance is necessary. Special formulations based on ethylene vinyl acetate and ethylene acrylic acid are available for application of low-temperature extrudates to aluminum foil.

In the case of plastic films, the ability of corona treatment to develop carbonyl and other oxidized sites on an otherwise nonpolar surface allows polyethyleneimine primers to fulfill many applications. High-demand applications, however, often require solvent-based polyurethanes. Another system for this use is melamine/formaldehyde primers. Note, however, that the usual relationship of extraordinary primer for extraordinary end use might be less rigid with film usage. At least one extruder uses polyurethanes in a broad spectrum of applications because he decided there was an economic advantage to keeping the primer in the pan for lengthier extruder runs rather than incurring downtime and more material loss with changeover to another primer. Alternately, polyethyleneimine based primers have qualified for a diversity of applications such as boil-in-bag (polyester to medium-density polyethylene), thermoformed meat packaging (nylon to low-density polyethylene to ethylene vinyl acetate copolymer), and ethylene oxide sterilizable surgical packaging (polyester to low density polyethylene). Extruders should consult primer suppliers regarding applicability of their products and carefully evaluate all new end uses or substrate combinations.

Polyester film can present a special case with some primer systems, especially those supplied in water, because even with a typical surface tension value as high as 44-46 dynes per cm on untreated film there might be poor wetting or bonding. Chemical modification or corona discharge treatment of the polyester film that raises the surface tension is sometimes necessary for proper adhesion of the primer.

When using coated substrates, one must consider the ability of the primer to wet on and bond to the particular coating. Furthermore, avoiding disruption of the anchorage or adhesion of the coating to the substrate (for example, by solvent attack) is equally important. Probably the most common coated surface encountered in extrusion processes is polyvinylidene chloride copolymers on films and papers. In this instance, polyethyleneimine primers have an affinity for the halogen portion of the polyvinylidene chloride molecule. One must, however, consider potential wetting problems caused by the offsetting of jaw release agents, slip additives, and other non-adherable materials and accommodate these problems through judicious use of co-diluents or corona treatment. Primer suppliers can often provide valuable assistance.

Lacquered aluminum foils, nitrocellulose coated cellophane, and acrylic coated films comprise other coated surfaces which require individual evaluation for wetting and adhesion performance.

OUTLOOK

A publication by Singer *(22)* describes the attributes of a perfect extrusion primer. While the primer manufacturer might never succeed in developing this perfect primer, two factors will influence his research efforts. The first of these will be to supply primers that conform to the various environmental concerns that are becoming more and more stringent as each day passes. Secondly, the customer will influence the primer manufacturer to provide him with products that offer better performance and/or better economics.

Primers in water or other environmentally agreeable forms will undoubtedly be the materials of the future. They will apply easily to the substrate and will offer adhesion to a wide variety of substrates with suitable heat, water, and chemical resistance. Finally, all of this will occur with low-cost materials that are readily run at rapid coating line speeds.

REFERENCES

1. Cromer, L.L., TAPPI 1973 Paper Synthetics Conference Proceedings, TAPPI PRESS, Atlanta, p. 169.

2. Bikerman, J.J., The Science of Adhesive Joints, Academic Press, New York, 1961.

3. Arbit, H.A., Griesser, E.E., Haine, W.A., Tappi 40(3):161 (1957).

4. Goslin, J.P. and Sweeney, H.F., Tappi 43(5):434 (1960).

5. Crolius, V.G., Ebeling, W.E., Parsons, R.C., Tappi 45(5):351 (1962).

6. Luongo, J.P., J. Polymer Sci. 42(139):139 (1960).

7. Hanes, R.A., American Chemical Society 17(2):69 (1957).

8. Twomey, B.J., TAPPI 1975 Paper Synthetics Conference Proceedings, TAPPI PRESS, Atlanta, p. 113.

9. Gregory, B.H., Adhesion In Extrusion Coating and Laminating In-company Seminar Notes, U.S.I. Europe NV., 1975.

10. Gray, C.L. Jr., MacCarthy, W.L., McLaughlin, T.F. Jr., Modern Packaging 34(6):143 (1961).

11. Siegel, D., Paper Film Foil Conv. 45(11):37 (1971).

12. Isbister, R.J., TAPPI 1983 Paper Synthetics Conference Proceedings, TAPPI PRESS, Atlanta, p. 117.

13. Havercroft, W.E., TAPPI 1975 Paper Synthetics Conference Proceedings, TAPPI PRESS, Atlanta, p. 157.

14. Henkle, R.N., Extrusion Coating Progress and Potential, Peacock Business Press, Park Ridge, IL, 1969, pp. 22-24.

15. Reddeman, N.G., TAPPI 1973 Paper Synthetics Conference Proceedings, TAPPI PRESS, Atlanta, p. 185.

16. "Extrusion Coating," E. I. duPont de Nemours & Co. (Inc.), Plastics Department, Polyolefins Division, Wilmington, DE, 1968.

17. Polyethylene Extrusion Coating-An Operating Manual, U.S. Industrial Chemicals Co., New York, 1962.

18. Grail, T.J. and Fingerman, S.M., Paper Film Foil Conv. 43(7):7 (1969) and 43(8):49 (1969).

19. Van Ness, R.T., Tappi 59(9):114 (1976).

20. Fuerholzer, J.J., Reddeman, N.G., and Zweig, S., U. S. pat. 3,616,191 (Oct. 26, 1971).

21. Tietje, A., Paper Film Foil Conv. 53(1):94 (1979).

22. Siegel, D. and Singer, F., Paper Film Foil Conv. 58(7):54 (1984).

CHAPTER 13

Coat Weight/Thickness Measurement and Control

William M. Kirtley

INTRODUCTION

This chapter describes the sensors normally used to measure extruded coatings. The objectives and special considerations related to measurement of coatings are described along with their solutions. The information discusses automatic control approaches, including automatic die control, and projects levels of quality improvement and cost savings which can be expected from coat weight measurement and control.

OBJECTIVE

The objective of coat weight/thickness measurement and control is improved profitability through materials savings, increases in net and gross production, and improved product quality. This is accomplished by giving the operator better product and process visibility and automatically controlling the coat weight or thickness to reduce variability and eliminate process bottlenecks. Typical results are as follows:

- 50-75% reduction in product variability
- 6-8% material savings
- 3-5% throughput increases
- 50-60% scrap reductions.

In addition, some systems incorporate linewide and plantwide control and information systems that typically increase corporate productivity 3-5%.

The savings generated by these results mean that an investment in these systems is commonly returned in one to two years or less.

SENSORS

Control of variability is limited by the accuracy of the sensors. Neither the operator nor automatic control can reduce upsets they cannot see. That is why the accuracy and reliability of the measurement provided by the sensors is so critical.

The two most common sensors used with extrusion coating applications are nuclear sensors and infrared sensors. Nuclear sensors are normally used where coatings are heavy relative to the substrate weight or where there is no need to differentiate multiple layers. Infrared is used to measure light coatings and, with new technology, will measure individual layers, for example, in coextruded coatings.

Nuclear sensors

These sensors are usually referred to as "beta gauges" because the nuclear source utilized in the sensor, normally Krypton 85 or Strontium 90, emits "beta" particles, or electrons. The source is on one side of the web, and the detector-typically an ion chamber-is on the other side. The mass between the source and receiver absorbs the electrons. The more mass there is, the more electrons are absorbed and do not reach the detector. In other words, the heavier the sheet is going through the sensor, the fewer electrons reaching the detector. The detector sends an electronic signal to the system proportional to the flow of electrons. Therefore, the sensor can be calibrated to measure the weight of the sheet.

Since the sensor "sees" mass, it can read out in units such as ounces per square yard, grams per square meter, or pounds per ream. If the specific gravity of the product is uniform, the system can be calibrated to read out in thickness, usually in mils or microns (Figs. 1 and 2).

Fig. 1. Nuclear sensor (beta gauge)

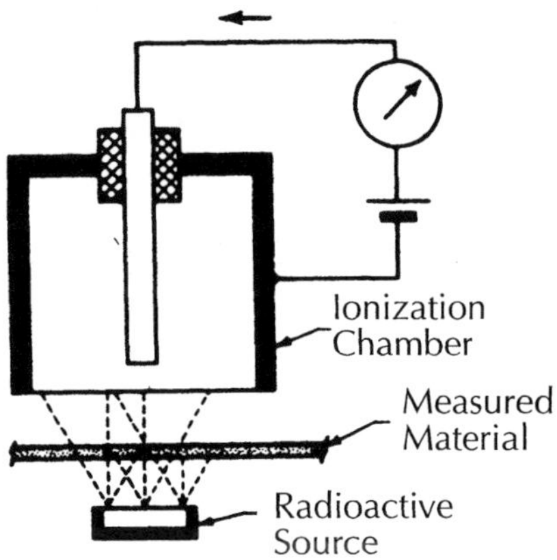

Fig. 2. Typical response of a nuclear sensor

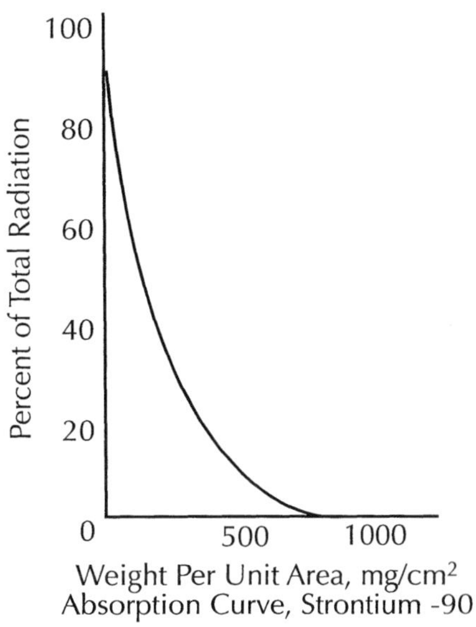

Nuclear sensors are very safe. The source is contained in a lead capsule that only allows particles to escape toward the receiver when the "shutter" is open. The shutter is automatically closed on power failure. Further, the sensors are inspected twice per year to comply with Nuclear Regulatory Commission requirements and to ensure that there is no danger. There has never been an accident reported relating to a nuclear sensor, and there have been hundreds in use worldwide for over 40 years.

Since the nuclear gauge reads the total weight or thickness of the product it is viewing, two sensors are needed to determine coating. (Sometimes one sensor is used where only total weight measurement is desired.) The first sensor scans the full width of the uncoated or "base" product, and a second sensor scans the product again after the extrusion coating is applied. This is referred to as the "gross" product (Fig. 3). In the simpler systems, the "base" reading is subtracted from the "gross" to calculate the weight or thickness of the coating or "net coat" measurement. The more sophisticated systems utilize a matched sensor configuration and calibrate directly for net coat. This significantly reduces the calibration effort necessary to ensure an accurate coat weight measurement in a multi-product environment.

Fig. 3. Extrusion coating

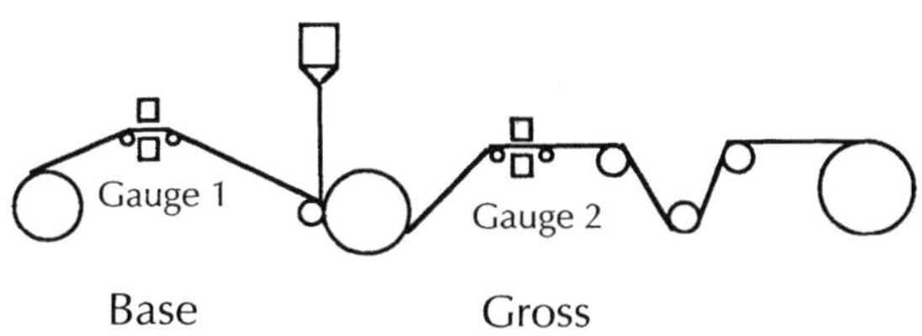

If an extrusion line has additional unwind stands or additional extrusion coating stations, additional nuclear sensors can be added as necessary. In each case, the sensor sees the total weight of the product, and various measurements must be subtracted to determine net coat weights.

Fig. 4. Extrusion coating and laminating

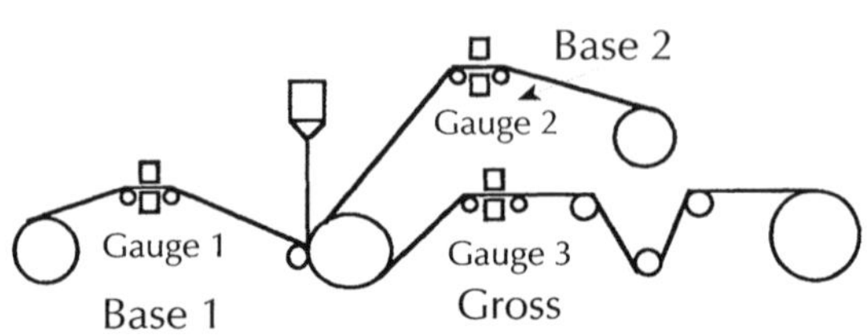

True same-spot measurement-changing line speed and sheet stretch

To maintain true same-spot measurement through a line speed change or sheet stretch, the speed that the sensors scan across the web must be slaved to the speed of the line. This can be achieved by driving each sensor with a dc motor and locking each motor to independent line speed tachometers (Fig. 5). For example, if the line is running at 100 ft/min, the sensor could be scanning at 10 in./s. With scan speed slaved to the line speed, the scanners would drop to 9 in./s when the line speed is slowed to 90 ft/min. System response time must be taken into account here. A slow response time would mean an error in same-spot alignment. On many applications, if the "same spot" is missed even by a small distance, same-spot measurement is of no value and, with a variable base sheet, coat weight error will be large.

Fig. 5. Same-spot measurement with scan speed slave to line speed

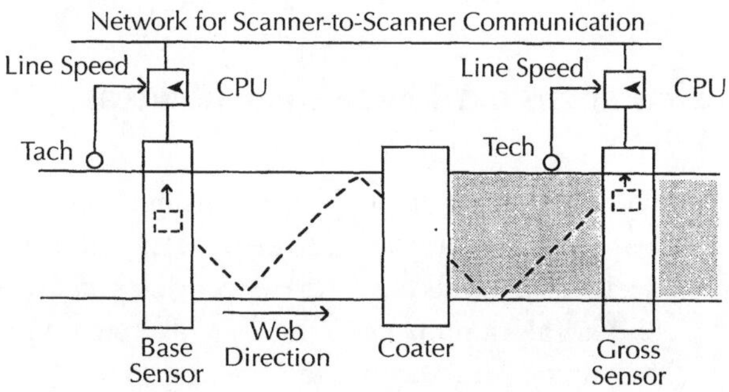

Without this feature of slaving scan speed to line speed, same-spot measurement is lost whenever the line speed changes, and coating measurement errors are introduced (Fig. 6). This occurs, for example, on product startup where reliable measurement is most critical.

Fig. 6. Same-spot head alignment with scan speed and line speed synchronization

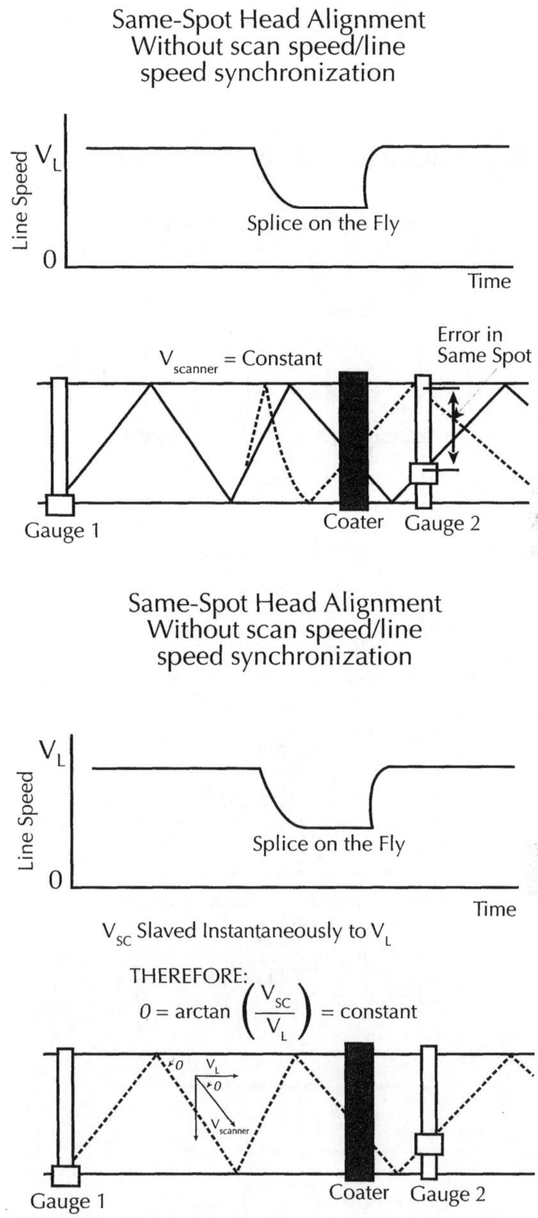

Infrared (IR) sensors

IR sensors use light in the infrared range of the spectrum to directly measure coating thickness. Conventional IR sensors use filters to focus on areas of the spectrum that are sensitive to thickness changes, a "measured" channel, and compare that changing signal to another selected area of the spectrum, a "reference" channel, where the signal is constant relative to thickness changes. The difference in these two signals can be calibrated to determine thickness. Rotating filter wheel, beam splitter, and full-spectrum technologies are now being utilized in the IR sensors commonly used for the extrusion coating application (Fig. 7).

Fig. 7. Alternative IR sensor designs

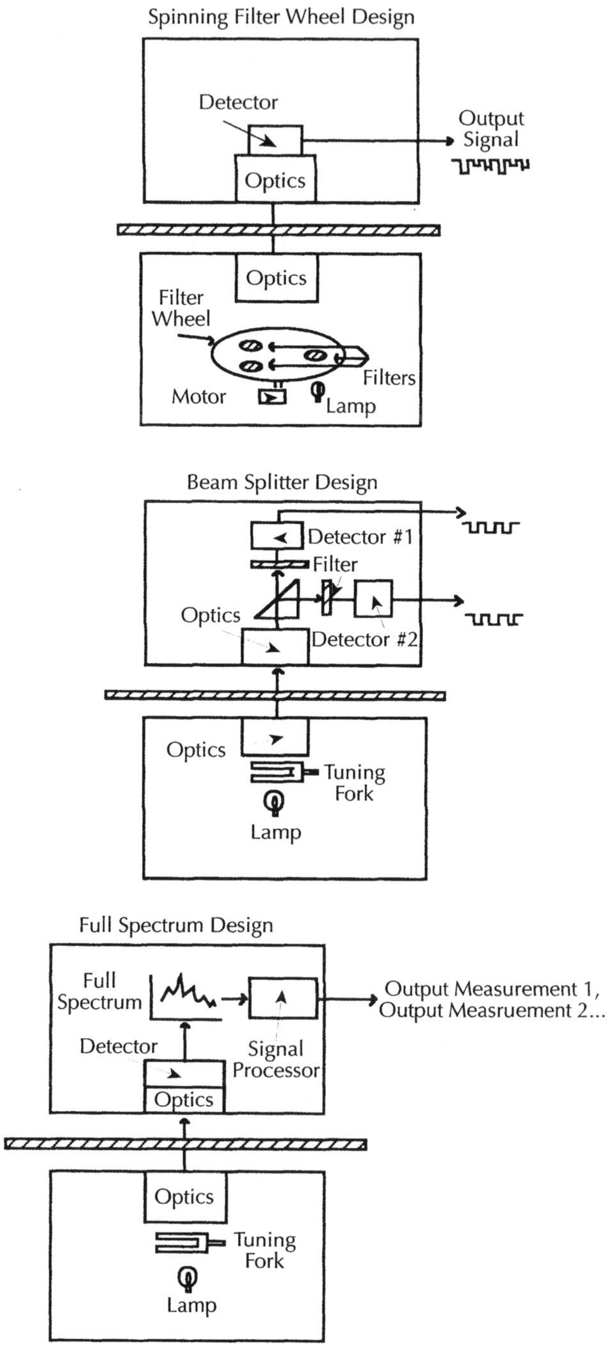

There are two modes of IR measurement:

1. **Transmission.** The source and detector are on opposite sides of the web. This obviously requires a transparent material that allows the light to pass through it.
2. **Reflectance.** The source and detector are on the same side of the web, and the light is reflected off of the base material back into the detector.

The infrared measurement technique has some inherent challenges in achieving a stable calibration. Such factors as stray light, changes in gloss, different base products, and pigments in the coating and/or the base typically can affect the gauge readings. Sensor technology attempts to compensate for these effects. New technology, looking at the full spectrum rather than isolated band widths, supported by powerful on-board computers, has introduced improved compensation for some of these effects as well as measurement of materials and layers which were previously impossible to see, such as silicone release coats, carbonless coatings, etc.

Limitations of IR measurement are as follows:

- The IR sensor cannot "see through" opaque materials, so if the extruded layer is "sandwiched" between two base products such as paper, aluminum foil, or heavily pigmented polymers, no measurement can be made.
- The IR sensor reads the coating directly and does not measure the base or total unless it is a transparent sheet where multiple layers can be measured. Therefore, in many applications, it is not possible to build a "flat" total product where "flat rolls" is the objective.
- In an A/B/A structure, an IR sensor cannot determine the individual thickness of each "A" layer. It will measure total "A" thickness (Fig. 8).

Fig. 8. An IR sensor will measure total "A" thickness in an A/B/A structure

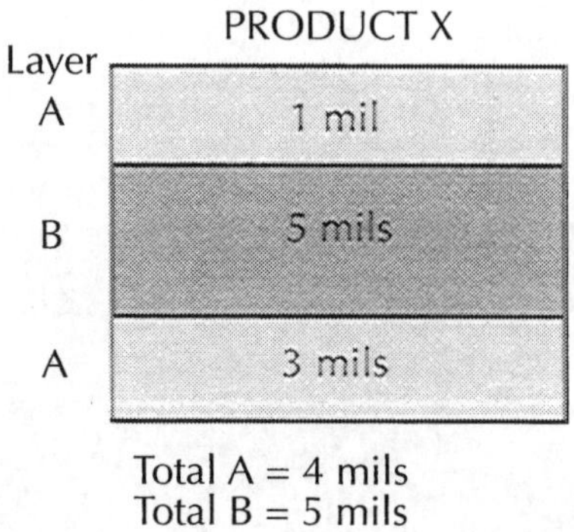

COEXTRUSION

The coextrusion process involves simultaneously extruding two or more polymer layers through a single die that is fed by two or more extruders. A nuclear gauge can measure only the total mass, so it is incapable of detecting discrete layers in a coextruded product. Since some infrared sensors measure a particular component, this technique is the logical choice for measuring individual layers in coextrusions (Fig. 9). In practice, however, with conventional sensors, there are some major factors that must be addressed.

Fig. 9. Measuring individual layers in coextrusions

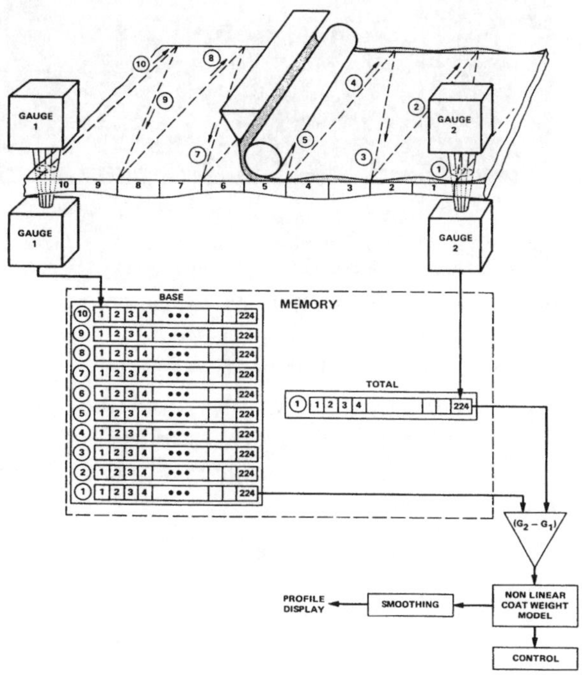

A separate filter must be selected for each different polymer to be measured. The IR sensor has difficulty separating the measurements of certain combinations of polymers because their IR attenuation signatures are quite similar. This can be overcome in many cases with a full-spectrum sensor.

TYPICAL SYSTEM CONFIGURATION

On an extrusion coater, the scanning frames are connected to an operator's console, which usually is located beside the line. This console traditionally has contained the computer uses a color video for product displays and a pushbutton panel for operating the measurement system. Newer systems have touch screens in place of push buttons for improved operator interface. The latest generation systems also distribute the computing power for scanning, measurement, and control to increase the power and reliability of the system.

Figure 10 shows an example of display normally viewed by the operator. The top two graphs show the variation in product weight or thickness in the cross-machine or transverse direction. These variations are referenced against the last scan average, and the die bolt positions are located across the display.

Fig. 10. Display of profiles/trends

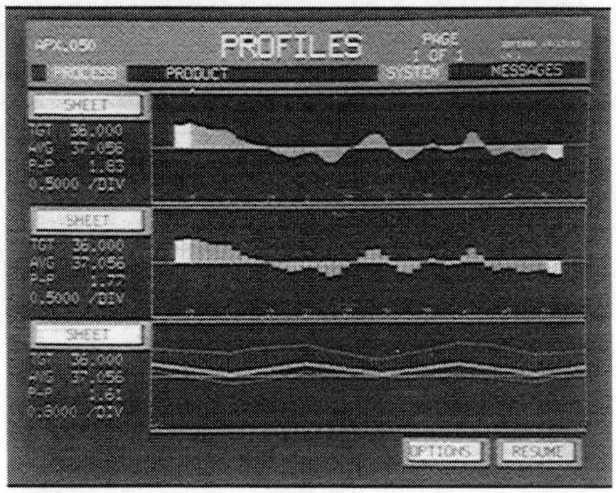

In the lower graph, the individual points show each scan average value for the last 80 scans. Superimposed over each bar is a narrow vertical line that represents the highest and lowest reading measured during the scan.

Other typical display pages include:

Recipe. Figure 11 shows a line setup condition. The currently running product is in the "Active" column, while the setup conditions for the upcoming product are shown in the "Preview" column. There can be multiple pages of these, and they can include not only sensor information but also temperatures, etc., from other systems on the line integrated into this system. All integrated systems can then be setup on a recipe change with one touch of the screen.

Fig. 11. Run specs for process setup

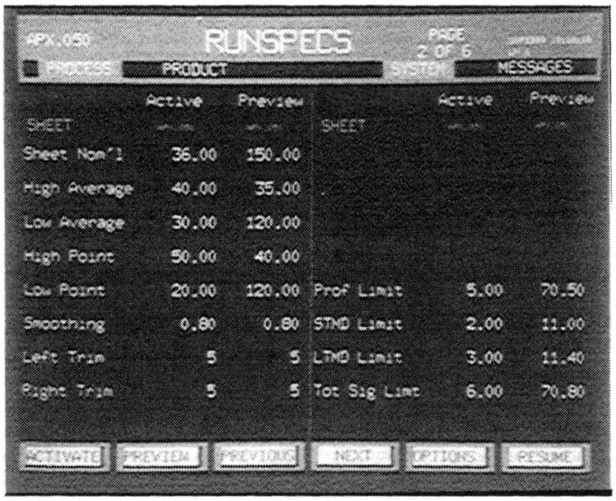

Process conditions and alarms. Figure 12 shows a display of current process conditions along with targets and high/low alarm points, whether from the gauging system or an integrated temperature system, chive system PLC, etc.

Fig. 12. Process conditions and alarms

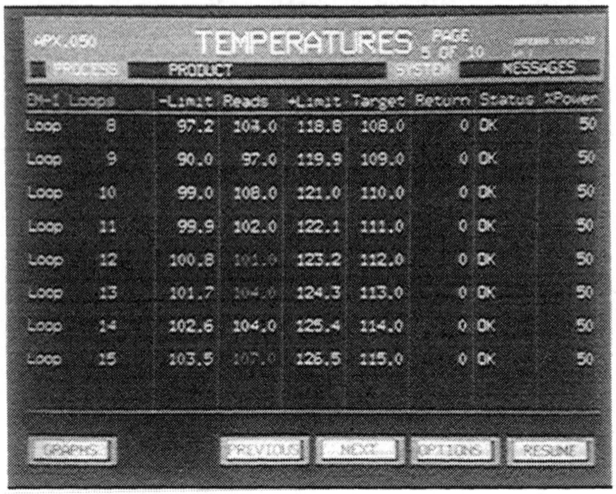

Special controls. Figure 13 shows a display used in conjunction with automatic profile control. (This feature is covered briefly later.)

Fig. 13. Die bolt control display

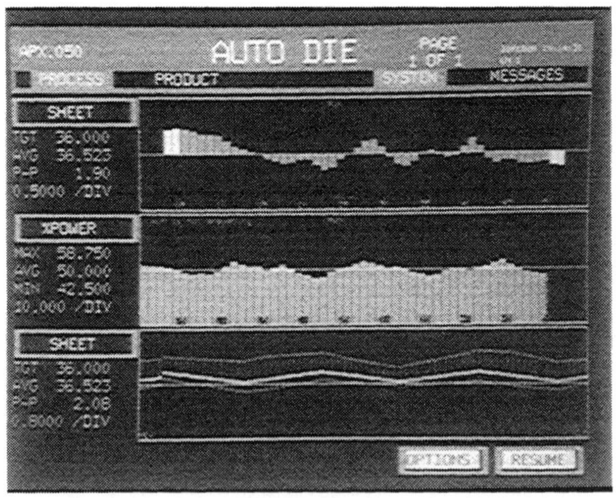

Trending. Figure 14 shows a typical process trend display. Some systems are capable of trending any variable being measured by the system and others can trend only a few selected variables. With some systems, variables from the gauging system, temperature control systems, PLC'S, etc., can be displayed numerically and graphically at a single operator console.

Fig. 14. Process trending

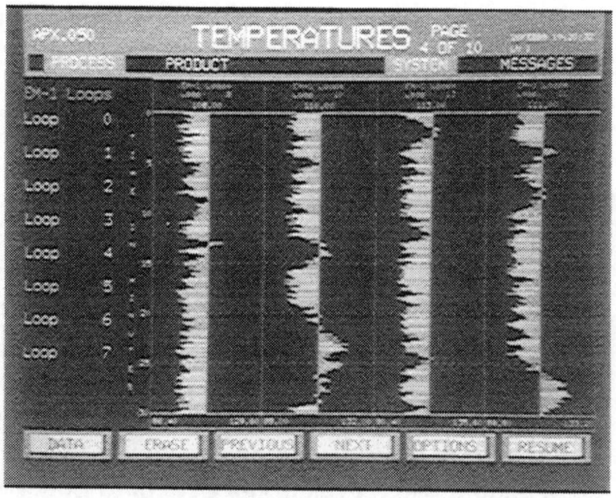

Statistical process control (SPC). Operators are becoming more and more involved in product quality improvement. However, a tool is needed to determine that the process is drifting out of control so that corrections can be made before poor quality and out-of-specification product is produced. Real-time SPC systems are available which alert operators when the process or product is drifting out of control (Fig. 15), show them the control limit and where the problem is occurring (Fig. 16), allow them to enter a cause if it is known (Fig. 17), allow them to enter tests that are not automatically recorded (Fig. 18), and show them current process capability information (Figs. 19 and 20).

Fig. 15. SPC/QC status

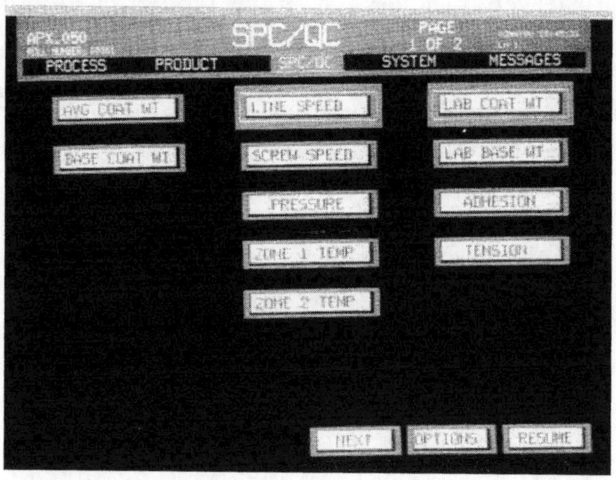

Fig. 16. X-bar and range charts

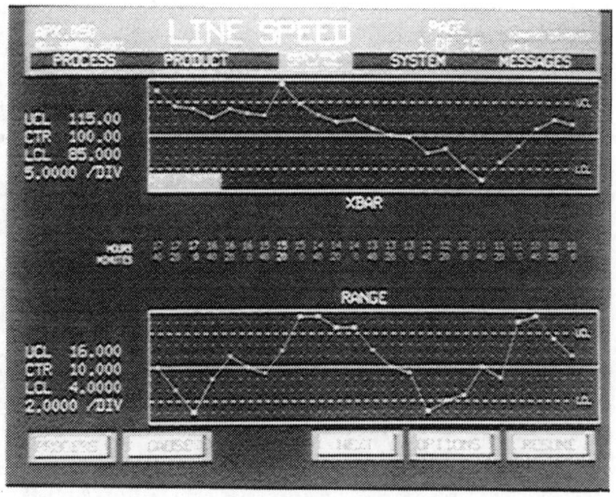

Fig. 17. Special cause of variation

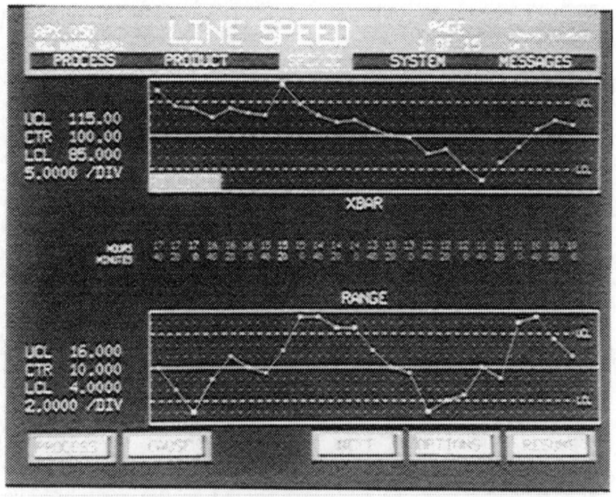

Fig. 18. Data entry

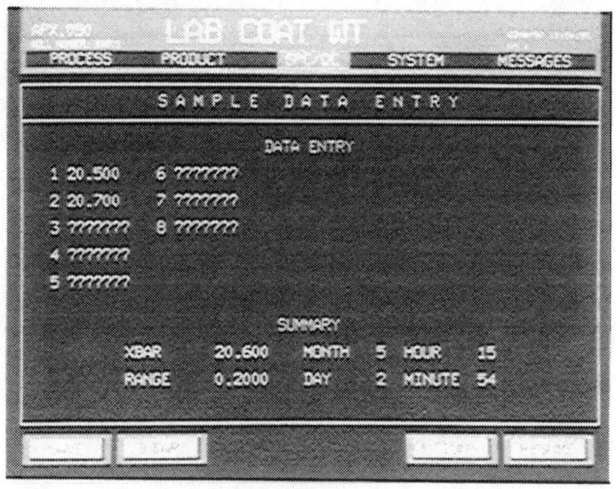

Fig. 19. Process capability

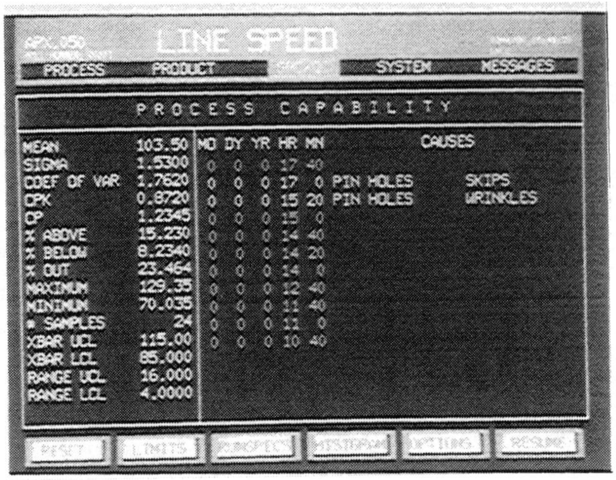

Fig. 20. Histogram

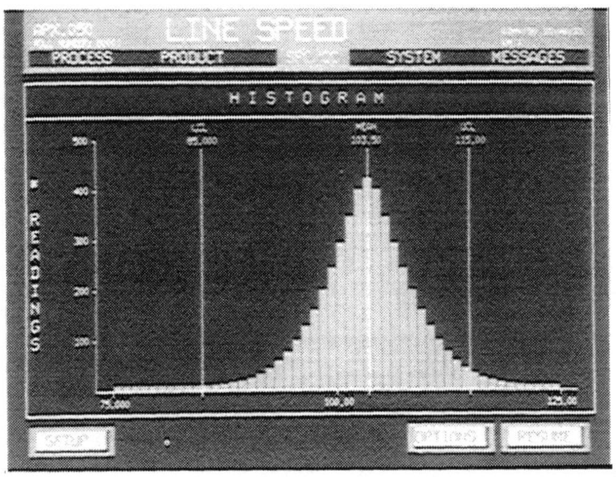

Information Systems. Typically, each system can be equipped with a printer to produce hard copies of the displays and canned "MIS" reports. Some systems incorporate personal computers and associated printers to handle data storage and printouts through computer links. Newer systems operate over local area networks and allow the user to configure custom reports from a wealth of data captured in a database. Data is made easily available at any level of the company. This is an area that should be thoroughly explored today, in light of computer integrated manufacturing (CIM) plans, before any investment is made in "islands of automation" which can be expensive to integrate and/or limited in functionality.

AUTOMATIC CONTROL

Every extrude can be characterized by some variation in the output. Only the magnitude and frequency of these variations differ from extrude to extrude. For example, there is a definite high frequency, or short-term, component of coat weight variation resulting from thermal or mechanical imperfections in the chill roll, melt instability, pressure fluctuations in the die, and slight fluctuations in line-speed or screw-speed drive units. Low frequency, or long-term, components of variation also exist in the machine direction. They are a result of batch-to-batch fluctuations in the coating resin, differences in operating conditions from shift to shift, and differences from operator to operator. A third component of variation occurring in the cross-web direction results from nonuniform flow through the die due to variations in gap opening or pressure drop variation across the die lip. The root mean square of these three components of variation (long-term, short-term, and profile) is called product sigma. It provides a measure of the overall capability of the process to hold a particular tolerance. A well-conceived process control system is intended to identify and reduce each component of variation. The overall process capability is improved, allowing both quality and economic benefits to be achieved.

Nominal control-machine-direction control to specification

The most basic type of machine-direction control consists of closed loop, proportional adjustment of either the haul-off speed or screw speed to regulate web thickness. With this type of control, the measured thickness is compared to a preset target and, should a deviation occur, an adjustment is made which is proportional to the magnitude of the thickness error and restores the thickness back to target (Fig. 21). This type of control is adequate for low-throughput extruders (less than 400 pounds per hour) where rapid control response is not an important consideration. After a correction is made, the system must wait until the effect of that correction has been seen by the gauging system before additional adjustments can be made. If the line speed is very slow, this can be a serious drawback because of the long time interval between control adjustments. Also, the responsiveness of the control system is fixed and is not sensitive to changes in process parameters. This can result in system over controlling at certain times and under controlling at other times.

Fig. 21. Nominal control maintains the product to target weight

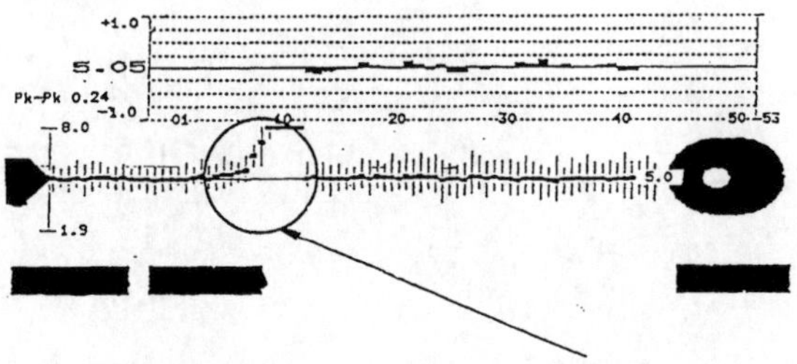

Some control systems execute more sophisticated control strategies, providing more rapid response to process upsets and better optimization of machine operating conditions. These systems utilize direct digital control (DDC) in which actual measured process parameters are fed into the system and compared with a process model stored in the computer. This makes possible feed-forward compensation for process changes that have not yet been seen by the gauge. DDC also self-tunes the response rate, or gain, of the control system based on changes that occur in other measured parameters.

Optimizing controls

The real economic benefit of more sophisticated control systems is their ability to carry out optimizing control strategies. Optimizing control provides a means of automatically selecting the optimum thickness target setting to meet product specifications while maximizing resin yield. The lower specification limit and percent below tolerance are programmed in for each product. These are used as inputs to the optimizing algorithms to continuously monitor the process variation against specifications. If a process upset occurs, the system senses the changes in process sigma and raises the target until the upset has cleared. When the upset has been corrected and process sigma has once again been restored to its normal operating lever, the target is reduced to the previous setting. In this mode of control, the process is no longer running to an average target; it is actually running against a low specification limit. This type of control is desirable when process economics favor minimizing raw material utilization.

Automatic control by the computer can also be used to manage the throughput rate of the extrusion coater. A targeted throughput rate is entered into the system along with maximum limits on key process parameters such as melt temperature, head pressure, motor current, line speed, screw speed, etc. When the system is put into automatic throughput control, it will slowly ramp the speed of the extrude. This continues as long as the system is not violating upper limits on key parameters. Once the targeted throughput rate has been achieved, the system will maintain the speed of the extrude at that rate. As the line speed of the machine is being increased toward the targeted rate, a corresponding adjustment is sent out to the drive system so that screw speed and other parameters are adjusted at the same relative rate. This means that a throughput increase can be implemented with no change in the basis weight of the product being produced. The real value of throughput maximization control is its ability to always run the machine at the optimum rate specified and eliminate changes in productivity from shift to shift and from operator to operator. This, on the average, increases the overall output of the extrude.

Automatic profile control (APC)

Nominal controls automatically adjust scan average coat weights, so the process is stabilized or optimized in the machine direction. The systems typically display cross-machine or profile variations but still rely on the operators to manually adjust die bolts to reduce variations in the profile. The automatic profile control solution, which was introduced in the 1970s, involved adding heaters to each die bolt so that bolt lengths could be changed thermally by automatically adjusting the level of heat to each bolt (Fig. 22). Since the economic return from the system can be limited by high-profile variation, the ability to flatten the product can yield high economic return.

Fig. 22. Die equipped for automatic profile control

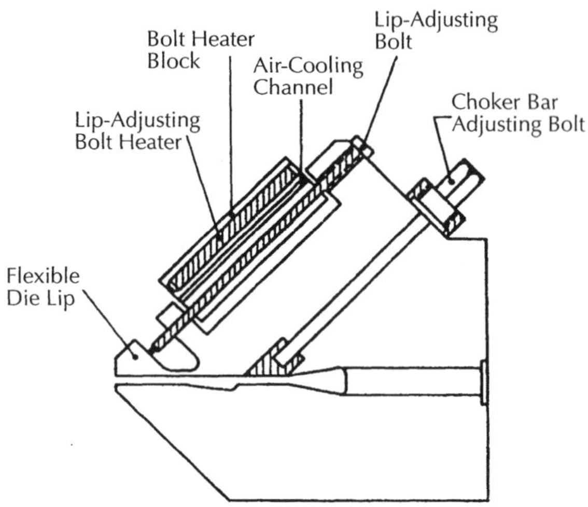

Fig. 23. Microprocessor interface to the automatic die

Fig. 24. Automatic profile control display

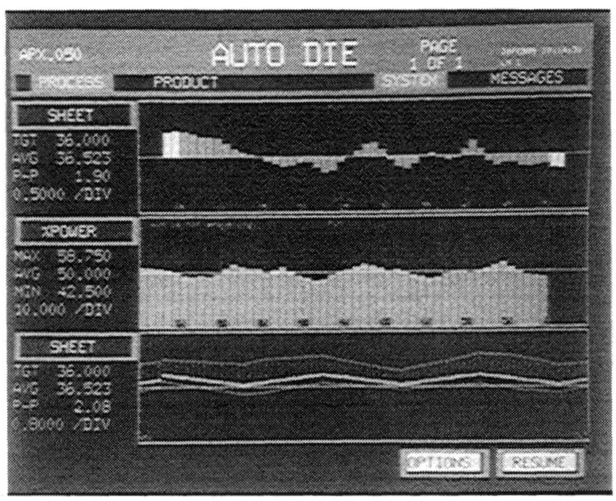

Control (APC) is typically implemented through a local microprocessor system communicating with a supervisory computer (Fig. 23). As each measurement scan has been completed, the new smoothed profile is compared to the target setting, and the computer sends incremental increase or decrease adjustments to the microprocessor for each die bolt position. The microprocessor implements these adjustments by raising or lowering the power level being maintained for each die bolt (Fig. 24). The control actuators are special die bolts that contain individual heaters. These heaters cause the bolts to increase or decrease in length as the power level is increased or decreased. Automatic profile control has the advantage of implementing profile adjustments across the full width of the die simultaneously. For proper operation, algorithms should be designed so that the net effect of all changes of the overall die gap is null, making it impossible to create a situation where, over time, the die lip would be over opened or over closed.

The more sophisticated versions of automatic profile control contain such features as PID and lead/lag control strategies. Also included are: accelerated time response algorithms, special measurement alignment to compensate for "neck-in" and die deckling, automatic control gains to deal with various die gaps and line speeds, line voltage compensation, and die bolt interaction models. Other important features include provisions to allow the system to be adjusted easily to produce shaped or "other-than-flat" product profiles and the capability to hold the die bolt heater values at their last setting incase of failure of the measurement or control system. This last feature ensures that the extrusion coater can continue to make a quality product and smoothly return to manual control in case of a system malfunction.

Scores of extrusion coaters are now utilizing automatic profile control (APC). The results are reductions in profile variability of up to 80%, with sustained variation less than ±1% in some cases, with the majority of the variation reduced in 10-20 minutes while the operator is attending to other duties on the line (Fig. 25).

Fig. 25. Actual reduction in profile variation with APC

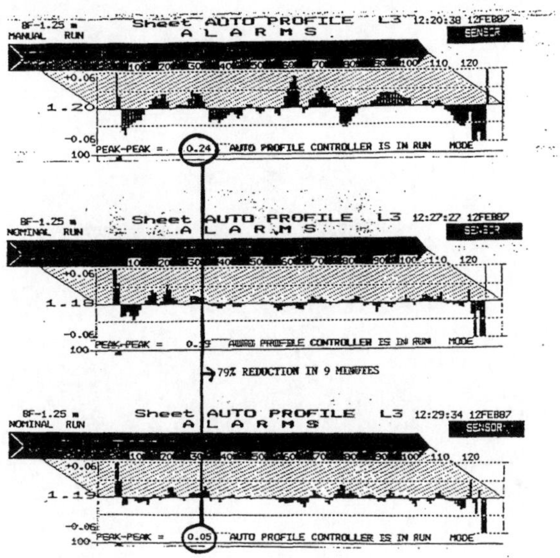

RESULTS OF COAT WEIGHT CONTROL

If the measurement and control system is chosen carefully, the initial objectives will be met and the investment will be justified through the cost savings of lower resin consumption, faster startups to produce on-specification products, greater throughput, or lower scrap rates. Another important result is the ability to penetrate new markets, which can frequently be achieved with a higher quality and more uniform product.

CHAPTER 14

Flame Plasma Surface Treatment

Joseph D. DiGiacomo, Flynn Burner Corporation

INTRODUCTION

Plastics and plastic-containing composites are currently utilized in a wide variety of packaging applications in the food, medical and pharmaceutical industries, and in many other consumer products, from container to automotive body components to geotextile materials.

In order to appeal to the sophisticated consumer of the 1990s, this packaging must be attractive and have eye-catching appeal. This means that it needs to be decorated. To obtain barrier properties, in the case of thin films, the substrate is laminated to at least one, but usually more, other material such as aluminum or other polyolefins. In automotive applications, the painted substrate must pass stringent quality control tests.

The very properties that make polyolefins so appealing for use in a multitude of applications are detrimental to converting processes such as painting, physical vapor deposition, printing/decorating, coating, and adhesion to other material. In order for polyolefins to be successfully converted, the reactivity of the polymer surface must be increased.

Surface tension, which is a measurement of surface energy, is the property—due to molecular forces—by which all liquids through contraction of the surface tend to bring the contained volume into a shape having the least area. A liquid placed on a flat, horizontal solid surface will either lay flat on the surface, called "wetting out" or "dewetting" to form a semispherical drop. An in-between state can also exist. The angle that the droplet or edge of the liquid makes with the solid is called the contact angle as shown in Fig. 1. The lower the contact angle is, the better the wettability. For solids, the nomenclature is surface energy; for liquids, it is surface tension.

Fig. 1. Surface contact angle

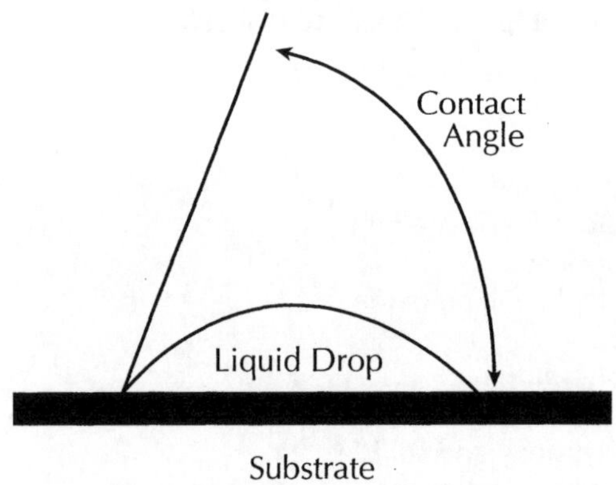

Table 1 shows surface energies of common substrates, and Table 2 shows the surface tension of liquids. The units are force/unit length (dynes/cm).

Table 1. Surface energy for typical polymer and substrates

HYDROCARBONS	Surface Tension
Polypropylene	29-31
Polyethylene	30-31
Polystyrene	38
Polystyrene (low ionomer)	33
Acrylonotrile butadiene styrene (ABS)	35-42
Polyamide	<36
Polymethyl methacrylate	<36
Polyvinyl acetate/polyethylene copolymer	33-44
Epoxy	<36
Polyester	41-44
Rigid Polyvinyl Chloride (PVC)	39
Plasticized polyvinyl chloride	33-38
Phenolic	N.D.
FLUOROCARBONS	
Chlorotrifluoroethylene (Aclar)	20-24
Polytetrafluoroethylene/ Polyterafluoroethylene (Teflon)	18.5
Polyethylene copolymer	20-24
Fluorinated ethylene propylene	18-22
Polyetrafluoroethylene (Teflon)	19-20
Polyvinylidone fluoride	25
Polyvinyl fluoride (Tedlar)	28
ENGINEERING THERMOPLASTICS	
Polyethylene Terephthalate	41-44
Polycarbonate	46
Polyamide	40
Polyaramid	N.D.
Polyaryl ether ketone	<36
Polyacetal	<36
Polyphenylene oxide	47
PBT	32
Polysultone	41
Polyethersultone	50
Polyphenylene sulfide	38
Polyhexamethylene Adipate (Nylon)	38-64
ELASTOMERS	
Silicone	24
Natural Rubber	
Latex	N.D.
Polyurethane	N.D.
Styrene buisdiene rubber	48
FLUOROELASTOMER	
Fluorocarbon copolymer elastomer	<36

N.D. = Not determined

Table 2. Surface energy for liquids

Liquid	Surface Tension dynes/cm
Alcohol	22
N-butanol	25
Toluene	29
Formamide	58
Glycerol	63
Water	72

As a general rule, acceptable adhesion is achieved when the surface energy of the substrate (measured in dynes/cm) is approximately 10 dynes/cm greater than or equal to the surface tension of the liquid. In this situation, the liquid is said to adequately "wet out" or adhere to the surface.

As long as solvent-based paints, inks, and adhesives were used in converting operations or substrate properties were such that paint adhesions was acceptable (as in the case of polyurethanes, as compared to polypropylene or TPO's) surface treatment, while important, was not critical because the surface energy of these materials is low. This all began to change with the Clean Air Act of 1979 and subsequent amendments. In order to be in compliance with the Act, converters began to use water-based coating, which-although helping to reduce volatile organic compound (VOC) emissions-created a criticality in surface preparation because water-based coatings have a much higher surface tension than solvent-based systems (see Table 2). Printing on films and non-absorbative substrates is considerably more difficult with water-based coatings.

TREATMENT METHODS

The primary objective of any surface modification or surface treatment is to improve the adhesion of a coating, ink, or substrate to the base substrate. Substrates can be polymers, metals such as aluminum, and glass. There are two approaches one can take: Decrease the surface tension of the liquid coating or add solvent to the surface energy of the liquid (but this will increase the amount of VOCs which will exacerbate the air pollution problem). The second approach, increasing the surface energy of the substrate, is more environmentally friendly.

Table 3 lists different methods used to treat polyolefins, metals, and glass to improve adhesion.

Table 3. Surface modification techniques

Direct flame plasma
Corona discharge
Glow discharge plasma
Chemical treatment
Coupling agents
Ozone
UV ozone

Direct flame plasma treatment consists of generating a plasma which results from the burning of a gaseous hydrocarbon fuel which reacts with the surface of the substrate. Flame plasma produces higher treatment levels (as shown Fig. 2) and longer lasting treatment levels (see Figs. 3 and 4). This technique does not produce ozone or other environmentally harmful by-products or use high voltage. Installation and operation is easy. Maintenance is low. Pinholing and backside treatment is not caused by flame treatment *(1, 2)*.

Fig. 2. Adhesion of LDPE to treated and untreated foils *(1)*

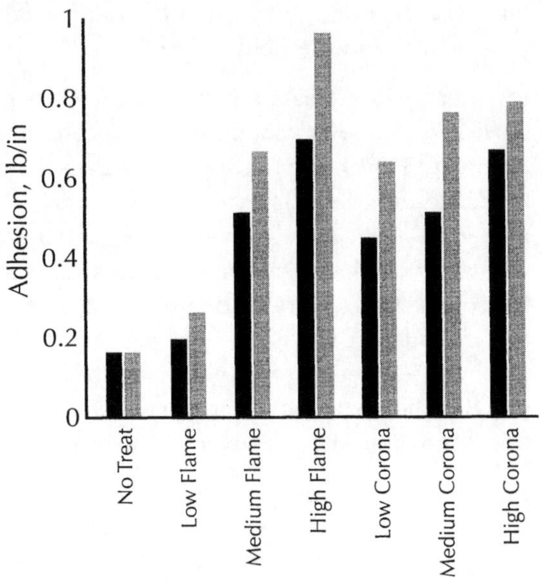

Fig. 3. Surface tension measurements *(2)*

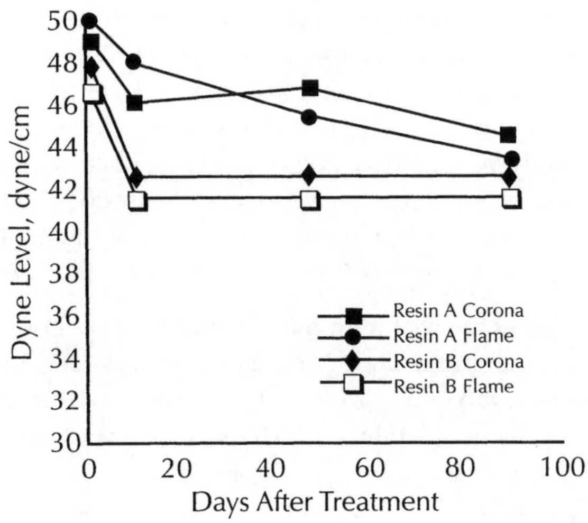

Fig. 4. XPS analysis of Resin A *(2)*

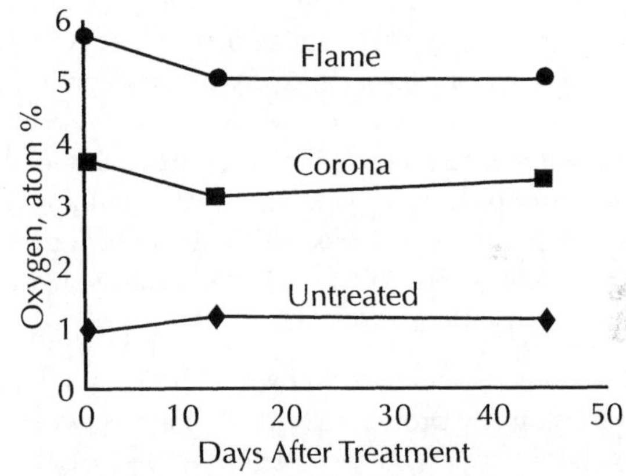

Corona discharge of a high voltage electrical discharge, a capacitor, in essence, using high voltage under atmospheric conditions its effectiveness is diminished under conditions of high humidity and produces an unwanted by-product, ozone, which must be eliminated by additional add-on equipment. Corona treatment does not easily conform to molded (three-dimensional) shapes, and the resulting treatment can be nonuniform. When corona treatment is applied to web or sheet-fed equipment, pinholing and backside treatment can result. Secondary treatment might be required by using water-based coatings or inks on hard-to-treat polymers such as polypropylene.

Glow discharge plasma is produced and controlled by ionizing a gas with an electromagnetic field.

Both radio frequency and microwave energy can be used. Glow discharge plasma generators are operated under vacuum in the range of 1-10 torr. Once ionized, the excited gas species react with the surface of the material exposed to the glow discharge plasma *(1)*. These systems are used primarily for molded polyolefin parts such as containers, catheters, etc. Operation is done on a batch or non-continuous basis. Cost of the equipment is higher than the other methods described.

Under chemical priming, the following methods of surface preparation are used: solvent cleaning, which removes surface contamination by dissolving the contaminant; detergent cleaning where oils, grease, and some mold release agents are emulsified in either hot or cold water solutions; and chemical etch using a highly reactive chemical such as an acid, a base, or an oxidizing or chlorinating agent, which change both the physical and chemical properties of the polymer surface. Most chemical etching methods require one or two water rinses and dry-off at higher than ambient temperatures. Many of the chemicals used are hazardous, corrosive, toxic, flammable, or poisonous. Safe handling, worker training, and appropriate control procedures are critical to minimize risk to the individual worker and harm to the environment *(3)*.

Coupling agents, also known as primers, improve adhesions by promoting cross-linking between the metal surface and polyolefin or between polyolefins and the organofunctional portion (of the primer) bonding or interacting with the polyolefin surface. The most widely used primers are salines. Their chemical structure is: $Y\text{-}R\text{-}Si\text{-}X_3$, where Y is the organofunctional group, R is the bulk polymer, Si is silicon, and X_3 is a CO, OH, OR group. Other chemicals used as primers are urethane polymers, isocyanates, nitrile phenolic, or vinyl. Cost, chemical hazard, non-responsive substrates and the requirement to apply the primer onto the surface almost immediately after cleaning are some disadvantages to this method. Primers are effective in promoting adhesion of metal to polyolefins as well as transferring strength from inorganic fillers to the resin *(3, 4)*.

Ozone, UV, and UV plus ozone provide functional groups that react with the polymer surface creating polarity (or functionality) at the surface. This is similar to the mechanism of flame, corona, and glow-discharge plasma methods. Ozone, UV, and UV plus ozone require much longer reaction times as compared to flame and corona but are effective in treating oddly shaped molded polyolefins. For example, a recent study (5) indicates that, compared to flame, the reaction time is 104 times longer.

A recent study comparing these methods shows that flame treatment most readily provides a highly wettable surface. Table 3 summarizes the contact angle measurements and ESCA (Electron Spectroscopy by Chemical Analysis), also called XPS (X-ray Photoelectron Spectroscopy) W/C (oxygen/carbon) atomic ratios, at a 38° electron takeoff angle, for polypropylene and PET films.

These data clearly indicate that-while corona, flame, and glow discharge plasma (referred to as remote air plasma) all attain approximately the same O/C ratios-the wettabilities are somewhat different. The flame treated films have the lowest receding contact angle and about the same advancing contact angels. The lower the contact angle is, the better the wettability is *(5)*.

SURFACE CHEMISTRY

The exact mechanism that occurs on the surface of a chemical inert, nonpolar polyolefin when direct flame is applied is unknown. There are several plausible theories advocated by knowledgeable people active in the field.

In direct flame treatment, the high temperature adiabatic flame temperature is approximately 1816°C (3300°F) combustion gases cause oxygen molecules to dissociate into free oxygen atoms and ions. Free electrons, carbon and nitrogen atoms as well as carbon-oxygen and oxygen-hydrogen groups, are also formed as a result of the combustion reaction. This gas is called plasma. These species react with each other as well as with the hydrocarbon polymer creating functionality within the top layer of the molecule.

Ether (R-O-R), ester (R-C=O-OR´), carbonyl (-COOH), carboxyl (O=C-O), and hydroxyl (-OH) groups and free radicals are also present in the flame plasma and are highly reactive.

Table 4. Surface properties of polypropylene films and treated PET films (5)

Treatment	Exposure Time(s)	ESCA Δ O/C Atomic Ratio[b]	θ_a (°)	θ_b (°)
None	-	0.0	70	53
Corona (1.7 J/cm^2)	0.5	0.13	42	<5
Corona (0.17 J/cm^2)	0.05	0.10	48	<5
Flame	0.04	0.12	29	<5
Glow discharge	0.1	0.11	44	<5
Ozone	3000	0.07	51	20
UV/air	300	0.15	48	15
UV/air plus ozone	300	0.17	41	14

a θ_a and θ_b are the advancing and receding water contact angles, respectively in air. Typical standard deviations are 2-3° for the contact-angle measurements and +0.03 O/C for the treated PET films.

b At a 38° electron take-off with respect to the surface

In studies of bonding, it was found that surfaces with high oxygen content and/or with oxy-carbon functional groups exhibited favorable results. Table 5 lists the species present in the flame plasma.

Table 5. Species present in flame plasma

Electrons
Oxygen ions/atoms
Ether and ester functional groups
Carbonyl, carboxyl, and hydroxyl functional groups
Free radicals

The flame plasma has an electron density of approximately 10^3/cm and electron energies of 0.5 ev (electron volts). While this is low compared to corona discharge, the mass flow rate of the plasma is much higher than corona discharge. The species contained in the flame plasma are highly reactive and affect the electron distribution and subsequently the electron density of the polyoelfin molecule. Oxidation is the transfer of electrons, and the species contained in the flame plasma are highly reactive and oxidize the surface of the polyolefin molecule. The electrons of the functional groups hold together atoms of different electronegativity, and these factors affect the electron distribution and shape of the electron cloud. The resulting asymmetry polarizes the surface of the polyolefin from a mono-molecular layer to several nonometers (10^{-9}m) *(5)*. By changing the polyolefin surface from a nonpolar to a polar surface, ink adhesion, coating, lamination, and metalizing are enhanced. The surface is much more receptive to subsequent converting operations.

EQUIPMENT SELECTION

The key in utilizing flame plasma surface treatment is control over key flame treating parameters such as:

- Flame geometry
- Energy output of the burner
- Boundary layer penetration
- Burner and web interface
- Dwell time
- Air/gas ratio control.

In order to set and control these variables, we need to review some basic information relating to combustion. Combustion is defined as a rapid chemical reaction between a fuel and oxidizing agent. In this case, the fuel is a gaseous hydrocarbon, and the oxidizing agent is the oxygen in the combustion air. The gas and combustion air must be supplied to the burner nozzle, to the reaction zone, or to the actual flame area, at the required rate. There are two basic types or flames involved in combustion: the diffusion (laminar flow) flame and the premixed (turbulent) flame. The diffusion zone is characterized by a gas-rich area into which oxygen must diffuse. The reaction kinetics of diffusion flames are slow compared to the premixed flames. The premixed flame is characterized with the fuel and air in close molecular proximity available for the combustion process and consists of an inner or primary core, a secondary or outer core, and the oxidizing portion of the flame.

For natural gas, the following reaction describes the combustion reaction:

$$CH_4 + 2O_2 + 8\ N_2 = CO_2 + 2\ H_2O + 8\ N_2$$

This reaction is exothermic, i.e., produces heat. The ideal air/gas ration is approximately 10 parts air to 1 part fuel. This is a stoichiometric air/gas ratio. A lower air/gas ratio is called sub-stoichiometric or "rich." A mixture with a higher air/gas ratio is "lean" and contains excess air. Best results are achieved when the air/gas ration is lean, or contains excess oxygen.

Flame Geometry

Flame geometry is controlled in a flame plasma treating system by the unique design of the burner which consists of a series of stainless steel "ribbons," the crimp and angle of which are designed to provide an extremely uniform flame front. This provides an even treatment across the web width. By changing the ribbon stack width and depth, an infinite variety of flame shapes can be achieved.

Energy Output

Energy output of the burner is controlled by the relationship between the air/gas mixer, combustion air pressure, burner discharge area, and burner flame retention features. The ribbon burner is designed for a wide range of burner firing rates. In addition, the Btu output of the burner must be programmed to follow line speed. As line increases, Btu output must increase correspondingly. Treatment level is a function of line speed as illustrated in Fig. 5. In the control system, an algorithm is programmed relating burner output to line speed.

Fig. 5. Treatment level vs. plasma value at different burner output

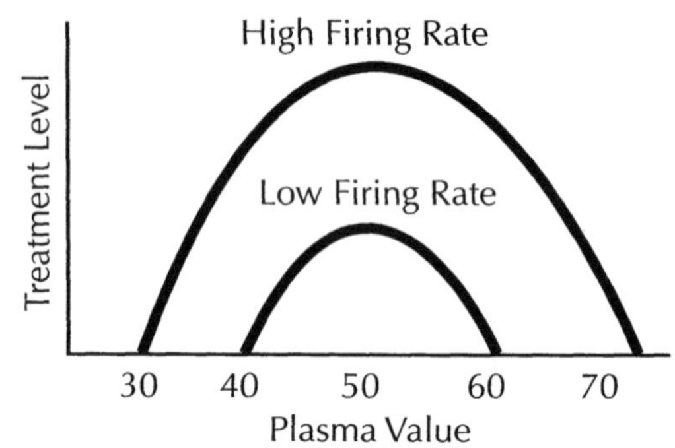

Boundary Layer

Boundary layer penetration is accomplished by producing a flame plasma with sufficient forward velocity to break though the boundary layer. A web which moves at typical operating speeds of 450-2500 ft/min induces a layer or air parallel to itself. Unless the flame plasma penetrates the boundary layer effective surface treatment cannot occur. The ribbon burner design provides enough velocity to the air/gas mixture that the boundary layer is penetrated at line speed of 2500 ft/min.

Burner/web interface

Burner/web interface-The flame plasma system for film, polymer coated paper, and aluminum foil requires a temperature-controlled backup roll and a water temperature control unit to provide water to the roll at approximately 45°C (105°F). This temperature assures that the film will be sufficiently flexible to have intimate contact with roll, but also eliminates any potential condensation. The water conditioning unit is also used to provide for cooling water to the burner.

The ribbon burner used for flame plasma surface treating has a cast-iron body with stainless steel ribbon ports designed to provide a very precise flame pattern. The flame is very smooth and continuous and will treat the surface evenly. A proprietary design minimizes burner deflection caused by induced thermal gradients in the burner.

The burner/web gap should be increased with increased burner output so that the plasma portion of the flame, which is just beyond the unburnt cones of air/gas mixture, is just at the web surface.

The burner should be designed so that little or no warping or bending due to heat or other factors occurs.

Dwell Time

Dwell time is an important factor in optimizing surface treatment. The moving web must be in contact with the flame plasma for sufficient time for the reaction kinetics to be maximized. This is accomplished by using multiple burner heads.

Air/gas ratio control is the single most important variable in achieving effective flame surface treatment. There are many methods that can be used to control air/gas ratio such as volumetric mass flow control, O_2 analysis, and flame treatment analysis. We investigated all of these techniques and found that none of them provided the close control required for effective flame treatment. We developed an air/gas ratio controller called the flame plasma analyzer, which controls air/gas ratio very precisely. Our work has indicated an oxygen concentration in the range of 0.1 to 0.5% excess oxygen in the flue gases is the optimum for flame plasma treatment. Figure 6 compares the scale for a typical O_2 analyzer and that for the Flynn flame plasma analyzer and illustrates the accuracy and the precise control provided by the analyzer.

Surface treating involves a plasma at the surface to be treated. Our approach is to simulate the reactions taking place at the surface by taking a 0.15 m_3/min continuous sample of air/gas mixture and producing a controlled flame in a closed chamber. The flame plasma produces an electrical signal, which is processed to produce the plasma value. The plasma value is an accurate, reproducible measure of the treating ability of an air/gas mixture and can be used in a control loop or displayed for information or manual control.

The Flynn plasma analyzer works as follows: A small sample of air/gas mixture is fed into the analyzer cell at a fixed rate. The cell and sensor are maintained at 812°C. The sample is ignited and the hot products of combustion flow passed the ceramic/platinum sensor and produce the electrical signal which is amplified, linearized, and processed to a voltage proportional to the percent excess oxygen. A discrete step function occurs when excess hydrocarbon (or unburnt gas) is present in the flue gas. This signal is in turn fed to a controller that has the desired PID tuning control loop constants necessary to operate the gas flow control valve. The air/gas mixer ensures a dynamically balanced mixture of air and gas.

A schematic representation of the Flame plasma analyzer is represented in Fig. 7.

Fig. 6. Typical oxygen analyzer scale compared to plasma value scale

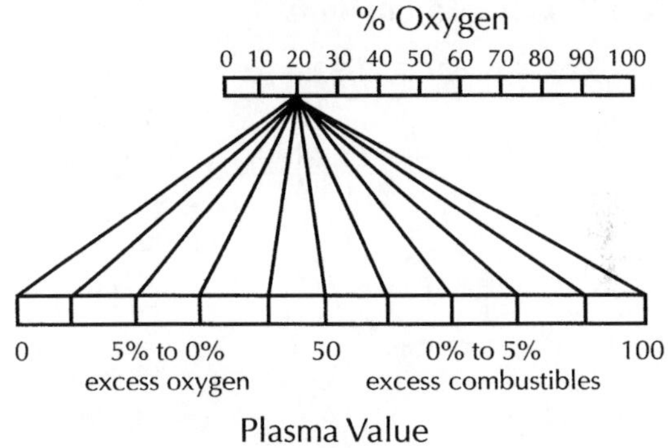

Fig. 7. Flame plasma analyzer

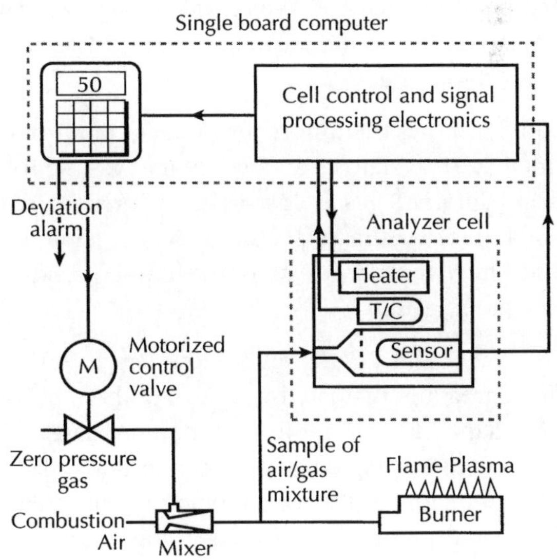

AIR/GAS RATIO

Combustion systems are designed on a volumetric basis. That is, approximately 10 parts of air are mixed with one part of natural gas to obtain the stoichiometric ration of air to gas. The combustion

Table 6. Effect of humidity and temperature on air density (6)

Temperature	Relative Humidity	Oxygen (O_2)	Nitrogen (N_2)	Other (Ar_2, CO_2, H_2)	Water (H_20)	Density (lb/cu.ft.)
60°	0	20.99	78.03	0.98	0	0.07632
60°	80	20.70	79.64	0.97	1.40	0.07592
60°	100	20.62	76.67	0.96	1.75	0.07581
90°	20	20.79	77.29	0.97	0.95	0.07189
90°	80	20.19	75.06	0.90	3.81	0.07111
90°	100	19.99	74.32	0.90	4.76	0.07086

Table 7. Gas composition

	Birmingham	Ohio	Pittsburgh	Alaska
Methane	90.0	94.1	83 .4	99.6
Ethane	5.0	3.01	15.8	-------
Propane	-------	0.42	-------	-------
Vol air/vol fuel (stoichiometric ratio)	10.47	10.72	11.70	10.52

reaction, however, depends on the mass of air to the mass of gas. Water vapor in air, expressed as % relative humidity will displace some of the oxygen molecules present, changing the air/gas ratio. Temperature changes, which affect air density will similarly affect in the air to gas ratio. Table 6 illustrates these affects.

Table 7 shows the difference in the amount of CH_2 (methane) available in the natural gas supplied from different gas fields. The difference in the amount of methane available in the natural gas will affect the air/gas ratio and consequently the treatment level.

Consistency in air/gas ratio is the key to effective flame treating; that is why control of the ratio is so important. For example, assume the air/gas ratio of combustion system is adjusted to maximize the surface treatment. If the temperature, humidity, or gas composition changes, then the air/gas ratio will change and, in all probability, will no longer be maximized. The treatment level achieved will not be as high as possible. This is the reason it is possible to produce, regardless of ambient conditions, the same air/gas ratio that was set to yield the highest possible surface treatment.

SAFETY AND OTHER CONSIDERATIONS

Many people unfamiliar with gas combustion systems are concerned about safety, as they should be. Industrial combustion systems have been utilized since the advent of significant gas usage for industrial processes beginning in the 1900s. These early systems were crude and were not supplied with the array of redundant safety controls available with current systems.

Modern industrial combustion systems are designed to standards specified by the Factory Mutual Insurance Association (FM) or Industrial Risk Insurance (IRI). Both of these organizations are umbrella companies, under which the majority of insurance coverage is written for industrial combustion systems.

The standards for the industrial combustion system include the following monitoring and flame and burner management equipment:

- High and Low Gas Pressure Switches—If the gas pressure is lower than or higher than the safe operating setpoints, the system cannot be started. If during operation, either of these or both pressures are outside of the setpoints, the system instantaneously shuts down.

- Low Combustion Air Pressure Switch—As above, if the combustion air pressure-at any time-falls below the operating setpoint, the system either not be able to start or will shut down if operating.
- Exhaust For Suction Switch—The system can be started only if the exhaust fan is operating and producing sufficient suction. This is sensed by the suction switch. If above setpoint, the system cannot be started or will shut down if operating.
- Minimum Speed Switch—The burner cannot be ignited or will shut down if the line speed does not reach the setpoint or falls below the setpoint.
- Burner Retraction Mechanism—If the line stops and/or minimum line speed is not attained, the burner will automatically rotate away from the web and gas will be shut-off to the air/gas mixer.
- IRI Double Block and Bleed Shut-off Valves—Two normally closed, zero leakage fuel gas shut-off valves with a normally open vent valve are provided in the gas train to positively and instantaneously shut-off the flow of gas to the system in the event that the starting and operating, interlocks (high, low gas pressure, etc.), flame failure, and customer interlocks are not at setpoint.
- Electronic Spark Ignition—This provides a 10-second trial for ignition of the burner. If flame is established and electronically verified, the shut-off valves remain open; if not, they instantaneously close in less than one second and the system must be purged before a relight is attempted.
- Electronic Flame Monitoring—This verifies by flame rectification, not heat, that a proper flame is always present when the gas valves are open. If flame is not present, the valves close instantaneously.

All of these interlocks, etc., are listed and/or approved by various testing agencies such as FM, UL, and CSA (which is the Canadian approval body). These listings/approvals are the result of very rigorous testing by these independent agencies.

We essentially utilize the same gas-fired system for all surface treating applications. This consists of a cast-iron ribbon burner, a plasma generator housing the combustion air blower, proportional mixer, motorized air and gas control valves, zero regulator, pressure transducers, pressure switches, and fuel shut-off valves. An operator control console, containing a single board computer with operator interface, message display, and keypad, are provided in the electronic control console. Electric relays, motor starters, and flame relay are housed in the electrical cabinet, which is integral to the plasma generator.

One last comment regarding safety, most of us are exposed to the combustion process by virtue of the fact that we drive automobiles. In an internal combustion engine, gasoline is volatized and burnt in the cylinder. Do any of us think it is dangerous to drive our cars? Not really, because of redundant controls. Modern industrial combustion systems have many more safety interlocks than today's automobiles.

CONCLUSION

With the successful development of the flame plasma analyzer, effective flame plasma surface treatment is now available. A formerly elusive variable in flame treatment has been identified and can now be measured and controlled. Modern flame plasma surface treatment provides high treatment levels without the deleterious side effects of other treatments, reliably, consistently, and efficiently. The use of water-based inks can now be actively pursued without concern for proper adhesion.

BIBLIOGRAPHY

1. Potts, M.W., Hansen, H.M., Kuettel, B.J., Goins, J.D. "Effect of Corona and Flame Treatments on Extrusion Coating Performance Properties," TAPPI 1993 Polymers, Laminations and Coatings Conference Proceedings, TAPPI PRESS, Atlanta, pp. 443.
2. Cooper, J.L., Cheatham, C.M., Hansen, H.M., "Surface Characteristics of LDPE Extrusion Coatings After Flame and Corona Treatments," TAPPI 1993 Polymers, Laminations and Coatings Conference Proceedings, TAPPI PRESS, Atlanta, pp. 321.

3. Harrington, W.F., in Coatings Technology Handbook (D. Saths, Ed.) Marcel Dekker, Inc., New York, 1991, pp. 279-285.

4. Waldman, B., in Adhesion Science and Technology Short Course Notes (K. Mittal, leader) Chapter J., The Center For Professional Advancement, New Brunswick, NJ, 1995.

5. Strobel, M., Walzak, M.J., Hill, J.M., Lin, A., Karbshenski, E., Lyons, C.S., "A Comparison of Gas Phase Methods of Modifying Polymer Surfaces," J. Adhesion Sci. Tech. 9(3):365 (1994).

6. Lindland, H.T. and Grandville, C., "New Developments in Flame Treating," TAPPI 1990 Polymers, Laminations and Coatings Conference Proceedings, TAPPI PRESS, Atlanta, p. 897.

CHAPTER 15

Winding

Jan Gronewold, Voith Sulzer Paper Technology

ABSTRACT

The following chapter covers a wide range of material properties, winding methods, types of operations, and winding plateaus, ranging from the basics to elaborate winding functions with the intention to benefit both the novice and expert through broad viewpoints and detailed insights to many applications. This is an update of the paper "Winding Tools and Variables," presented at the TAPPI 1995 Finishing and Converting Conference, and excerpts from the 1998 TAPPI PRESS book *Winders, the Complete Guide for Paper Mills and Converters*. From center to center-surface winding, from duplex to differential, gap and contact winding, this paper relates to all types of winders, ranging from reels to dual turret winders with in-line slitting.

INTRODUCTION

Winders are found in almost all manufacturing processes of "raw" web materials (such as textiles, rubbers, steel, paper, film, nonwovens, and composites such as film/aluminum foil laminates, countertops, wallpapers, and carpets) that are produced in a "continuous" process, that is, in an uninterrupted manufacturing cycle without stopping. These processes are continuous because capital and energy-intense machinery do not produce any return on investment during downtimes. Frequent stops (due to maintenance, machinery changeovers, or technical incapability) result in substantial money losses, often creating hours and/or days of labor to get the process running again. Thus, the less production interruptions there are, the better it is. This explains the need for continuity in a raw material process.

Following a crude windup that often focuses on "continuous running" and quantity rather than on quality, we can find subsequent slitter-rewinder systems that produce smaller rolls (in width and diameter) of better quality (straight sides, roundness, consistent density) *(1)* not only for shipping and storage reasons, but also for subsequent processes. As "web-type materials" are converted into sheets, strips, ribbons, cups, or plates, they are unwound and rewound another 2-6 times in other converting processes *(2)* such as coating, printing, and extrusion lines that do not only have continuous turret unwinds and rewinders, but their own subsequent slitter-rewinder systems.

The objectives of paper mill production and finishing departments (as well as paper, film, aluminum foil, and nonwovens converters) are to take an unfinished product, turn it into a finished product, and pass it on to a customer in a way that it provides a return on investment, fulfills and motivates employees, and makes customers happy so they will come back for more. Apart from having to run an economically and ergonomically efficient operation with a respectable return on investment, the main focus is normally placed on providing customer satisfaction through high quality, low cost and quick delivery.

When establishing, changing, or improving a winding operation, we must keep the same objectives in mind. In pursuit of excellence, the task is to understand the basics, consider given parameters, fulfill production or process requirements, and put out paper and rolls of good structure and quality. When doing so in a competitive environment, we must not only analyze our current application and flaws, customer satisfaction, and our competitor's and customer's winding methods, but we should also look at other industries that might handle much broader winding challenges or that possibly serve their customers much better in other ways.

When we want to learn about reasons for one particular winding application, we ought to consider other industries and winding methods as well, including other types of operations and materials.

FUNDAMENTALS

Materials and their properties that effect winding

When looking at "all" winding aspects, we need to consider and differentiate as many web-type materials and their properties as possible. While web-type materials are usually differentiated by basis weight/grammage, thickness/caliper, density, sheet formation and surface, breaking strength, coefficient of friction (COF), compressibility (z-direction)/K2 stiffness coefficient, elasticity, bending stiffness, moisture contents, and porosity, *(2, 3, 4, 5)* more practical examples will enable us to better understand which winding method should be applied:

- Elastic/extensible materials (some films such as stretch-wrap, PE, and thin Teflon; tissue; most nonwovens; creped paper)

  versus: more rigid goods (paper, board, a few nonwovens such as Tyvek, some rigid films such as OPP or PET)
- Dense products (supercalendered papers, fax paper, LWC paper, glassine)

  versus: high loft/bulky materials (most nonwovens, tissue, filter media, most boards)
- Heavy weights (metalized papers, some laminates such as film-aluminum foil-paper)

  versus: lightweights (board, nonwovens, tissue)
- Smooth/slippery surfaces (silicone/glassine coated)

  versus: rough surfaces (encapsulated papers, sandpaper, tipping, uncalendered grocery bag)
- Low tensile strengths (aluminum foil, most tissue, ceramic nonwovens, some glass fibers)

  versus: high tensile strengths (corrugated medium, kraft, board, some nonwovens such as Tyvek)
- Surface/nip/gravity sensitive (carbonless papers; materials that are required to maintain high loft like some nonwovens; very dense materials that become very heavy as the roll increases in diameter, thus being prone to gravity induced defects, such as supercalendered papers, thermal papers, metalized papers)
- High caliper/gauge variations (most boards, films, P/S laminates)

  versus: low caliper variations (aluminum foil, a few papers)
- Recycled fibers (many boards, some papers such as kraft and newsprint)

  versus: virgin fibers.

The process of winding

If the above web-type materials are to be wound into rolls (for easy handling, less costly shipping, and to fit the next process such as a printing line) and if those rolls are to be of good quality while trying to meet ever-increasing demands from customers (who require to run higher speeds and larger diameter rolls, maintain closer tolerances, and obtain better print quality than ever), each material and its property requires a different winding method with varying tools such as tension, nip, or torque or a combination of tension, nip, and torque (commonly referred to as the TNT *(4)* tools of winding). (See Figures 1 and 2)

Fig. 1. The TNT tools of winding *(2)*

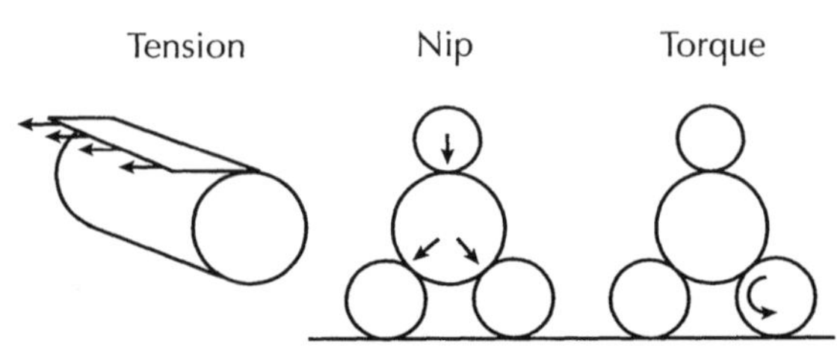

Fig. 2. Tension, nip, and torque *(2)*

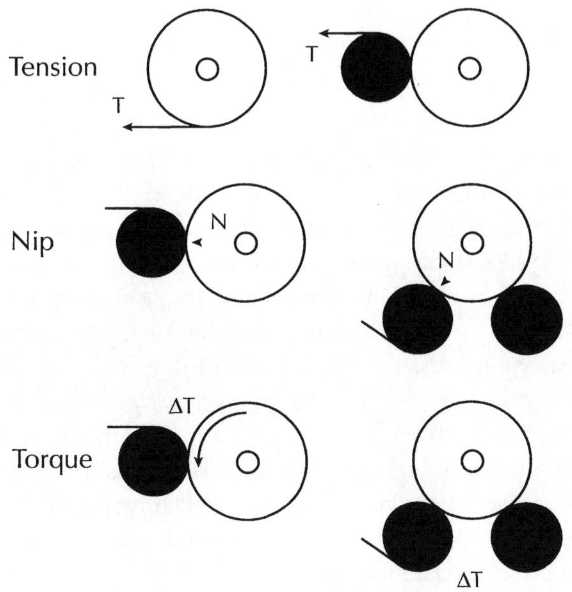

Roll quality *(1, 2, 4)*

A good quality roll can be summarized as having a roll of:

- Good edge quality (no dished or ripped edges)
- Good slit quality and tolerances (no dust, no slitter rings, no fuzzy edges)
- Good roll density, i.e., uniform roll hardness (as levels of wound-in tension)
- Tight starts (as a solid roll structure foundation)
- Proper outside dimensions (diameter, width, web footage inside the roll)
- Roundness (to run at high speed without vibration and to avoid cyclic tension surges)
- Roll must also be kept round during handling and storage
- Concentricity (meaning that the core is in the roll's center)
- Aesthetically and cosmetically well in appearance
- Absence of roll and web defects (such as internal bursts, wrinkles, slippage during winding, and rope marks)
- Runnability in subsequent processes
- Good furnish/material quality
- Density/fiber distribution/surface form (few cross machine variations)
- Good core quality (must support the weight of the roll, must transmit braking and winding torque, must have very little dimensional variations, must be less compressible than the roll (must be stiffer), the stiffness of cores must withstand speed of subsequent converting operations without vibration (consider natural frequency of cores).

Basic roll structure

Basic roll structure comprise the following:

- The TNT tools of winding

  Drives, brakes, drums, rollers, and tension controls that produce tension, nip, and torque

  Drum size relation, rolling nip theory, J-line/slip measurement

- The forces that hold a roll together *(4)*

  Inwound, i.e., residual tension (derived from tension, nip, and torque)

  Radial compressive forces (derived from tension, nip, and torque)

  Centrifugal forces during high-speed winding

  Material properties and behaviors

  Coefficient of friction

  Compressibility

  Tensile strength

  Modulus of elasticity.

Understanding tension

Tension, or rather "web tension" based on material tensile strength and modulus of elasticity, is applied to the web to: *(6)* avoid lateral web movement, avoid slack areas, avoid wrinkles, assist in spreading, and assist in producing inwound tension as the web moves from a process line, unwind, or slitting system to the rewind area. (Fig. 3)

Fig. 3. Web tension and control

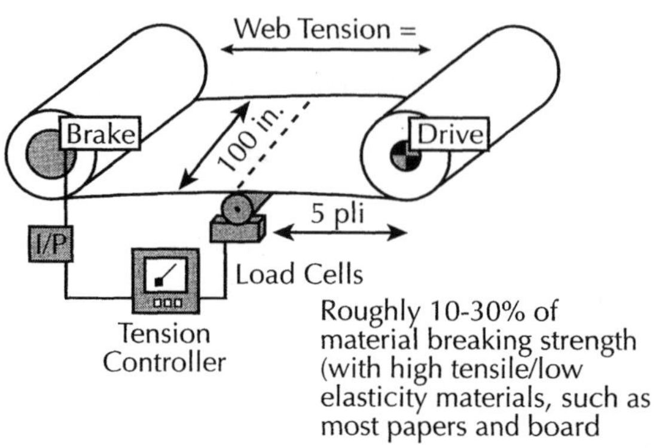

However, most high elasticity materials (such as creped tissue, elastic film, or nonwovens) need to be handled with different care and often with much lighter tension despite their breaking points. This is sometimes due to their inability of elastic recovery, but most often to prevent deformation during slitting and winding.

Web tension has a direct effect on rewind roll quality, unless unwind or process tension is totally isolated from the rewind area. Tension isolation can purposely be manipulated with drum diameter (amount of wrap), drum coating (for more or less friction, e.g., with tungsten carbide), center-surface drive features, and completely separate tension isolation rolls/pull rolls prior to the windup. *(1)*

In most cases, however, web tension is not completely isolated from the rewind area through capstan effects (friction, wrap around a drum), *(4)* nip, or torque. On two-drum winders, it is estimated that approximately 20-30% of web tension is carried into the rewind area, having a direct effect on roll hardness and roll quality outcomes. The remaining tension is changed/isolated through drum wraps (friction), nips, and torque. Even though not all roll hardness/quality relates directly to web tension, it is one of the most important factors in winding because inwound tension and radial compressive forces derive from it. See Table 1.

Further tension charts are listed in the TAPPI PRESS book *Winders,* and in the *Tappi Journal* article "Selecting Proper Winding Tensions" *(6)* by Gary DuBois (see references).

Web tension can be further segmented into tension creation, tension measurement, tension control/adjustment, tension isolation, inwound tension, and tension tapers.

Understanding nip

Nip is a pinching force between two surfaces. *(2)* (Fig. 4) In winding terms, it usually means that two rollers are forced together or that a winding roll of material is forced against another surface (often by its own weight) such as a winder drum. Nip-like stress areas can also be found inside a roll during unwinding and centerwinding without forcing a roll against another surface/roller. Internal pressure areas and nip-like effects are formed because the roll's own weight must be supported by a core and by already-wound web layers the weight of which causes pressure areas. (Fig. 5) Nip is omnipresent in all unwinding and winding operations, unless we find a way to eliminate gravitational forces.

Fig. 4. Nips with pull-rolls and winders *(7)*

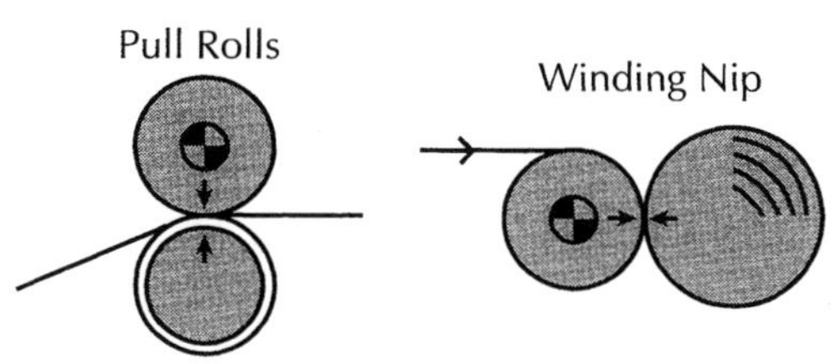

Fig. 5. Nips inside a roll *(8)*

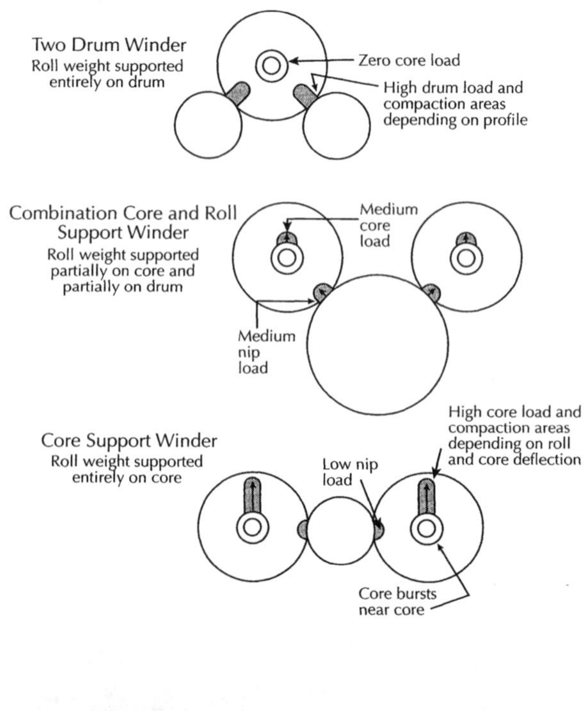

Table 1. TAPPI reference tension data for typical converting materials

Material	Tension (lb./in./mil)	Paper and Laminations	Tension pli
Aluminum foils	0.5 to 1.5 (1.0 aver.)	20 lb/R-32.54 gm/m^2	0.50 to 1.0
Cellophanes	0.5 to 1.0	40 lb/R-65.08 gm/m^2	1.0 to 2.0
Acetate	0.5	60 lb/R-97.62 gm/m^2	1.5 to 3.0
Mylar (Polyester)	0.5 to 1.0 (0.75 aver.)	80 lb/R-130.1 gm/m^2	2.0 to 4.0
Polyethylene	0.25 to 0.30		
Polyproylene	0.25 to 0.30		
Polystyrene	1.0		
Saran	0.05 to 0.20 (0.10 aver.)		
Vinyl	0.25 to 0.20 (0.10 aver.)		

Substrate*	Approximate Tension (lb/in.)
Mylar and oriented propylene	
0.0005 in.	0.25
0.001 in.	0.5
0.002 in.	1.0
Cellophane	
0.00075 in.	0.5
0.001 in.	0.75
0.002 in.	1.0
Nylon and cast propylene (non-oriented)	
0.00075 in.	0.15
0.001 in.	0.25
0.002 in.	0.5
Paper	
15 lb/ream (3000 ft^2)	0.5
20 lb/ream	0.75
30 lb/ream	1.0
40 lb/ream	1.5
60 lb/ream	2.0
80 lb/ream	2.5
Paperboard	
8 pt.	3.0
12 pt.	4.0
15 pt.	5.0
20 pt.	7.0
25 pt.	9.0
30 pt.	11.0
Laminations	
25 lb paper/0.0005 in. PE/0.00035 in. foil/0.001 in PE	3.0
0.001 in. cello/0.0005 in. PE/0.001 in. cello	1.5

* When these substrates are coated with polyethylene, nylon, polypropylene, EVA, EAA, and EEA, add the following tension values to the values listed here for the substrate only.
Coating thickness = Approx. tension lbs/inch: 0.0005 in. to 0.001 = 0.12 lb/in., 0.0011 in. to 0.002 = 0.25 lb/in.

Nip can induce tension on the surface of a winding roll that is significantly higher than the tension that is applied to the web up to the winding station.

"Surface" winders require nip to drive a roll (nip eliminates slippage between the driven drum and the roll of material) and to provide more winding hardness than tension or torque can accomplish alone. It is also used to prevent air from entering the rewind roll. Furthermore, nip is helpful in isolating web tension areas, as found with pull rolls. However, if nip is not administered adequately, nip can be the culprit for wrinkles and creases due to speed mismatches and uneven roller surfaces. It can also cause vibration problems with materials of high caliper variations and high coefficient of friction. *(4)* Overall, it can be said that nip-especially nip nonuniformity- induces more defects into a winding roll than any other winding tool. A winding nip often brings out the worst in a web and the list of nip-induced defects (NIDS) is quite long (see Roisum's book *Mechanics of Winding,* Smith's book *Roll and Web Defect Terminology*, or Frye's book *Winding*), which should make us appreciate torque and tension factors even more.

However, we must recognize that it is rarely "nip alone" that causes defects, but it is usually "nip nonuniformity," based on nonuniform paper and nonuniform control of roll structure. *(4)*

Nip-induced defects (NIDS) *(2)*

Nip can induce tension, compression, and shear stresses that will cause bursts near the roll's periphery (from MD tension), crepe wrinkles (from MD compression), corrugations (from MD shear), and bursts near core (combination of tensile, compression, and shear). Nip can also cause vibration (when uneven or out-of-round, driven rollers are nipped against each other), dishing, interweaving, and loose cores.

A more detailed discussion about nip in the book Winders includes the following subjects:

- Mechanics of the "rolling nip" (nip and drum diameter relationship, J-line)
- Nip and drum diameter relationship, and first nip
- Slippage (J-line measurement).

Understanding torque

Torque applied to a winding roll should be viewed in conjunction with the source (drive) and how it is applied. *(9)* Torque can be induced through outer web layers of a roll or directly applied to the center of a roll (true torque). In order to comprehend torque, we must first look at electric drives.

Drives provide speed, torque, and horsepower in order to *(10)*:

- Pull or push a load (total web tension = pli × width)

  Pull the load into the windup (motoring)

  Hold it back at the unwind (regenerating)
- Web tension while unwinding (regenerative braking)
- Provide derived roll hardness at the windup
- Drive auxiliary rolls (spreaders, slitter rolls, etc.)
- Provide speed match to the web (with auxiliary drives)
- Provide accelerating, decelerating, and E-stop functions (at desired rates, depending on material grades)
- Provide start, jog, thread, coast to stop, and s-curve functions.

Understanding the speed, torque, and horsepower relationship

In winding terms, horsepower (work delivered) and torque (rotational force) are required to pull a certain web tension (load) and transport the web over a distance at a certain speed. Horsepower, torque, and speed work in relation to each other: hp = torque (ft·lb) × speed (rpm) ÷ 5252.

Horsepower

Horsepower is a given amount of work performed within a unit of time. It is based on James Watt's finding that a strong horse could raise a weight of 550 pounds to 1 foot in 1 second. Thus 1 hp = 550 foot-pounds per second. Since a more common term in winding is "feet per minute" (not per second), the equivalent of 33000 foot-pounds "per minute" is used instead (550 pounds × 60 seconds). 1 hp = 33000 foot-pounds of work per minute.

Horsepower is "exerting a force over a distance at speed," but in simplified terms you could look at delivered horsepower as "work being done at a particular rate."

The "work" could be moving a 500-pound object (100 in. wide web at 5 pli) over a distance of 2000 feet. Work is calculated by multiplying the item's weight (500) by the distance it moves (2000), resulting in this case in 10,000 foot-pounds.

"Pure surface winders" pull a web over a distance with an external member, i.e., winder drum of a given diameter by means of tension horsepower. With a fixed winder drum diameter, the rotational speed (rpm) is directly proportional to web speed. The drive usually works in the constant torque range of the motor because the load (web tension) usually stays the same, but speed and horsepower vary as the winder accelerates and decelerates.

The following is a simplified tension horsepower calculation for "speed-regulated" surface winders/drives *(10)*:

$$\frac{\text{Web tension} \times \text{Web width} \times \text{Speed}}{\text{Constant}}$$

or

$$\frac{5\text{ pli} \times 100\text{ in.} \times 2000\text{ ft/min}}{33{,}000} =$$

30.3 hp (plus approximately 10% for friction losses).

Because of the fixed winder drum diameter, we can also calculate the circumferencial drum speed:

$$\text{rpm} = (\text{ft/min} \times 12) \div (\text{diameter} \times \text{pi})$$

$$= (2000 \times 12) \div (20\text{ in.} \times 3.14).$$

A 30.3-hp motor with 382 rpm base speed would be required, if such a motor existed (motors come in standard frame sizes with standard base speeds, e.g., 350 rpm, 850 rpm, and 1150 rpm). The hp will vary from zero to base speed of a motor; thus, we need to ensure that the drive develops sufficient hp at the determined speed and that the proper "base speed motor" is selected.

Horsepower is directly proportional to the speed of the winder and, because machinery is expected to reach a certain speed, most winder drives are considered "speed-controlled" drive applications, even though the motor itself might actually be torque-regulated. Despite the ultimate drive control, all web-type/production machinery must have a speed reference.

This is where we need to distinguish the "technique of drive control" versus "what the drive delivers." Speed control is not always equal to drive control. Torque and speed control have less to do with center vs. surface drive than with the function that needs to be accomplished and where the power is delivered to. For instance, most (surface winding) reels in the industry are torque controlled according to Pete Werner of Rockwell Automation. This way, the drive can control the sheet tension in a more controlled fashion. Werner further indicates that the drive control often depends on the material to be run. For example, in a tissue application-where no additional tension should be induced into the sheet-all drives and rollers are speed controlled, based on one reference point.

The different torque and speed control solutions are further listed in the book *Winders*.

Torque

Torque is a rotational force measured in engineering units such as foot-pounds or Newton meters. (Fig. 6) In the case of an electric drive, it is proportional to current flow. *(4, 10)* Power divided by speed is directly proportional to torque (Power ÷ speed = torque). Example: A fast spinning motor delivers less torque than a slower spinning motor of equal power.

Torque is responsible for pulling a load or web tension (Total tension = pli × width), whereas horsepower is responsible for moving the web over a distance at speed.

Torque is proportional to force but, in simplified terms, you could say that torque is "potential work to be done."

Fig. 6. Torque = rotational force, pulling a load.

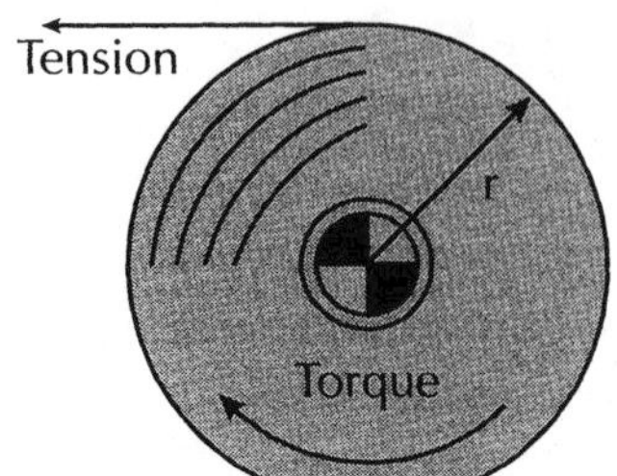

Torque = radius × web tension (20 × 5 × 100).

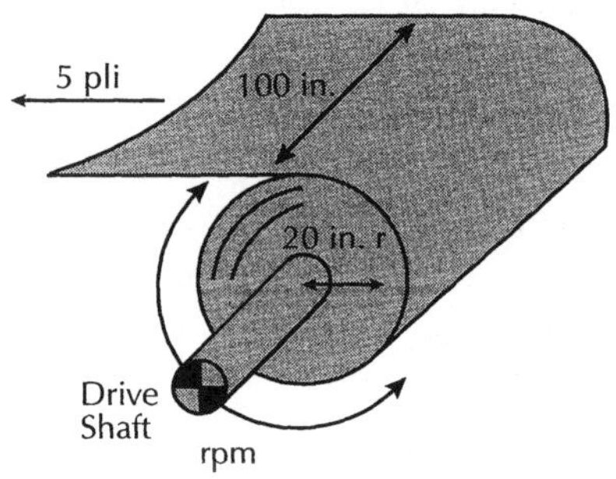

Unwinding (and center winding) is a typical "torque" application, as we are dealing with a "load" to be pulled (web tension) and with varying "rotational forces." Again we want to remember that horsepower, torque, and speed work in relation to each other (hp = torque (ft·lb) × speed (rpm) ÷ 5252).

As an unwinding roll turns down in diameter, rotational speed increases and torque requirements decrease at equal web speed and web tension. The torque produced by a motor must not only hold back web tension, but it must also overcome the roll's inertia (remaining web layers that are still wound). In other words, the total load consists of web tension plus the remaining web layers in a roll (apart from frictional forces). As the unwinding roll turns down in diameter, less and less torque is required because of the decreasing roll diameter (even though web tension remains the same). Thus, increased rotational speed and decreased torque (at a decreasing roll diameter) works out naturally fine in a torque-controlled environment. We can find the combination "high torque at low speed" and "low torque at high speed." Therefore, it is generally said that unwinds (and center winders) which deliver reasonable levels of torque (force) to the sheet use "torque-regulated drives." The drive works at constant horsepower (variable torque) over the speed range of the motor. This is accomplished by varying the field current and related excitation (field) flux to the motor.

The same "torque" principle also applies to pure center-drive winders, just reverse: As the rewinding roll grows larger in diameter, rotational speed decreases and torque must increase as the load increases (web tension plus already wound web-layers, i.e., the roll itself) at equal horsepower. Therefore, the torque requirements are directly proportional to the roll's buildup ratio.

Noteworthy with pure center winds is that increasing torque can be tapered off as the roll grows larger in diameter, so outer layers are wound less tight than inner layers with very slippery materials, preventing telescoping or dishing. *(1)* However, this principle depends very much on the material's coefficient of friction (see the section "The holding forces that affect roll structure"). If the material is very slippery (low coefficient of friction) and stretches (high elasticity), the statement about "tension/torque taper" holds true, as torque is directly proportional to in-wound tension times roll radius. But the less slippage there is between layers, the less effect will torque have on already-wound web layers because compressive forces will dominate inside the roll (see the section "The holding forces that affect roll structure").

Torque needs to be controlled in relation to the decreasing (unwind) or increasing (rewind) roll diameter, based on maintaining equal web tension (total tension = pli × width of sheet). Torque (often measured in foot-pounds) can be manipulated by increasing, decreasing, or reversing the electric current flow to the field or to the armature. The field current is often proportional to roll diameter, the armature current is then a function of tension setpoints, acceleration, deceleration, friction, and windage. With many dc drive winder applications, "armature current" is a direct proportional substitute for tension.

As web speed changes during acceleration and deceleration, armature voltage increases as winder speed increases.

The tension related torque required can be calculated by multiplying the roll's radius with the total web tension. For instance, a 6-in.-diameter roll (outer core diameter = start of run) under 500 pounds of tension requires 125 foot-pounds of torque (3-in. radius × 500 lb ÷ 12 in./ft), while a 40-in.-diameter roll (finished roll = end of run) under 500 pounds of tension requires 833 foot-pounds of torque (20-in. radius × 500 lb ÷ 12 in./ft).

Horsepower for a "center drive" is calculated as follows:

$$\text{hp} = \frac{\text{rpm} \times \text{torque (in lb·ft)}}{= 5252 \text{ constant}}$$

For this 2000-ft/min winder speed example, the required tension related power is the same at the core OD of 6 in., as it is at the full roll diameter of 40 in.:

$$\text{hp (linear)} = \frac{\text{500-lb tension} \times \text{2000 ft/min}}{\text{33,000 lb-ft/min/hp}} = 30.3 \text{ hp}$$

The resultant roll speed with the 6-in.-diameter core at 2000 fpm will be:

$$\text{Speed (rpm)} = \frac{\text{2000 ft/min} \times \text{12 in./ft}}{\text{6-in.-diam.} \times \text{pi}} = 1273 \text{ rpm}$$

The required torque at the 6-in.-diameter core to develop 500 pounds of tension would be:

Torque (lb·ft) = 125 lb·ft (as calculated above)

Therefore the rotational horsepower required at the roll for 500 pounds tension, at 2000 ft/min, when at the outer core diameter of 6 in. would be:

$$\text{hp} = \frac{\text{1273 rpm x 125 ft-lb}}{5252} = 30.3 \text{ hp}$$

The resultant roll speed at 40-in. full roll diameter at 2000 ft/min would be:

$$\text{Speed (rpm)} = \frac{\text{2000 ft/min x 12 in./ft}}{\text{40 in. dia. x pi}} = 191 \text{ rpm}$$

Per the above, the torque required at full roll diameter of 40 in. to develop 500 pounds of tension would be 833.3 ft·lb (as calculated above). As well then, the rotational horsepower required at the full roll diameter of 40 in., for the same 500-pound tension, at 2000 ft/min, would again be the same:

$$\text{hp} = \frac{\text{191 rpm} \times \text{833.3 ft-lb}}{5252} = 30.3 \text{ hp}$$

This is why center winding and unwinding at constant tension and surface speeds is considered a "constant horsepower application." At full speed, the motor works in the "constant horsepower" range during roll buildup or build-down.

The buildup (winders) or build-down (unwinds) ratios are important relative to selecting appropriate drive motor applications. In this example, the buildup ratio is 40:6 which translates to 6.666:1. This is a relatively wide ratio and is impractical to satisfy with a motor having a variable flux range (torque constant range) equal to this 6.666:1 range. Generally, 4:1 to 5:1 are maximum available field ranges (also referred to as flux and/or torque constant ranges). As such, there are two common solutions to this example:

- Fixed (field) flux motor
- Variable (field) flux range motor that can be field weakened to pick a lower hp, lower rpm motor size.

Further details would become too intricate for the purpose of this particular chapter, and the ultimate drive and motor size should always be selected by the drive manufacturer who must consider other functions as well.

In our example, the center drive ended up being a 200-hp, 1750-rpm fixed flux motor, and alternatively a 100-hp drive, 850-1750 rpm variable flux motor (field weakened to 1750 rpm).

The holding forces that affect roll structure *(4)*

What happens inside a roll while it is being wound, and what are the different forces that can make a roll collapse or hold together? This is important to know because the stresses in a wound roll determine the quality of a roll. Roll density is proportional to the levels of inwound tension. Inwound tension originates not only from web tension that is wound with the web onto a core (or shaft) and into the roll, but it is also induced through nip and torque. With paper, once a couple of layers have been built, inwound tension creates radial compressive forces that compact the paper below it. This compression of the layers causes loss of inwound tension 2-3 in. below the surface of a roll. (Fig. 7)

Fig. 7. Compressive forces and diameter reductions (courtesy of David Pfeiffer, Ken Frye, Alex Olshansky, and David Roisum)

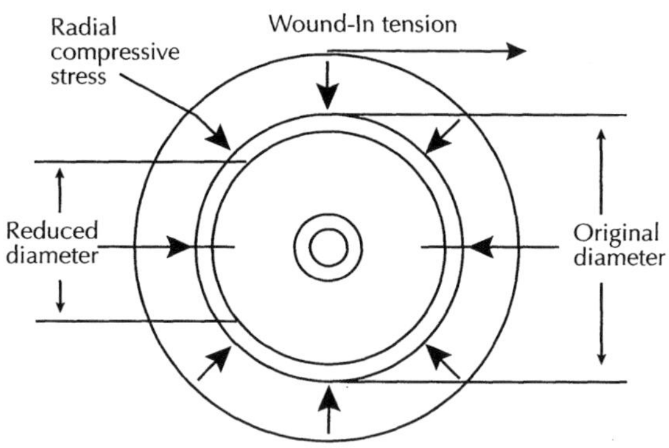

Compressive forces, in conjunction with the coefficient of friction of paper, create a friction force that resists layer-to-layer slippage inside a roll and that hold a roll together. Another way to visualize roll structure is to represent the levels of wound-in tension locked inside a roll during the process of winding. This can be measured by Beloit's WIT-WOT test (wound-in tension, wound-out-tension lab winder) or by the Cameron "gap test" which you can perform yourself, if you follow detailed instructions. pli = (gap width ÷ roll circumference) × caliper × modulus of elasticity. Residual strain (%) = (gap width × 100) ÷ circumference.

The higher the friction coefficient of a material is and the higher the compressibility is, the faster a roll goes into radial compression (vs. tension) and the less layer-to-layer slippage occurs. Less web tension is necessary to wind this roll. *(11)*

On the other hand, extremely slippery materials will hardly go into compression and inwound tension is most proportional to web tension. In addition, highly elastic materials will build up more inwound tension. Therefore inwound tension is often tapered off (through torque) as the roll grows in diameter in order to prevent stresses that causes slippage inside the roll.

Slippage between layers can cause the roll to dish (in extreme cases to collapse) as it is being wound, or to telescope in a subsequent unwind processes. The more layer-to-layer movement occurs, the more web and roll defects (such as crepe wrinkles) are found inside a roll. *(12)* However, this is most often due to "nonuniform" inter-layer movement.

With drum-supported winders, materials of very high coefficient of friction and high caliper variations are prone to cause vibration and roll throwouts, whereas materials of very low COF and low tensile strength will cause slippage and web defects such as crepe wrinkles and bursts.

When building a roll, it is of the utmost importance to have a good foundation, that is a tight start near the core. Loose cores, i.e., a loose start, are the culprit for many roll defects. A good start can be accomplished by temporarily increasing the unwind tension, by inducing torque and having adequate rider roll pressure after a few layers have been wound, and by utilizing adequate cores (with close tolerances, high stiffness, etc.) and web-to-core attachment methods.

TYPES OF OPERATIONS

Let's look at the different types of operations found at mills and converters, and their specific requirements. In winding terms, operations can be distinguished by:

1.) What a roll is wound on *(2)*

- Web to spool/coreshaft operation
- Web to core operation

- Spool to core operation
- Core to core operation
- Core to spool (traverse wound)
- Core to sheet operation

2.) Process

- Continuous operation (a paper machine or process line such as nonwovens or coating lines)

  versus: discontinuous operation ("stop-and-go" printing or coating lines with frequent changeovers, or a typical slitter-rewinder)
- In-line winding (in-line with a paper machine; or other process line via turret winder)

  versus: off-line winding (off paper machine; more focused and accurate winding such as most slitter-rewinders)
- Continuous off-line winding (e.g., with an off-machine coater; or an automatic slitter-rewinder that converts large diameter unwind rolls (60 in.) to small-diameter rewind rolls (8-12 in.) for drafting paper, butcher paper, or gummed tape that is not "in line" with a paper machine)

  versus: discontinuous off-line winding (typical slitter-rewinder)
- Continuous in-line winding (paper machine)

  versus: discontinuous in-line winding (stop-and-go process line)
- Shafted winding (paper machine reels, re-reelers, off-line-duplex slitter-rewinders, off-line-center-surface simplex slitter-rewinders, some coating and printing lines)

  versus: shaftless winding (higher production winders, e.g., two-drum winders, single-drum duplex winders; high output and more costly converting lines)
- Winding on cores (very typical for shipping and subsequent handling purposes)

  versus: winding without cores (often with pulp; less often with some nonwovens, some board, counter rolls of gummed tape, kraft paper, food wrap)
- Winding on cores with glue/adhesive tape (to attach the web to the core; very typical)

  versus: winding on cores or shafts without glue (less often found: some heavy boards; some nonwovens; miscellaneous applications where subsequent processes disallow glue such as photographic or medical substrates, and some printing presses)
- Varying widths (shaft deflection must be considered; large width rewind shafts are often cumbersome for handling; narrow webs can encounter a hurdle with cost justification, as a narrow winder must have the same features and functions as a large web winders.)
- Varying diameters (large-diameter, heavy rolls adversely affect nip mechanics with two-drum winders. Large-diameter rolls require a high torque buildup ratio with center-drive winders. Also, large-diameter and heavy rolls might not be handled well by a relatively small-diameter shaft/core that deflects excessively when the rolls are slit.)
- Varying speeds (the higher the web speed, the more we need to consider critical speeds of idler rolls and drums, vibration, deflection, possible web floating (over idler rolls), air-entrapment, type of slitting, particularly for narrow cuts over wide widths; and centrifugal forces.)
- Varying tensions (low tension materials might call for positive tension isolation from the rewind section and they might dictate "center-surface" winding; heavy tension web materials allow a lesser degree of tension isolation and allow higher torque and nip forces to take effect.

3.) Slitting operation

- Slitting off-line
- Slitting in-line
- Slitting, allowing shafted operation
- Slitting in a shaftless winding operation (disallowing rewind shafts)
- Slitting, allowing alternate rewind arbors = duplex winding
- Slitting materials with high caliper variations such as board

- Slitting materials with low caliper variations such as aluminum foil.

In-line vs. off-line slitting during winding in a converting operation is discussed in more detail in the TAPPI PRESS book Winders.

THE DIFFERENT PLATEAUS OF WINDING

Pure center winding

The simplest form of winding, yet the best for some type of materials is "center winding." (Fig. 8) Torque seems to result in less web and roll defects than nip. Finer torque control and torque-tapers allow winding of very soft rolls which is desirable for some elastic films and flexible packaging laminations with high caliper variations. However, it can be detrimental to other materials because soft rolls can cause dishing. Torque applications become limited with larger roll diameters as buildup ratios become extreme, and high torque needs to be transmitted through small diameter shafts, chucks, and cores.

Fig. 8. Center winder

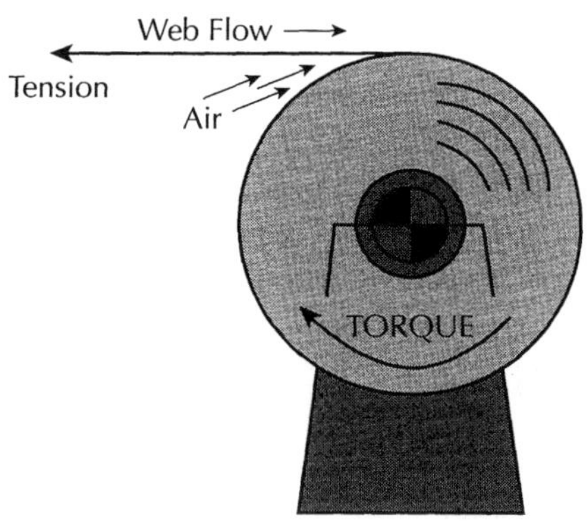

As torque is transmitted from the drive through chucks, cores, shaft, and web layers that have already been wound, the most important fact to know about this type of winder/drive is that torque-being applied to the center of the roll-cannot exceed the web tension applied during unwinding, process, or slitting. Nor can it exceed the tensile limits of a specific material or the torque limitations of core chucks (particularly during accel and deceleration). However, inwound web tension often needs to be higher during the start of the run in order to prevent dishing of a roll as it grows larger in diameter or to prevent telescoping when the roll is unwound in subsequent processes. Again, outer web layers need to be wound less hard then inner layers. Therefore, many center-winders follow tension isolation rolls, so additional torque can be applied at the start to wind the roll tight near the center.

As inwound tension needs to be reduced, some torque-controlled drives feature a potentiometer or a mechanical cam control or a computer programmable curve to taper-off increasing-torque as the roll increases in diameter.

Very simple drives that run at low speeds feature constant torque. As the roll diameter increases, more torque is required to maintain web tension. Thus, if the drive puts out the same torque, its force becomes weaker in relationship to the growing roll, providing a self-induced taper, i.e., less tight peripheral web layers. However, the buildup ratio is very limited (up to 30 in.). Otherwise, the roll would become too soft and would be prone to collapsing.

Pure center-drives work well with some films, film laminates, rubbers, textiles, nonwovens, and papers at fairly slow speeds. (100-300 m/min) Some materials stretch more than others (without damage) and are therefore more forgiving during winding; others have high caliper variations that can utilize "slip-core" winding (when slitting), where individual rolls are allowed to slip on an overspeed shaft in order to compensate for different roll diameters and different peripheral speeds. If several narrow slits are taken in the same rewind position, surface winders often have an adverse effect with dense off-gauge materials, thus, center winders are used; especially in the converting industry. However, heavy off-gauge materials–such as low density, high tensile board–rarely lend themselves to center winding due to torque limitations of drives (i.e., high hp requirements) based on roll buildup ratios.

The other most significant feature of a pure center winder (drive) is that air is allowed to enter the rewinding roll, which can prevent blocking with some films (positive), but often causes dishing with most other materials (negative), especially with materials of smooth finish/surface such as slippery silicone coated papers. The air reduces the effectiveness of the layer-to-layer friction that holds the layers in place and prevents slippage.

Center drives can be shafted or shaftless. However, a shaft is required when slitting is involved.

Center winding with gap (gap winding)

Most film laminates (pressure-sensitive materials with release liner) require a center-drive due to the slippery surface of the release liner (typically silicone coated) that necessitates a "high torque application" and "high torque tapers." However, when many slits are taken, a guide roll becomes necessary to eliminate any long and unsupported web runs to the windup, and to feed a "flat" wrinkle-free web to the windup; thus, the next level of winding includes either an "idler roll" or a "driven feed-roll" that is kept close to, yet not in touch with, the winding roll. (Fig. 9)

Fig. 9. Gap winding (short draw/advance system)

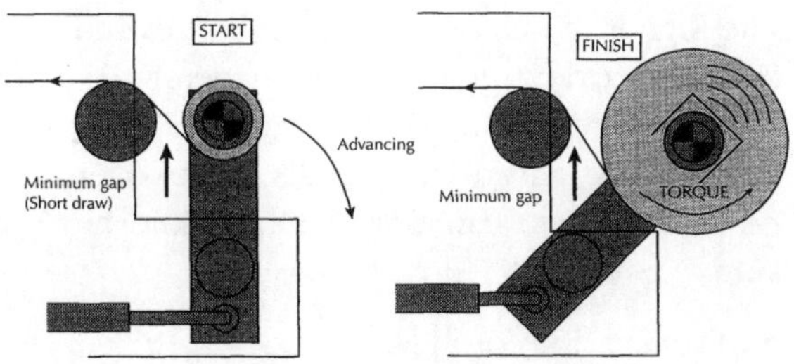

As the roll grows larger in diameter, either the "feed-roll" or the "rewind arm" moves out and away to maintain a controlled gap to the winding roll based on a signal from a photo-eye/sensor. This type of winding is called "gap winding" or "short draw/advance system" and works well with a large variety of materials (from film, aluminum foil, and nonwovens to pressure-sensitive papers and laminates) at low speeds (e.g., 100-300 ft/min).

Again it can be shaftless or shafted, but slitting requires a shafted operation.

Center winding with nipped lay-on roll (contact winding)

The next step is to nip a "lay-on roll" or "idler roll" (a roll that is not driven) against a center-driven rewind roll. (Fig. 10) It can be done with a lay-on roll that produces a nip by its own weight/by gravity, or a nip-roll that is loaded via pneumatic or hydraulic cylinders (see below). The result is that air is ironed out, i.e., that most air is prevented from entering the rewind roll because of nip. The harder the roll is supposed to get, the less air should be allowed to enter a roll. Nip, or the amount of pressure applied to the rewind roll (without damaging the material) via lay-on-roll, is a determining factor in building roll hardness. The harder the nip is, the harder the rewind roll will become.

Fig. 10. Center winding with lay-on/nip roll

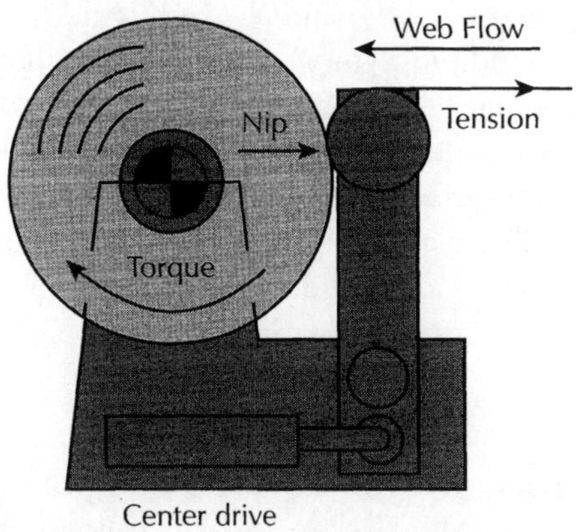

In most cases, this is the minimum level of winding sophistication in a continuous process line where maintenance is kept to a minimum, yet reasonable roll quality outputs are desired. It works well with paper at slow speeds and with rigid or slippery films that require center winding, yet it necessitates a nip-roll to prevent air entrapment.

Simplex center-surface winding

If a nip or lay-on roll is driven (and possibly friction-coated), rewind tension can be isolated from the process or slitting tension, enabling the center rewind drive to apply additional torque at the beginning of the windup to ensure a tight start. This is the next plateau of winding. (Fig. 11)

Fig. 11. Center-surface winder

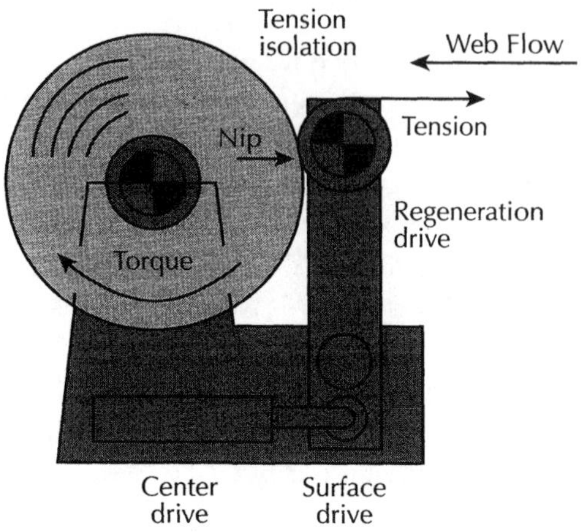

Fig. 12. Duplex center-surface winder

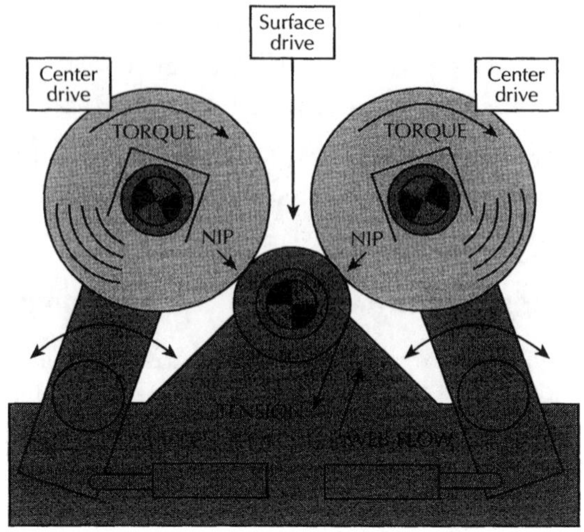

Torque is now allowed to exceed the web, slitting, or process tension for a specific material, and nip allows tighter winding of a roll.

This system also allows different load shares depending on the material. Light grades are usually wound with more torque from the center drive, while heavier grades require more nip and therefore a good amount of surface drive. The larger the surface drive roller diameter is, the more wrap will isolate web tension; and the larger the roller's footprint is due to larger diameter, the less severe are impacts from nip.

Duplex center-surface winding

If slitting is part of the center-surface winder operation and materials are of high caliper variation, individual strips can be alternately guided to two separate rewind shafts. (Fig. 12) This not only allows load shares by a center and a surface drive, but the total load can be shared by two separate rewind shafts. Because slitting is involved, it must be a shafted operation, but automated shaft-handling systems can make this winder as productive as a simplex or shaftless winder.

The tools of center-surface winding:

- Nip pressure control:

Nip pressure can be adjusted to taper off as the roll grows larger in diameter, and/or be adjusted for different material types. With center-surface winding, nip usually remains constant while the center-drive provides a tension taper (through increasing-torque taper).

The higher the nip is, the harder the rewind roll is. Heavy and rigid grades (most papers) allow high nip pressure, while lighter, high bulk, pressure- and surface-sensitive materials (nonwovens, carbonless papers) can only tolerate little nip pressure.

- Tension and torque:

A surface roll is often speed-controlled and it typically provides most of the web pulling, but it can also fight a center-drive through regeneration (hold-back). It can also provide partial tension isolation.

A center drive is often torque-controlled and usually assists in overcoming roll inertia and in providing additional roll hardness during the start. High torque is often applied at the beginning of the run and then tapered off. (Fig. 13) The higher the torque is (from the center drive), the more tension isolation or "hold-back" is required by the surface drive. The surface drive therefore must be regenerative. High torque is particularly important with very slippery materials.

Fig. 13. Increasing-torque taper with center drives and winders

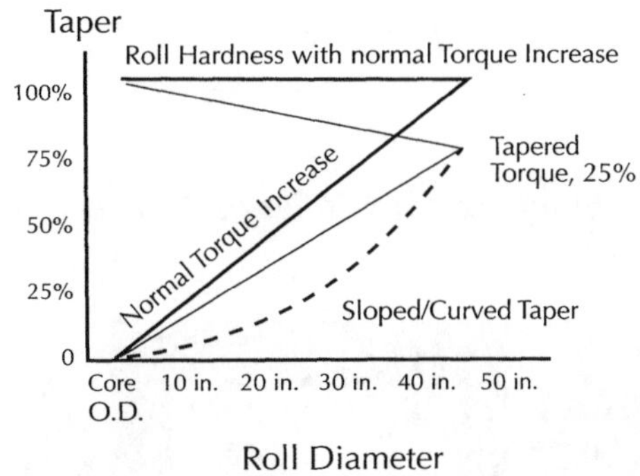

Fig. 14. Load share of center-surface winder

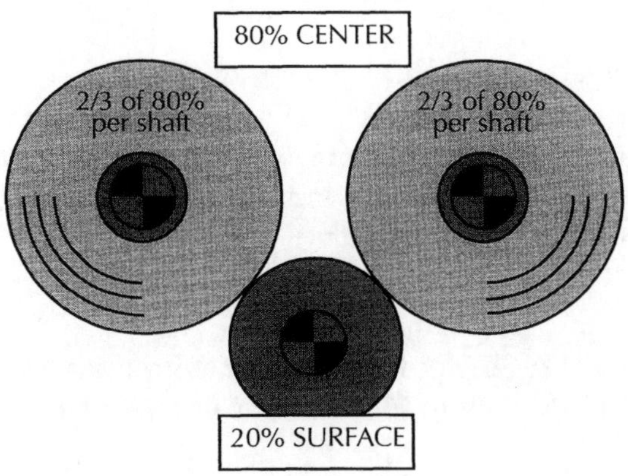

Load share and tapers: Some manufacturers differentiate "center-surface" winders (where the center winder is the dominator in producing roll hardness) from "surface-center" winders (where the surface winder is mostly accountable for providing roll hardness while the center-drive simply overcomes the growing roll's inertia). In either case, the load is shared by the two drives (see also the information under headings "load share" and "drives"). (Fig. 14) Torque and nip can be tapered off via straight taper or slope taper, and load share can be programmed to change as the roll grows larger in diameter.

Light grades are best wound with little nip and high torque, whereas heavy materials allow higher nip forces, making torque a less important (yet helpful) factor. Slippery materials require high torque tapers.

Center-surface winding abstract: A center-surface type winder with positive tension isolation is one of the highest forms of winding, as it utilizes the most tools found in the winding arena: torque, nip, tension, load shares, torque tapers, torque differential and tension isolation.

The importance might not be so much the TNT itself as it is the way each tool is applied. Each tool can be controlled individually or adjusted differently in the same application with load share programming, straight or sloped taper cams, curves, etc. It can be a converting duplex slitter-rewinder or a large single drum mill winder with individual rewind stations.

Differential winding

Differential winding (also called "slip-core winding" and "end-core-loading") is a sub-form of duplex winding where multiple individual rolls (also called "pancakes" or "donuts") are allowed to slip while being wound on a center-driven shaft that is driven at slight overspeed.

When high tensile strength materials of high caliper variations are slit into multiple webs, each rewinding roll will grow to a slightly different diameter, therefore having a different circumferencial speed that results in different tension. If the rolls were locked on a driven center shaft (as with "lock-core winding"), the largest diameter roll would take a dictating role (pull more web or lift the smaller rolls away from the drum) and become much harder, while the small diameter rolls would be wound too loose and eventually collapse.

A differential shaft features a keyway to accept spacers with keys and cores that are alternatively mounted (one core, one spacer, one core, etc.). As the shaft is driven, only the spacers will positively turn with the shaft (due to the keyway). (Fig. 15) The shaft is driven at a slight overspeed and the cores are loaded against the spacers from one or two sides of the winder to provide the torque to turn the rolls. The higher the loading pressure is from the side, the more simultaneous each roll will turn with the shaft. The less loading pressure there is, the more each rewind core/roll is allowed to slip independently. The slippage acts to equalize individual strip tension variations due to caliper variations in the web, i.e., roll diameter variations during winding. With more tension-sensitive materials, low friction core adapters (or low friction shafts) are used that slip easier on the shaft than the core. In that case, the core will sit on an adapter.

Fig. 15. Differential winding (courtesy of Cameron)

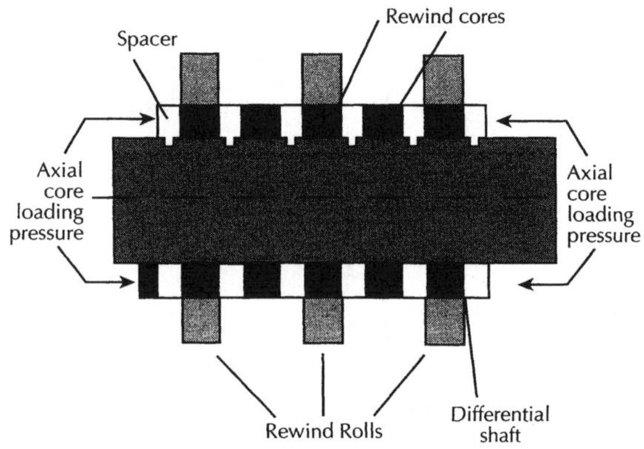

Some differential winders drive both, the rewind arms (duplex) and the main pull roll with a single drive. With "lock-core winding," clutches allow each rewind shaft to reduce rotational speed as the rolls grow larger in diameter. However, with a differential winding one-motor drive system, all rolls will slip more and more as they grow in diameter. The cores will ultimately heat up due to the high friction from slippage as the rolls approach final diameters. This application therefore has speed (500 ft/min) and diameter (24 in.) limitations. Also, as the roll weight increases, the effectiveness of the thrust to control the winding tightness is reduced.

Single-drum winding with multiple individual rewind stations

As the slit web is separated to two separate rewind areas (with duplex winding), each roll can be held against a large-diameter (24-60 in./600-1500 mm) center drum (surface drive) by individual rewind arms (with or without center drive). (Fig. 16) This allows each individual rewind roll to build up independently without regard to the neighboring roll (usually controlled with nip and torque). The large-diameter center drum provides a large footprint for low nip impact, and increasing roll weight variation has no or limited impact on nip between roll and drum. Also, it allows a shaftless operation. This winder features many individual rewinders within one frame.

Fig. 16. Single-drum winder with overhead web lead

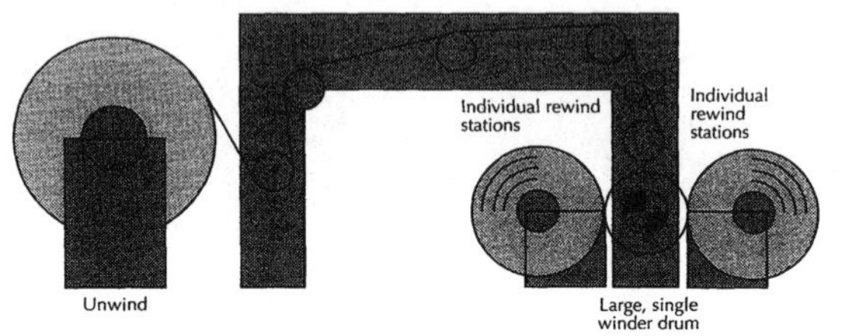

Large single-drum winders can have the so-called "9 o'clock/3 o'clock" roll positions for zero roll gravity effects on drum nip (Fig. 16), or the "11 o'clock/1 o'clock" rewind roll positions with only little roll gravity effects (Fig. 17).

Fig. 17. Single-drum winder with underground web path (courtesy of Jagenberg)

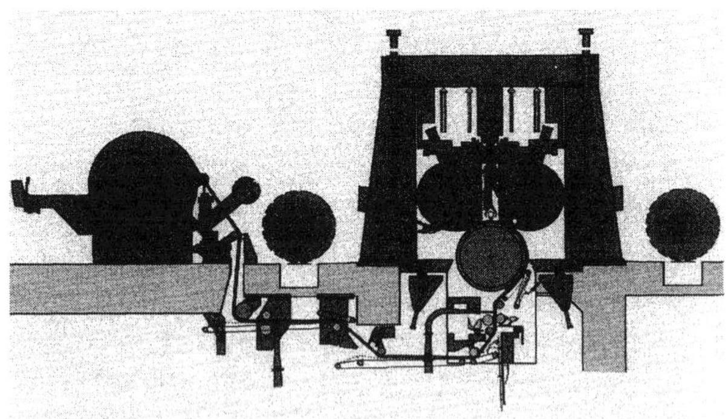

However, as we are getting more sophisticated in winding, it also becomes more complex, even though "full automation" of such winders help overcome its complexity and setup times.

Some of the large-diameter single-drum winders have the drum located at floor level (so the drum pocket can be accessed by operators) with a preceding slitter section either way up in the air or way down in the pit or lower floor level. This system usually requires higher initial investment cost, a lot of space, and longer installation periods when installed over two floor levels (earlier winders also have longer downtimes during set changes). All this is in comparison with the two-drum winder but, if a single-drum winder features duplex center-surface winding with individual roll buildup, it is considered one of the highest forms of winding.

Pure surface winders

Two-drum winders

When we talk about two-drum winders, we mean surface winders because the roll is accelerated, driven, and decelerated at its periphery/surface by the surface of external rollers (winder drums) plus a third roller, namely the rider roll that applies additional nip during the start of a run. (Fig. 18) A two-drum winder is predominantly a "nip" winder (not only because it would not work without nip, but because nip is a predominant factor in building roll hardness), even though roll structure is controlled by a combination of TNT tools. Because of the nip, it can produce much harder rolls than a pure center winder, which is desirable for board and most paper products.

Fig. 18. Two-drum winder

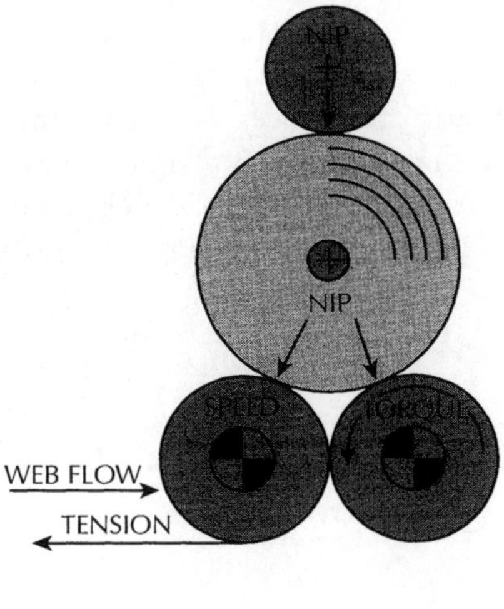

The reason for a two-drum winder's long and continued existence is high efficiency (easy setup, easy handling without shafts, good visibility and accessibility of components, and self-contained roll handling via integral roll pickup and lowering devices, often referred to as a "one-man operation"), productivity, versatility, and the ability to slit without use of rewind shafts.

Materials handled on a two-drum winder are normally neither surface- nor nip-sensitive (such as carbonless papers or some nonwovens which need to maintain bulk). On the other hand, most tissues and nonwovens can easily be handled on a two-drum winder because the material is not very dense or heavy, reducing nip forces caused by their own weight. Neither are they extremely slippery (such as silicon papers) or high in caliper variation (such as film and laminates) except board.

Winding tools of a two-drum winder are tension, nip and torque, which are explained below. Figure 19 is a graphical representation of tension, nip and torque in a two-drum winder system.

- Tension from the unwind is partially isolated through the large amount of wrap around the back drum (capstan affect) and through nips between winder drums, rider roll, and rewind roll. Therefore web tension has a partial effect on rewind roll quality. Consequently, a good tension control system is not only important for a good sheet run but also for good rewind quality. Also, programmed S-curves from the drive provide gradual speed changes that allow the tension controllers to keep up.
- Nip is initially produced by the rider roll to provide added winding hardness during the start, but also to prevent slippage between drums and rewind roll. As the rewinding roll increases in diameter, the roll will produce its own nip and the rider roll pressure to the roll must be reduced. (Fig. 19) With tissue winders, nip is reduced through the shaft, as most tissue winders do not have a rider roll.
- Torque can be applied to the second winder drum that is traction coated. It is connected to a separate drive which follows the first drum drive with an override control. The torque applied to the drum results in additional tension on the outer wrap/web layer of the roll, which is particularly important at the start of a run.

Fig. 19. TNT taper chart

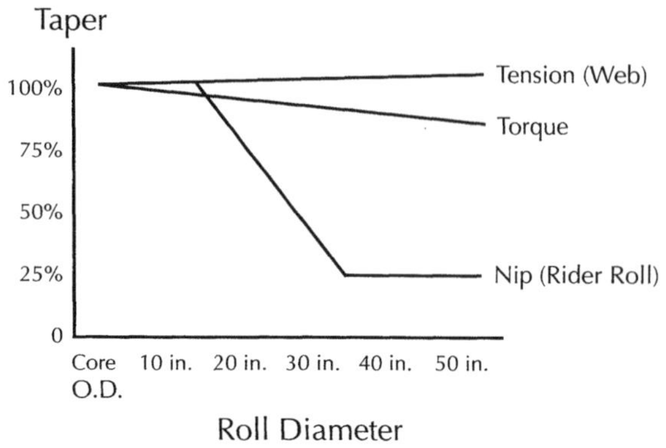

Large single-drum winders, vertical two-drum winders, belted three-drum winders, air-injection winders

As a roll of material grows larger in diameter, its weight increases, sometimes causing excessive nip and induction of excessive residual stresses and compression. This in turn can cause bursts near the core and other roll defects. When winding very dense papers, the gravity forces become even more intolerable, particularly after the roll has grown above 40 in. in diameter. In that case, vertical two-drum winders and large single-drum winders are beneficial, since the roll's gravity has less of an impact on roll quality.

For instance, a vertical two-drum winder (dated design) features two winder drums stacked up vertically with the roll of material moving horizontally away from the drums. The roll is only in touch with both drums up to a diameter of 7-10 in. to ensure a tight start. Increasing roll weight therefore has no impact on nip coming from the drums. (Figs. 20 and 21)

Fig. 20. Vertical two-drum winder (dated design)

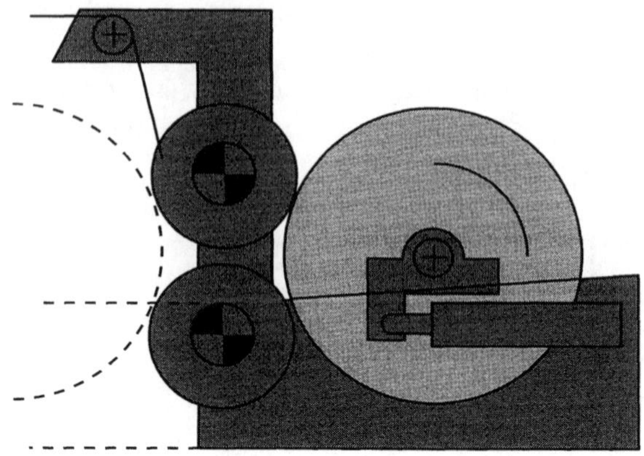

Fig. 21. Large single-drum winder or MIR (multiple individual rewinds)

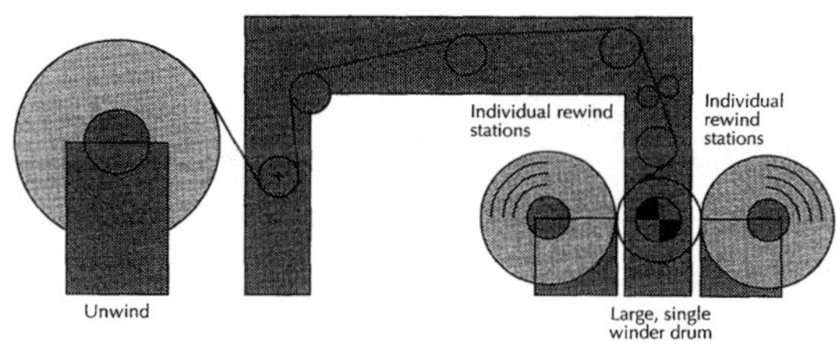

Because increasing roll gravity can have negative effects on winding and because single-drum winders are usually connected to a higher initial investment cost, various winder manufacturers have come up with alternatives for two-drum winders that feature near single-drum roll quality through soft-nips, articulating rider rolls, belts, and gravity-compensating air-cushions. (Figs. 22, 23, 24)

Fig. 22. Soft nip winder

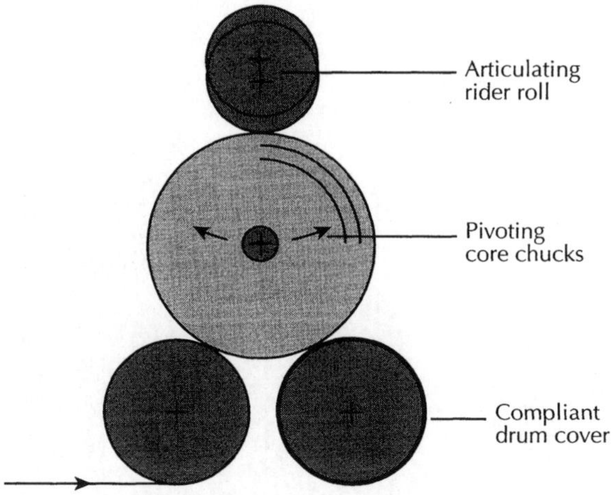

Fig. 23. Belted winder

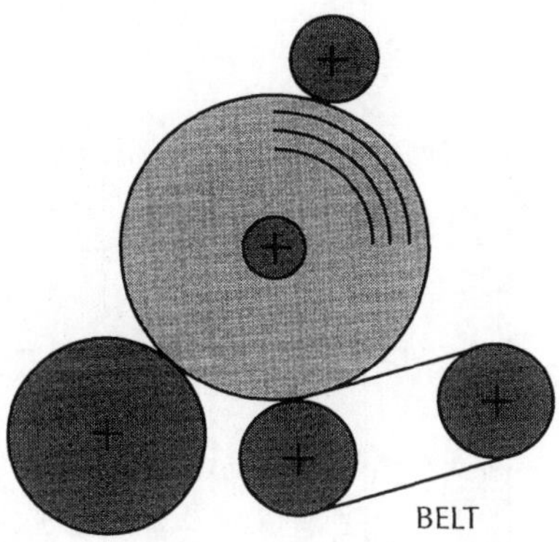

Fig. 24. Air-flotation winder

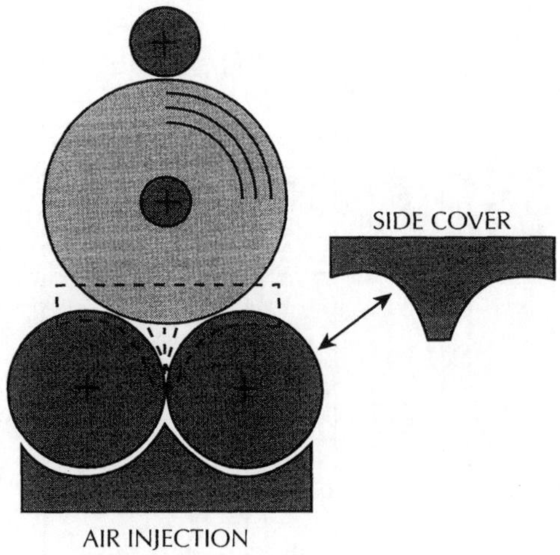

All of above systems combat nip-induced defects.

Conventional reels

The last category of pure surface winders includes "conventional" paper machine reels (the latest reels also feature a center-drive assist).

Fig. 25.

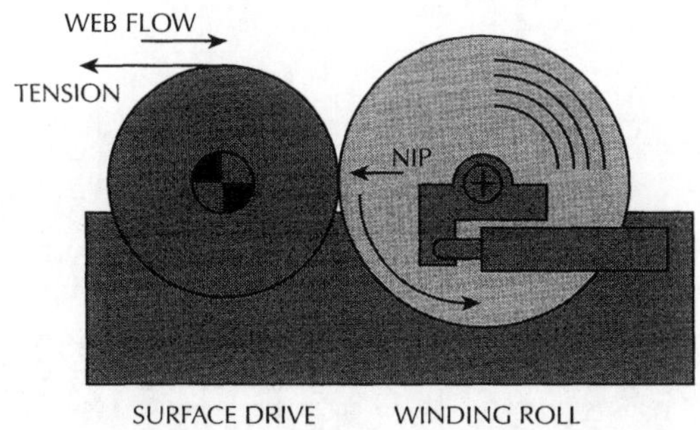

The following, conclusive chart indicates which winder, winding tool and roll support is typically* used for specific materials and properties.

** Before selecting a winder, you should always check with an original equipment manufacturer first.*

Table 2. Summary of material properties and their winding tools

Material Property	Example	Winder Type/Roll Support	Drive Application	Main Tool	Assist. Tool	Taper	Tension	Tension Isolation	Torque Differential (Load Share)
Low elasticity, high tensile, high compr.	Kraft, board	Drum supported (two-drum)	Surface	Nip and tension	Torque and tension	Nip and torque taper	High tension	Not necessary	Optional by two-motor drive for tight starts
Low elasticity, low tensile, low compressible	Aluminum foil, some nonwovens	Core/shaft suspended or drum support	Center or surface	Torque or nip		High torque and nip taper	Fine tension control very important	Low tension throughout	N/A
High elasticity, high tensile, high compr.	Some nonwovens	Core/shaft and drum supported	Center-surace or surface-center	Torque, tension and nip		Little torque and nip taper	Medium tension	Good isolation is important	Depends on output: bulky vs. tight rolls
High elasticity, low tensile	Thin tissue, films, some non-wovens	Core/shaft suspended	Center	Torque	Low tension and little nip	High torque taper	Fine tension control very important	Low tension throughout	Usually N/A
High density, low compressible	Supercal., bond, C_2S, fax, offset	Core/shaft susp. or combination of core/shaft and Drum	Center or center-surface	High torque	Uniform nip and tension	Yes	Medium tension	Limited tension isolation	Mainly center; little surface drive
High bulk/ high compressible	Tissue, non-wovens, board	Core/shaft or drum supported	Surface or combined center-surface	Tension, nip and torque		High nip taper, little torque	Depends on tensile strength	Partial isolation is okay	Mainly surface, little center
Heavy roll weight	Metallized papers, laminates	Combination core/shaft and drum supported	Center-surface	Tension, nip and torque		High nip taper,little torque	Medium tension	Not important	Mainly center, some surface
Light roll weight	Tissue, non-wovens, board	Drum or core/shaft supported	Surface	Nip and tension	Torque	High nip taper	Depends on tensile strength	Not important	Optional by two-motor drive
Low coefficient of friction	Silicone/ glassine, C_2S	Combination core/shaft and drum supported	Center-surface	Tension, nip and torque	Torque	High nip, high torque	Medium tension	Good isolation is important	Center and surface
High coefficient of friction	Sandpaper	Combination core/shaft and drum supported	Center-surface	Tension, nip and torque	Torque	High nip, high torque	Medium tension	Not important	N/A
Surface/nip/ gravity sensitive	Carbonless thermal, some nonwovens	Individual rewind stations	Center or (center-surface)	Torque and tension	(Rarely nip)	High torque taper	Medium tension		High center (low surface if applicable)
High caliper variations	Some films, P/S laminates	Duplex (differential)	Center or center-surface	Torque and tension	Nip	Yes	Medium tension	Depends on specific material	Depends on specific material
Low caliper varations	Aluminum foil, some papers	Simplex	Surface or center	Nip and tension or torque	Torque or nip	Yes	Depends on specific material	Depends on specific material	N/A

REFERENCES

1. Smith, R.D., "The art of winding good rolls," Paper, Film & Foil Converter (4):(1991).
2. Roisum, D.R., "The Mechanics of Winding," Tappi Press, Atlanta, 1994.
3. Scott, W.E., Abbott, J.C., Trosset, S., "Properties of Paper: An Introduction," TAPPI PRESS, Atlanta, 1995.
4. Olshansky, A., "Understanding Winding," Beloit Finishing Technologies Seminar Notes, 1996, internal document.
5. Pfeiffer, J.D., "Using paper compressibility measurements to enhance roll quality," *TAPPI 1996 Finishing and Converting Conference Proceedings,* TAPPI PRESS, Atlanta.
6. DuBois, G.B., "Selecting Proper Winding Tension for Various Web Substrates" Tappi J. 76(5):91 (1993).
7. Roisum, D.R., "The profile of a nip," Converting Mag. (6):(1994).
8. Frye, K.G., "Winding," TAPPI PRESS, Atlanta, 1990.
9. Schmid, G. and Werner, P., "Advantages of the variable frequency vector drive applied to PM winder backstands," *TAPPI 1993 Finishing and Converting Conference Proceedings,* TAPPI PRESS, Atlanta.
10. Duncan, D., "ABB Drive Course," Beloit 1996 Finishing Technologies Seminar Notes, Beloit.
11. Pfeiffer, J.D., "Nip forces and their effect on wound-in tension," Tappi 60(2):(1977).
12. Smith, R.D., *Roll and Web Defect Terminology,* TAPPI PRESS, Atlanta, 1995.

SUGGESTED OTHER READING

Eddens, G.R., "Web Tension and Tension Control," CEMA 1993 Seminar Notes, CEMA

Frye, K.G., "Winding," TAPPI PRESS, Atlanta, 1990.

Greiner, T.S., "Winders with individually programmed stations and slitters," Tappi 63(10):(1980).

Gronewold, J., "Computer-aided winding can give competitive edge," Paper, Film & Foil Converter (12):(1991).

Gronewold, J., "WinderNews," Newsletter Factfinders, 1995, internal document.

Gronewold, J., "Winding Tools and Variables: A Look Beyond Paper and TNT," TAPPI 1995 Finishing and Converting Conference Proceedings, TAPPI PRESS, Atlanta.

Gronewold, J., Paper, Film Foil Converter Magazine (12):(1991); "Computer-aided Winding Can Give Competitive Edge."

Gronewold, J., "WinderNews," Newsletter Factfinders, 1995 internal document.

Jones, J.P., "Determining tension values," Converting Mag. (11):(1989).

Kirchner, R.F., "Slitting and winding," Modern Plastics Encyclopedia, 1989, McGraw Hill, New York.

Lucas, R.G., "Internal gearing in a roll of paper," TAPPI 1981 Finishing and Converting Conference Proceedings, TAPPI PRESS, Atlanta.

Lucas, R.G., "Winder crepe wrinkles their causes and cures," TAPPI 1981 Finishing and Converting Conference Proceedings, TAPPI PRESS, Atlanta.

Lucas, R.G., "Dishing in winding rolls of paper," Tappi 60(7):121(1977).

Mykytiuk, A., "Examining the trends in drives and motors," Paper, Film & Foil Converter (1):(1996).

Mykytiuk, A.R., "Electronics help common slitter reach maturity," Paper, Film & Foil Converter (3):(1996).

Mykytiuk, A.R., "Slitters are finally getting the respect they deserve," *Paper, Film & Foil Converter* (11):(1995).

Olshansky, A., "Introduction to roll structure," *Tech '94 Theory and Practice of Papermaking Course Notes,* CPPA, Montreal.

Olshansky, A., "The process of winding," *Tech '94 Theory and Practice of Papermaking Course Notes,* CPPA, Montreal.

Pfeiffer, J.D., "Mechanics of a rolling nip on paper webs," Tappi 51(8):(1968).

Pfeiffer, J.D., "Wound-off tension measurement in paper rolls," *Tappi* 60(3):(1977).

Pfeiffer, J.D., "Nip forces and their effect on wound-in tension," Tappi 60(2):(1977).

Pfeiffer, J.D., "Understanding rolling nip action through video technology," TAPPI 1993 Finishing and Converting Conference Proceedings, TAPPI PRESS, Atlanta.

Pfeiffer, J.D., "Pursuing the elusive crepe wrinkle," TAPPI 1997 Finishing and Converting Conference Proceedings, TAPPI PRESS, Atlanta.

Pfeiffer, J.D., "Internal pressure in a wound roll of paper," Tappi 51(8):(1968).

Pfeiffer, J.D., "Measurement of the K2 factor for paper," Tappi 64(4):(1981).

Roisum, D.R., "The mechanics of web spreading," TAPPI 1993 Finishing and Converting Conference Proceedings, TAPPI PRESS, Atlanta.

Roisum, D.R., "The profile of a nip," Converting Mag. (6):(1994).

Smith, R.D., "Converting to shaftless winding boosts efficiency at slitter-winder," Pulp and Paper (9):(1977).

CHAPTER 16

Machinery Upgrades and Rebuilds

Louis W. Piffer, Egan Davis-Standard

INTRODUCTION

Requirements for extrusion coating lines have changed dramatically since 1960. These changes include higher speed, new resins, new films, new primers, and an increased environmental and workplace awareness. Due to today's competitive environment, adapting to these changes by purchasing new equipment is not always economically feasible. This paper will investigate options open to extrusion coating converters, enabling them to respond to the challenges of today by upgrading existing older equipment.

DESIGN CRITERIA

The world of extrusion coating was much simpler before 1970. The coating was LDPE, the substrate was normally paper, and the line could look as shown in Fig. 1. Since then, technology has introduced a multitude of new resins, substrates, and primers to assist the extrusion coating converter in making new and more varied products. (See Table 1.)

Fig. 1. Pre-1970 E/C line

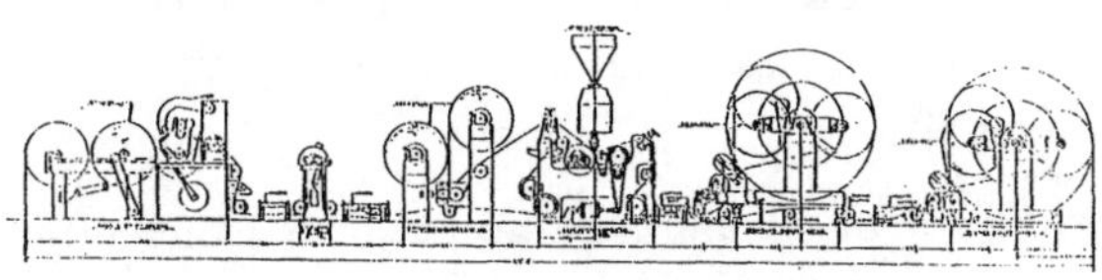

Table 1. Post 1970 developments

Resins	EAA, EVA, EVOH, Nylon, EMA and Ionomers
Substrates	BOPP, PET, thinner aluminum foils
Primers	Commerical use of water based primers

The commercial use of these new materials places more rigorous demands on the extrusion coating line. To run today's structures competitively you need higher speeds, better tension and temperature control, more corrosion resistance, and a lot more versatility. This task would be simple if extrusion coating equipment was replaced as more complex product structures were introduced. Unfortunately, the longevity (+30 years) of extrusion coating line and their high costs can make this option impractical.

The following pages will address how a pre-1970 extrusion coating line can effectively run the sophisticated product structures of today through the use of strategic upgrades, as opposed to simply replacing equipment. The question of how the upgrade will be accomplished, why the upgrade needs to be done, and (if possible) the level of improvement achievable is discussed. It is important to note that these suggestions are meant only to assist in expanding the capability of the older line; they will not allow all product structures to be run, nor are all pre-1970s extrusion coating lines unable to run today's product structures.

PROCESSING INFORMATION

Unwinds and winders

Single to double direction unwinds: Today's converters need to squeeze the greatest possible amount of flexibility out of their equipment. One way to do this is by changing an unwind from single to double direction. A general list of changes needed to accomplish this include: A modification in the stroke of the cylinder that controls the splicing arm indexer, replacement of the clutch system with one that allows spindle rotation in both directions, modification of the indexer drive so it can go in either direction, and replacement of the single-position knife with a dual-position arrangement. Also, the entire unit might have to be raised to provide clearance for the double direction indexing of the full roll diameters. (See Fig. 2.)

Fig. 2. Single-double direction

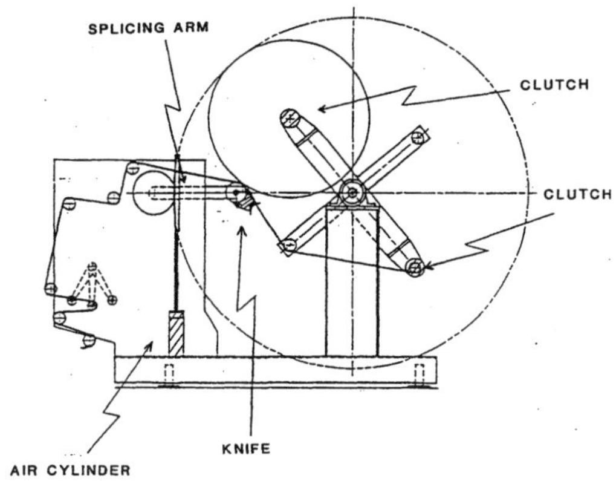

Manual to flying splice: Unwinds and winders that already have a turret base can possibly be upgraded from manual to automatic by installing an independent splicing/transfer assembly. The cost effectiveness of this type of upgrade might be in question because of the amount of equipment added.

Fig. 3. Manual splice winder

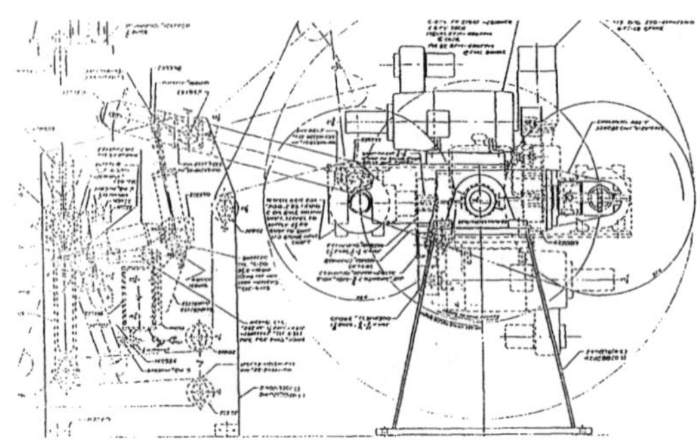

Manual splice winders already equipped with a lay-on roll assembly are more readily upgraded to a flying splice by the addition of a new splicing arm. (See Fig. 3.)

Note that in both cases, idler rolls might have to be installed to ensure that the web is perpendicular to the firing knife.

PLC programming. Unwinds and winders that are currently under relay control can be upgraded by the installation of a dedicated PLC. This upgrade will not improve the performance of the unwind or winder, but it will make troubleshooting and maintenance easier. The PLC will also help maintain a consistent operation during line speed changes.

Brakes to D.C. drives. Replacing unwind mechanical brakes with drives will provide better tension control at lower tension levels. This might be needed when the frictional losses in the spindle brakes tension required for the product. Before attempting this upgrade, consider the possibility that unwind spindle arms designed for brakes might not be able to support the weight of a heavier spindle drive, meaning that the drives would have to be floor mounted, creating the further consideration of whether there is enough room on the existing shaft to "quill" the drive to the shaft.

An alternate upgrade to allow an existing unwind to handle a lower tension range is to install an infeed holdback after the unwind. The unwind brakes could be replaced with more sensitive brakes and the infeed holdback would then handle the excess tension requirements. Additionally, the drive cost might be lower since the higher tensions would be created by one holdback drive instead of two spindle drives.

Running lower tensions can require adjustment of parts other than brakes and/or drives. Air cylinders controlling dancers might have to be modified or replaced to control at lower tension ranges. Similarly, dead shaft idlers with greased bearings might have to be replaced by easier to turn live shaft idlers with oiled bearings.

Winder tension regulation: Most winders of the 1950s and '60s have two motors with a speed match for transfer. However, earlier drive technology was limited to a current regulator with dancer trim. This system has the disadvantage of a constant gain dancer, causing a more unstable response during the roll buildup and requiring the drive to be retuned and the gain averaged over the entire buildup ratio. The other disadvantage with a current regulator is that frictional losses will show up as an additional current load, limiting these winders from running very light tensions.

Today's winders are speed controlled with dancer trim. The current (torque) goes to a value that satisfies the dancer. This allows a wider range of products as well as lighter products to run. The gain of the dancer is decreased proportional to buildup; thus, dancer stability can be easily maintained without retuning the drive.

Infeed holdback

Older, higher tension systems often had only one tension zone between the unwind and the laminator. When running lighter more extensible films that might already be printed, this can cause a stretching of the film as the laminator drive has to overcome the friction of the idlers in the web path, causing a distortion of the print. The addition of a holdback will create a more uniform pulling force, less stressful to the web. The other advantage of using a holdback is that the laminator is isolated from tension fluctuations in the web caused by unwind transfers.

Prime coaters

One of the more notable changes in the last 30 years in extrusion coating prime coaters has been the gradual replacement of solvent-based coatings with water-based ones. The coater originally designed for solvent-based coatings can have reduced effectiveness when running water-based coatings. In this case, there are some possible modifications to the coating unit. They include:

Gravure cylinders: Solvent-based primers often require application of three or more wet pounds per ream in order to achieve a 0.75-1 dry pound per ream layer. This is accomplished with relatively coarse etched cylinders. Using aqueous systems, one often needs only 0.01-0.1 dry pound per ream. This range can be most effectively applied with finer cylinders laying down only one wet pound per ream. Expense of a new cylinder usually is repaid through reduced primer cost.

Enclosed doctor blades: Solvent-based coatings containing such VOCs as MEK or alcohols generally do not tend to foam, even at comparatively high speeds. Due to differences in wetting tension, pH, and other characteristics, primer formulations based in water often are prone to develop foam, given sufficient agitation. The entrained air can interfere in a number of ways and the ultimate result can be a faulty application. An enclosed applicator system reduces the amount of primer exposed to turbulence or agitation at any given time. This tends to ensure smooth, consistent laydowns.

Positive displacement pumps: Where circulating systems are set up for feeding a priming station, a common tendency by converters who also have gravure or flexographic printing equipment is to use a spare centrifugal ink pump for primer service. By design, centrifugal pumps create turbulent movement and would provide an excellent condition for foam development in aqueous liquids. Positive displacement pumps such as piston, diaphragm, and peristaltic are recommended as being much less abusive.

Larger gravure roll: On equipment involving an applicator cylinder lifting primer from a fountain of liquid, an additional modification might be the use of a larger diameter roll. The larger diameter will reduce the rpms and the amount of primer cascading off the doctor blade. Both foaming and splashing could be more easily controlled as line speed requirements increase.

Corrosion prevention: Some coatings are based on polyvinylidene chloride (PVDC). As a group, these have very low pH. The more acidic environment usually requires all metallic components of the applicator station to be fabricated of stainless steel or other material capable of resisting the corrosive effects.

Tension control: One final modification would center around the drive system of the coater. Older coaters would simply have a single drive on the gravure cylinder that would act as a line speed follower. This system would be used when the coater nip was opened or closed. With today's more extensible materials (such as 45-gauge OPP and 25-gauge PET), an additional tension control zone can be implemented by adding a transducer between the coater and the line lead section. In this manner, the coater drive can be trimmed by the transducer when the nip is closed and acts as a line speed follower when the nip is open. The original motor on the coater can often be saved during this modification.

Dryers

Drying capacity is often thought of as the last limiting factor in an extrusion coating line. Increased requirements in line speed, along with the advent of water-based coatings can often tax this drying capacity.

Drying capacity and efficiency can be increased by the replacement of parts of the drying system.

Dryer nozzles: Dryer nozzle designs have developed over the years such that a nozzle replacement can provide as much as a 25% increase in dryer efficiency.

Fans: Air flow increases are possible by providing larger fans. (Note that ductwork might have to be modified.) Also larger burners or heating coils can allow dryer temperatures to increase for a given air flow or to be maintained for increased air flow.

IR dryers: An additional way to increase dryer capacity is to install IR dryers before the main dryers. The IR units will do some of the initial drying, raising the temperature of the main substrate, with the main unit providing the final drying. This combination can significantly increase the drying capacity of the complete system.

Process control: Modern process control techniques can be adapted to existing dryers in many ways.

Dampers: Supply, exhaust, and makeup air flows can be controlled via automatically actuated damper systems. Alternatively, dampers can be eliminated by replacing the drives on the blowers with variable-frequency drives.

Web temperature: Web temperatures can be read through optical pyrometers and placed in closed loop control. This can be very helpful when dealing with certain webs like PET that can curl when exposed to too high of a web temperature. The automatic system has the advantage of responding to variations in coating weight or line speed and still maintain web temperature.

A different consideration for dryer upgrades are adherence to environmental laws.

As VOC environmental limits are reduced, older dryers that had a direct supply/exhaust air system can be modified to provide recirculation, reducing the amount of exhaust into the atmosphere.

Extruders

The heart of an extrusion coating line is the extruder. As attempts are made to increase line speed, additional output is needed from the extruder. Increasing an extruders output is probably the most common form of upgrade done to an extrusion coating line. The following are various ways this can be accomplished, tabulated by increasing cost and complexity.

Increase screw rpm: The screw rpm can be increased by a change in sheaves, by field weakening an extruder drive, or by changing the gearbox ratio. Since an extruder screw is basically a volumetric device, this will increase output. The three methods can be described as follows:

Sheave changes: This upgrade, limited to belted extruders, is not normally meant to allow for great increases in rpm, but rather as a tool by screw designers to make fine adjustments in screw rpm. The increase in rpm is limited by the amount of torque that can be transmitted by the belts and pulleys to the gearbox and by the amount of bending moment in the gearbox shaft as the "overhung load" increases with decreased sheave ratio.

Field weakening: Through the addition of a field regulator, a drive can increase its input rpm to a gearbox. This will help increase the screw rpm, but at the new field range (base speed to maximum rpm) the motor is running at a constant hp and a reduced torque. Care must be taken that this reduced torque is enough to run the resin at the higher rpm and that the motor frame is sized to allow dissipation of heat at the higher field weakened range.

Gear set change: A gearbox ratio can be changed by the replacement of high- and low-speed pinions, shafts, and gears. Each gearbox has a range of gear ratios available. These limits are dictated by the strength of the gear teeth and the physical size of the gearbox.

As stated before, increasing screw rpm will increase output, but possibly at the expense of melt control. As rpms increase, the heating or cooling contributed from the barrel will decrease. This will cause melt temperatures to either rise or fall depending on the contributions from the barrel. This change in melt temperature can be corrected by changing barrel temperatures, but at a price. Correcting for melt temperature changes in this manner can cause melt temperatures problems at lower screw speeds. This is because as the barrel heating and cooling contributes more to the melt temperature, the melt will gain more heat or chill from the barrel, drastically lowering or increasing its value. To avoid these temperature fluctuations, a new screw could be designed for the higher rpms.

New screw designs: Screw designs have come a long way since the 1960s. Computer-aided designs with more efficient mixing sections can allow the screw to process more resin but maintain melt quality.

Note that all of the above modifications can only be provided if there is enough torque in the existing drives. If this does not occur, a larger drive must be installed.

Larger drives: In the previous examples, the modifications can be limited by the size of the drive. There is a direct correlation between the horsepower (hp) of the motor and the theoretical total output of the extruder. (See Fig. 4.) Doing a basic energy balance around the extruder,

Fig. 4. Hp vs. rate and diameter extruder and motor sizing

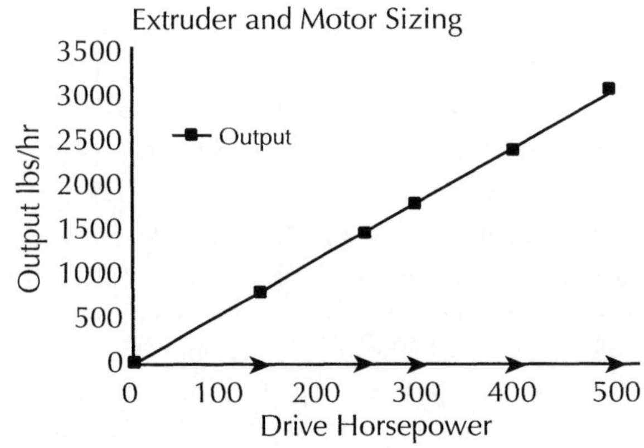

mechanical energy is imputed into the system by the motor. Total energy coming out of the system will be by way of frictional losses and enthalpy of the resin. (Note this is a very general balance; constants such as energy from heaters are excluded.) Therefore, in most cases, the more hp there is, the more resin there is. During this retrofit, care must be taken to make sure the original isolation transformer is rated for the higher hp. Also there might be gearbox horsepower and torsional torque limitations with which to contend.

Increase barrel lengths: Extruders of today commonly are provided in 28:1 or 32:1 L/D lengths. Extruders of the pre-1970 era were often provided at 24:1 or even 20:1 L/D lengths. At these short lengths, there is a finite output that can be provided despite the mechanical energy (hp) that enters the system. When this happens, a larger L/D barrel and screw could be provided. Increasing barrel L/Ds from 24:1 to 28:1 can provide as much as a 10% increase in output for LDPE with everything else remaining the same. It must be ensured when providing this type of retrofit that the power panel and control system provides for the additional wattage and/or zoning. Also, a die support might be needed to prevent excessive barrel deflection with the increased over-hung load off the new barrel. Finally, there must be enough room behind the extruder during retraction to account for the increased length of the barrel.

As extrusion coating operators struggle to come up with ways to increase the output of their extruders, they must also, of course, keep a watchful eye on the quality of resins that those extruders produce. Again, resin and product changes in the past 30 years can require certain modifications to a pre-1970 extruder. They include:

Water- and air-cooled heaters: Many older extruders were designed simply for LDPE. The heaters on these extruders were often ceramic band heaters, which provide heat but not cooling. Resins that require lower melt temperatures such as EVA, EVOH, and EAA might not be able to be run as efficiently on a "heat-only" extruder. Even when running higher melt temperature resins, the addition of water or air cooling will help in providing a more consistent melt temperature by maintaining a more constant barrel profile.

Note that in both examples, new controllers must be used, and barrel covers must be altered to account for either the blower and shrouds of the air cooled system or the solenoid valves, water inlet and return manifolds of the water-cooled system. Installers of water-cooled retrofit must additionally ensure there is enough cooling capacity to handle the additional Btu load of the extruder.

Ionomer barrel: Extruders of ionomer are often plagued with the problem of resin sticking to the root of the screw in the feed section. Many different methods of screw design have been employed to solve this problem, but one tactic that is universally accepted is to keep the feed section of the barrel cool. To accomplish this, the extruder could be upgraded by providing a barrel that has an integral feed section. This will eliminate the heavy walled feed section casting and will provide better heat transfer and cooling capability in the feed zone.

Alternatively, the initial feed zone can be modified by piping cooling water directly to each heater half, instead of having one solenoid valve for the main water flow, then splitting into two lines for each heater half. In the retrofit, there would be a solenoid valve for each heater half that would be wired in parallel.

Adapters

The basic design of the side access adapter has remained relatively consistent throughout the years. Any modification to this design would not be to improve on its functionality, but to extend its life. With the introduction of acidic co-polymers, corrosion has continued to be a problem with adapters because the point where resin hits the valve stem and the surrounding adapter body is a perfect place for resin to "hang up," causing corrosion.

If corrosion presents itself at the tip of the valve stem and surrounding body, the valve stem can be replaced and the body can be placed on a lathe and bored out to the depth of the corrosion. Then a corrosion resistance sleeve can be placed into the adapter body with a hole bored in it to accommodate the new valve stem. This modification can extend the life of the adapter or fill in the time gap from when the problem is first noticed until the new adapter is fabricated.

Coextrusion

With the development of such specialty resins as EVOH, EAA, and EVA, coextrusion came about as an economical means of applying them. The advantages of coextrusion are well documented and beyond the scope of this chapter. The information discussed here involves the problems facing a converter when attempting a coextruder retrofit.

The first thing that must be looked at is finding a place for the "satellite" extruder. Today, most extrusion coating heads are sold with a three-way coextruder carriage, but this was not always the case between 1960 and 1970. Many of the carriages held only one extruder and had automatic movement in only one direction. If there is only a mono carriage in place, there are two options available: removing the existing extruder and placing it on a new coextruder carriage, or modifying the existing carriage to accept the additional extruder.

Removing extruder: The original extruder is removed from the carriage and installed on a new arrangement consisting of a separate extruder and new carriage. This option has the advantage of utilizing a new three-way extrusion coating carriage.

The disadvantage is that the coating line will be down as the new extruder is being installed.

Modifying the carriage: One modification would be to build a sidecar alongside the original carriage. This method has the advantage of allowing the line to continue running during most of the construction of the platform modifications and as the satellite extruder is installed. The disadvantage is that there is a lot more work involved, including building a sidecar and pouring a foundation for additional tracks. Also, the completed assembly will still consist of the original extruder carriage. Note that this upgrade is limited to a single-direction carriage if downtime is to be avoided. A three-way carriage could be modified, but the modification would be more complex and require additional extruder downtime.

The other modification method would be to build a smaller platform cantilevered off the main unit. Not all carriages have enough room to provide a support for the cantilevered support. This option is more practical when a very small satellite extruder (2 ½ in. or less) is added.

The converter must decide which of these two methods is preferable.

After a place has been found for the extruder, the utilities and controls for this new extruder must be provided for. A new power panel for the new barrel, coextrusion piping, and feedblock zones must be installed. The isolation transformer needs to be checked to see if it is still sufficient for the new load and augmented if it is found lacking. Water (if water cooling is used) should be piped to a new water circulation system for the satellite extruder, or to a new larger system that will be shared by both extruders. Finally, control changes for the additional zones must be implemented. This would involve new control modules for a stand-alone system (which would also take care of the power panel requirements) or software changes in a supervisory system.

Other coextrusion considerations are as follows:

Air gap: With the installation of a coextrusion feedblock, modifications need to be made at the adapter/downspout area to ensure that there is the same die to nip relationship. This can result in a reduction in the length of the installed static mixer. Depending upon the individual products made, this might not be an acceptable choice.

If the carriage can raise the extruder, the extruder could be raised to a point where a full-length static mixer can be installed. However, with this option, the maximum air gap attainable will be reduced. If this is not desirable, the mixer can be replaced with one that will give similar mixing in a shorter length or the entire carriage can be raised by placing shims between the carriage and the caster bearing blocks.

Cooling: When adding a satellite extruder to an operation, the extrusion coating line can be limited by cooling capacity. The additional output will put a higher Btu load on the chiller unit, requiring increases in chiller tonnage, chill roll diameter, and/or water flow rate to maintain the temperature drop across the roll and stripping temperatures. This indicates why designing a laminator and chilling system for future output increases is a good idea.

Dies: Coextrusion dies have special design considerations in the manifold and the die inlet. Depending upon the die, the die could be upgraded from mono to coextrusion at the time of the retrofit. Note that these changes are based on a T-slot design since, for simplicity, it is assumed a coat-hanger design would not be used for mono extrusion.

A rectangular rather than circular die entrance is used for coextrusion applications. This type of entrance provides minimum flow disturbance to the coextruded layers by matching the rectangular discharge end of the coextruder block. This change is easily accomplished when the original die is furnished with a die inlet plug.

The manifold of the T-slot die would need to be changed from a round "gun-drill" arrangement to a "teardrop design."

Tests indicate that the "gun-drill" manifold is not acceptable for any coextrusion structure other than 50%/50%.

Finally, software and hardware changes are needed for the control system during a coextrusion retrofit. The extent and ease of these changes depend upon the specific system used.

Laminators

Increases in cooling requirements and general age might require a converter to modify an existing lamination system. Scale buildup will reduce heat transfer coefficients, limiting the total cooling capacity and, possibly, line speed.

A converter can check the cooling efficiency of his chill roll by reading the inlet and outlet water temperature coming from the chill roll and comparing the two. As this number starts to lower, less heat is being pulled away by the chill roll, indicating an old or plugged water path. A loss in cooling efficiency can also occur as the internal baffles corrode and disintegrate. This can be seen as a hot spot on the film at a consistent location or by bumps on the roll face as you run your hand across it when the roll is at rest.

Upgrades of coextrusion will have the same effect on a chilling system as an extruder output increase described earlier. If the size of a laminator chill roll needs to be increased, there might be a mechanical interference with the other rolls such as the stripper roll, bow roll, and idler rolls.

In older systems, water cooling was normally only used in the chill roll in a three- or two-roll laminator. Today, pressure and backing rolls are cored and cooled to help reduce chances of delamination of the rubber roll.

Earlier laminator system transferred power for the motor via a chain system. The drives supplied did not have the degree of speed regulation that is available today. To maintain a consistent torque, the drives would have to be de-tuned to prevent them from responding too aggressively to the perceived change in load caused by the movement of the chain links. If the chains were replaced with timing belts and if speed regulation and drift are important, the drive could be replaced, which could give 0.01% regulation and 0.1% drift.

Automatic dies

Perhaps the biggest technological breakthrough in extrusion coating systems since 1970 is the development of the automatic die. When looking at an extrusion coating line of today and comparing it to one of pre-1970 vintage, it is by far the biggest difference. Resin savings associated with faster start-up times and reduced cross machine variation can make this upgrade well worth the cost. Additionally, converters benefit from both the systems ability to provide enhanced statistical data on rolls being produced and that reduced cross machine variation make for a web that will run easier in downstream printing or form, fill, and seal equipment. These capabilities are often being requested by the more demanding converting customer of today.

The converter must be aware of the following when looking at an automatic die upgrade:

Die length: Depending upon the type of die purchased, there is a chance that the overall length of the die is greater than the die it is replacing. Techniques like those described in the section on coextrusion will need to be used to raise the die to maintain the same die exit to nip relationship.

Die weight: A new automatic die might be heavier than the die it is replacing, causing a barrel deflection problem where there was none before. Therefore, it is wise to check barrel deflection after an upgrade of this type and install a die support system if needed.

Power panel modification: Die body zones on a new automatic die might be at a higher watt density and more numerous than with the original die. The power panel would have to be modified to accommodate these changes.

Control systems

Discussions on control systems will be limited to the developments of the main components in a typical temperature closed-looped system, the controller, power control switch or valve, and sensor.

Sensors: The standard temperature sensor in extrusion coating lines is the type "J" thermocouple. Developments associated with this piece of equipment have been more along the lines of better reliability than better sensor accuracy.

Controllers: Today's controllers offer faster response times and greater accuracy than earlier models. These earlier models were limited to proportional control only until the late 1960s when proportional, derivative, and integral control was developed. The greater use of digital technology also offered a significant advantage over its analog counterpart. Circuitry of older systems could be susceptible to error or a drift in current measurement as current goes through the analog components, resulting in a maximum control offset. Digital technology allows the accuracy of the controller to be dependent on the resolution of the particular math package as opposed to the accuracy of the controller.

When dealing with a digital signal, it is much easier to maintain repeatable control. Setting up a potentiometer in the exact same position might prove difficult; this is not a problem for a digital signal.

Controllers are available today that will act in a "self tune" mode, viewing the history of the response of the controlled variable, and recalculating the controlling parameters (i.e., time constant, reset rate) based upon that response. If required, specialized algorithms can be downloaded through an RS-232 cable to the individual controller.

Power control: The first power control systems available were mechanical contactors. Still available today, mechanical contactors now offer different metals at the arcing point, offering better heat dissipation and greater reliability. One possible concern with mechanical contactors is their longer cycle times, 1 cycle per 20 seconds, which can lead to greater thermal shock in the heater, limiting its life. In answer to this possible problem, mercury mechanical contactors were developed with a cycle time of 1 cycle per 3.5 seconds.

Another commonly used power system is the silicone-controlled rectifier (SCR). Developed in the late 1960s, the SCR has the advantage of higher reliability and faster cycle time, 1 cycle every 1/2-1 second.

With the development of master operating systems that provide logic signal outputs, an SCR assembly with a solid-state relay was created. This unit is fired with a much lower level output signal, creating a safer system since no signal at line voltage is brought to the front panel controller.

Gauging

Gauging systems developments throughout the years closely follow the improvements found in most electrical systems. Gauges built before 1967 were of vacuum tube design. Between 1967 and the mid-'80s, solid-state relays were used. After the mid-'80s, a complete digital system was employed. With these changes, there has come a decrease in required maintenance and an increase in sensitivity with gauging systems.

Unlike most line equipment, the decision to upgrade a gauge can be a difficult one because of the shorter life of the gauge and the higher upgrade costs as a percentage of the cost of a new gauge.

Treaters

Treater designs have also benefited from improvements described in the gauge section of this paper. Power supplies are smaller and systems are more efficient. Like gauges, due to their lower costs, complete replacements are more the rule than upgrades.

SAFETY

When modifying equipment, often safety devices such as guarding are removed. These safety items often must be modified to allow them to retain their function. The desire to "get back in production" before all safety items are installed must be resisted.

When upgrading a system for higher speeds or output, safety devices used on the original system might not be adequate for the upgrading equipment. A safety evaluation for the upgraded line should be performed as if it was a brand new piece of equipment.

FOR FURTHER READING

1. Keller, R., "Evaluating Alternative Processes for Plastic Film Production," TAPPI 1985 PLC Conference Proceedings, TAPPI PRESS, Atlanta.
2. Bayless, R., "Selection of Optimal Die Design for Coextrusion Coating," TAPPI 1986 PLC Conference Proceedings, TAPPI PRESS, Atlanta.
3. Helmy, H., "Aspects of the Design of Coathanger Dies for Cast Film and Sheet Applications," TAPPI 1987 PLC Conference Proceedings, TAPPI PRESS, Atlanta.

4. Singer, F., personal communication, MICA Corporation, June 12, 1994.

ACKNOWLEDGMENT

The author would like to thank Fred Singer of Mica Corporation, Dick Palmer of Wollock & Lott, and everyone else that assisted in the research of this paper.

CHAPTER 17

Safety in Extrusion Coating Design

Louis W. Piffer, Egan Davis Standard

INTRODUCTION

Accidents in the factory can be painful in terms of both worker injuries and reduced company profit margins. This chapter discusses safety options available in extrusion equipment for both new equipment buyers and older equipment owners. The information also describes individual pieces of extrusion coating equipment and the more recent advancements in their safety features.

DESIGN CRITERIA

Safety in extrusion coating design is no longer just a "good ideal"; it's a cost-effective way to do business. Across all industries, billions of dollars and millions of production days are lost due to safety-related injuries and deaths. Despite greater safety consciousness in the industry, resulting in a reduction of safety related events, the cost of these events have skyrocketed.

Many factors contribute to this apparent paradox. Litigation, which is responsible for as much as 50% of workers' compensation costs, is becoming more common. Medical expenses, accounting for 40% of workers' compensation benefit payments, have been rising at an alarming rate. This rise has been further exacerbated by the apparent disparity in workers' compensation versus group health insurance claims.

According to *"The Issues Report,"* published by the National Council on Insurance, workers' compensation medical costs per claim have been rising almost 50% faster than per capital costs in the United States since 1980.

However, despite this environment of rising insurance rates, equipment owners do have options. Savings can be realized by reducing money expended and lost worker time.

PROCESS INFORMATION

Safety advances by equipment manufacturers have been assisted through discussions with customers and safety organizations, and indirectly by technical equipment innovations. These upgrades and innovations have presented the new equipment buyer and older equipment owner with many safety options. This chapter will now review these options in the individual pieces of equipment of a typical extrusion coating line.

Guarding systems

The most visible safety devices on an extrusion coating line are the guards placed around the equipment. Examples of this guarding can be seen in Figs. 1 and 2. This chapter does not describe every area where a guard needs to be supplied. When new equipment is supplied, the equipment manufacturer will provide this service in concurrence with OSHA requirements. The guards have the dual purpose of limiting operator access to dangerous areas but, at the same time, not severely limiting the equipment's operations.

Fig. 1. Equipment guards

Fig. 2. Equipment guards (laminator shown without chill roll)

Many safety-related incidents occur not because the guards were built incorrectly, but because they were not there at all. Guards are often removed during maintenance because of an equipment modification. If a guard is being removed because it "makes the machine easier to run," then the design of the guard should be discussed with the equipment manufacturer.

COATERS AND DRYERS

Applying a coating onto a moving web and then drying it can be a potentially hazardous situation due to mechanical movement and, in the case of organic solvent-based coating, possible solvent explosions.

Mechanical movement

Coaters: Operators often change coating rolls during color or product changes. The introduction of the cartridge coater has helped eliminate a possible area for operator injury. This coater design allows the bottom roll(s) to be removed with the doctor blade and pan in one assembly, allowing the operator to remove the rolls off-line in a safer manner.

Dryers: Movement of a clamshell-style dryer during cleaning can create another potential safety problem. Areas of concern and their safety solutions are described as follows:

A stop bolt assembly is attached to the corners of a planer retraction dryer. This assembly consists of a plate attached to the upper dryer half and an adjustable "stop" bolt connected to the lower half, which function together as an adjustable mechanical stop for the closing movement of the dryer. When the bolt and plate meet, a pinch point develops; therefore, a stop bolt guard should be included. This is easy to retrofit by installing a telescoping sleeve that surrounds and extends beyond the plate. Upon lowering the dryer half, the sleeve contacts the operator first. The telescoping feature prevents any pinch point from occurring and gives the operator time to remove his hand. (See Fig. 3.)

Fig. 3. Telescoping sleeve

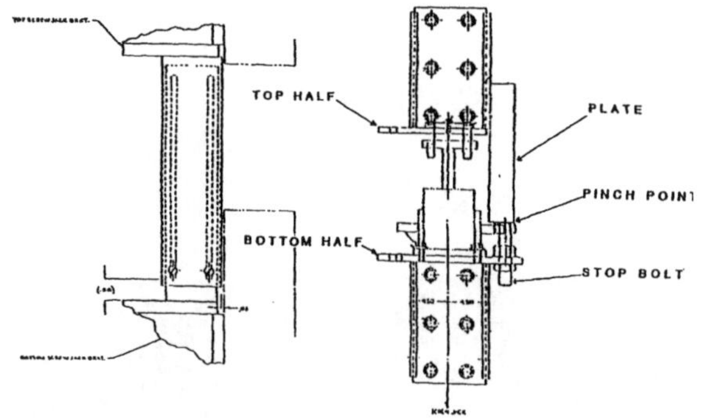

Life-line safety switches are often used between the two halves of a clamshell dryer. These lines, similar to the type used before an extruder carriage, will be tripped if an object gets trapped between the dryer halves, putting pressure on the line and stopping the dryer movement. (See Fig. 4.)

Fig. 4. Lifeline

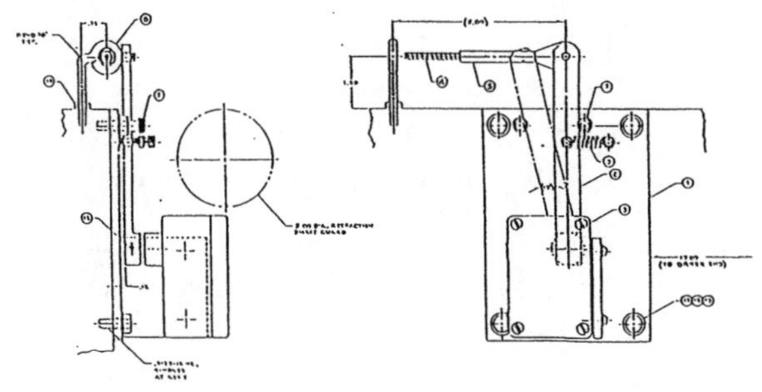

In a clamshell dryer, the two dryer halves are hinged in the back, with two screw jacks opening the two halves in the front. Newer designs have been created where the screw jacks are not placed between the clamshell halves, pushing them open; instead the dryer is pulled open by an air cylinder attached to an above structure. This creates better access to the dryer's internals for the worker, but it also creates a potential safety problem. If there is a failure in the lifting mechanism, the top dryer half will quickly close without warning. This can be prevented by either mechanically holding the cylinder in the open position, installing additional cylinders to retract upon failure of the primary system, or by installing a cable system to work in conjunction with the air cylinders. In the cable system, during retraction, the cable is wound up on a drum connected to a ratchet clutch. When the retraction has stopped, a rod clevis is inserted into the clutch arrangement, preventing the cable drum from unwrapping even if the lifting mechanism fails. (See Figs. 5 and 6.)

Fig. 5. Clamshell dryer with overhead lifting

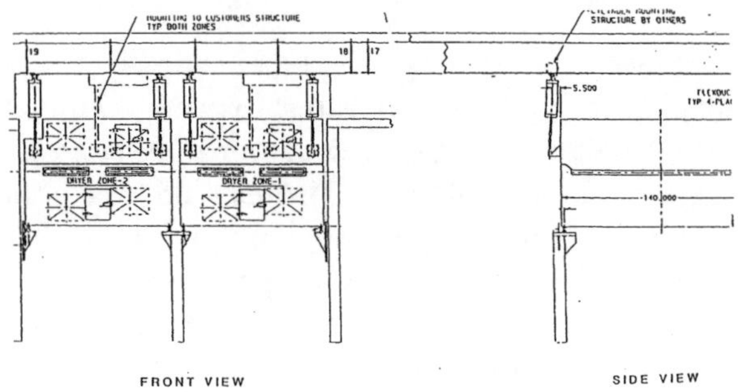

Fig. 6. Retraction cable assembly

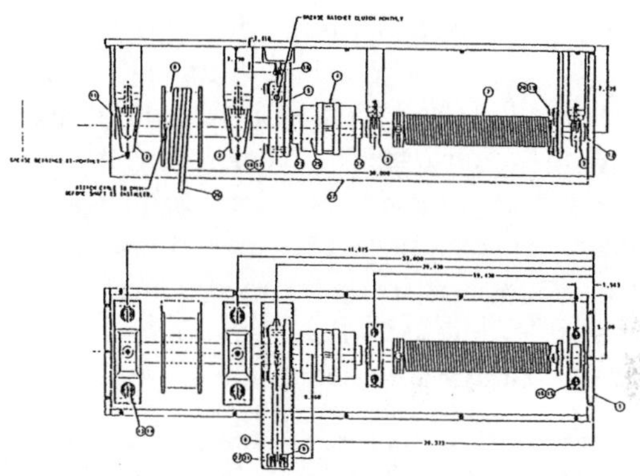

Explosions

Coaters: Care must be taken around coaters that are applying a solvent-laden coating. Figure 7 shows the current NEC specifications for Class 1, Division 1 and 2 explosion proof area boundaries for a given solvent source. Class 1, Division 1 requires all electrical devices to be explosion-proof or purged from an external air source. Class 1, Division 2 requires all arcing and heating devices to be explosion-proof, with other devices being of a standard design. It should be noted that the specifications shown in Fig. 7 should also be applied to the coated web leaving the coater and entering the dryer when determining the explosion-proof boundaries in an extrusion coating line.

Fig. 7. Explosion proof area boundaries

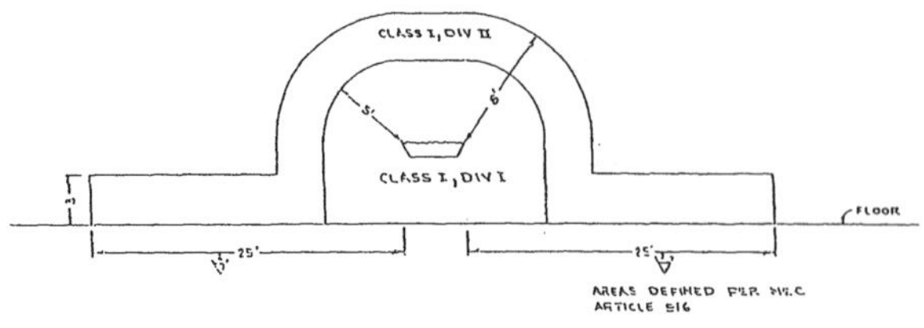

Dryers. As with coaters, dryers run the risk of explosion due to solvent vapor. Since it is impractical to eliminate this ignition source in its entirety, the vapor is diluted with fresh air to a point below its lower explosion limit (LEL). OSHA requires that when the LEL is not monitored, the dryer should to be designed to run at 25% of the LEL. If the LEL is monitored, blowers and exhaust ducting in the dryers can be designed for 50% of the LEL.

When a dryer is designed to dry both a solvent- and water-based coating, the above guidelines might easily be exceeded. Running a high percentage of recirculation air is often the case when running water-based systems. But if the same damper arrangement is used in the water-based system, an excessive LEL level could result. For this reason, it is recommended that an LEL monitor should be used when both water- and solvent-based coatings are used in the same dryer.

Further safety features attempt to reduce the impact of a dryer explosion. Safety relief ports are built into dryers' designs. (See Fig. 8.) These ports vent the explosive force in a controlled manner. The ratio of the area of these explosion vents is 1 $ft^2/15\ ft^3$ of dryer volume; the minimum acceptable ratio is 1:20. Explosion vents have the following features:

- They will be secured by cables. These cables allow the panels to blow off of the dryer without turning them into projectiles.
- They will have an explosion latch. This latch will be preset to release at a given pressure, which will be less than the pressure created by an explosion or achieved by the kick from a trapped worker.
- Explosion vents are placed in areas where workers are not normally present; for example, they are frequently located on the drive side or on the top of a dryer.

Fig. 8. Safety relief port

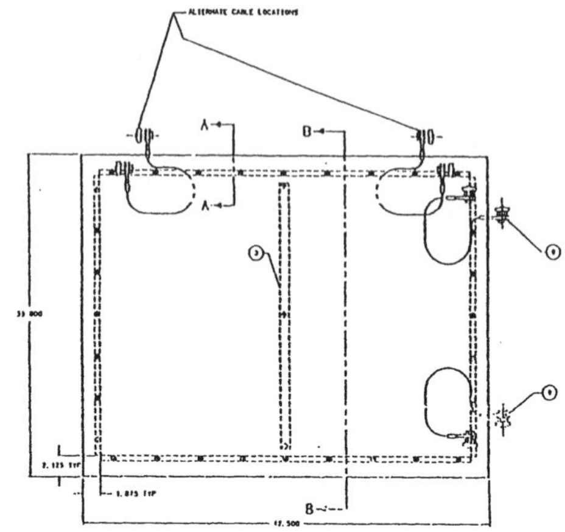

Line equipment

Unwinds and winders are the most consistent areas for operator interfacing. Shafts and rolls must be installed at the unwinder and removed at the winder. The lifting and moving of these items can risk injury. The most prominent type of injury associated with this lifting is a back injury.

Back injuries can create a significant dollar loss. Recently, a flexible packaging company with 210 employees on the floor investigated its workers' compensation records, revealing between 4000 and 5000 hours lost, due only to back injuries. This time loss translated into a dollar loss of over US$ 80,000/year.

Because of this high potential for injury, the following alternatives are needed:

Shaftless designs: Line equipment of this type eliminates the need for an operator to install shafts.

Shafted/shaftless design: This design has a chuck that will accept both shafts and cores; cores will be used on the wider width webs and shafts will be used for smaller widths. The shorter and lighter shafts allow for safer handling by the operator.

Composite shafts: As roll diameter requirements increase, shaft diameters and weight increase.

OSHA requires any shaft over 22 kg (50 lb) to be lifted by two people and any shaft over 45 kg (100 lb) to be lifted by a hoist. Unfortunately, operators do not always follow these guidelines. Composite shafts can reduce the weight classification of a given shaft, creating less chance of breaking OSHA rules and operators backs! Weight reductions of 50-75% have been reported.

Lower weight shafts: These shafts have up to a 35% less weight then standard shafts. This weight reduction is created by the use of multiple lower weight internal bladders instead of a single, heavier bladder.

Roll handling and shaft pulling systems: These systems can eliminate manual lifting of shafts, reducing potential mishaps caused by manually moving rolls.

Control systems

With the development of automatic control systems, extrusion coating safety has benefited. If a variable can be represented by a properly conditioned voltage or milli-amp signal, it can be monitored. Monitoring variables such as extruder drive amps, extruder pressure, or barrel temperatures allows you to set warning flags when predetermined upper hazardous limits are reached. The control system can also respond when warning flags are generated. These responses range from slowing an extruder down to shutting off a barrel heater.

Another safety feature is a cold start function. This function requires an operator to wait a predetermined amount of time before the extruder drive begins to turn the screw. The delay ensures that, when the machine is turned on, the hardened resin inside is given a chance to heat up and melt. This prevents the development of extreme high pressures from a motor trying to turn a screw through unmelted resin.

Temperature ramping, like a cold start function, is another safety feature that the control system provides. Tests have shown that transfer piping requires a slower heat-up process than the rest of the extruder with explosive results if they are heated up too quickly. The control system can give you the capability of ramping individual temperature zones at a reduced rate, in a safer fashion.

An automatic die was not created because of safety concerns, but it is a very fortunate indirect benefit. The die laminator nip is one of the most dangerous places in an extrusion coating line. Most auto dies do not require the operator to be adjusting die bolts during normal operation, creating a safer environment.

Extruders

The main variables of concern in an extruder are temperature and pressure. Some important safety features are:

- Barrel covers are insulated to prevent burning the operator.
- Guards are placed over heater terminal points to prevent injury due to electrical shock.
- Thermocouples are threaded into a heavy bushing that is mounted on the barrel. Older designs threaded the thermocouple adapter directly into the barrel without the added support of the bushing. During heat-up of the barrel heaters, thermal expansion of the heaters could possibly shear off the thermocouple, creating a runaway heater situation.
- Rupture discs are placed in the barrel. Older extruders used a 1/411 × 20 tap. Unfortunately, this tap was the same as a PMTC, and operators would use a handy PMTC plug in the event there was a lack of rupture discs. This tap was eventually changed to prevent the health hazard. Note that the outlet point of the rupture disc should be directed downward so that the plastic melt can be discharged in a controlled, safe manner.

Extrusion coating dies

The outlet of an extrusion coating die is the one area in an extrusion coating line that an operator is most susceptible to coming in contact with the heated resin. Unfortunately, the die is a high maintenance item, requiring periodic cleaning by the operator. Important points to remember during this operation are:

1. Wear protective clothing during cleaning. Gauntlet gloves are commonly used during die cleanup. These gloves cover the hands and arms of the operator. They are tied in the back, preventing the gloves from slipping off during use.
2. Only clean die lips when the extruder is drooling off line.
3. Die curtains during purging can be affected by random gusts of air. Opening an outside door could create gusts that, left unimpeded, could cause the curtain to blow onto an unsuspecting operator. Dampening out these gusts before they reach the die curtain would reduce this health hazard.
4. Water is often placed in a drool stick or pan to quickly cool the incoming melt and prevent sticking. Care must be taken when employing this practice because water heated up by the incoming melt could turn to steam under the melt, causing the melt to blow onto the operator.
5. Never walk under a die even when the melt flow appears to have stopped.

Slitters

Today's slitting units are designed to give a clean cut of the incoming web. The necessary sharpness of the slitter blades can create health problems when the operator is exposed to the blades during setup and maintenance procedures. Limiting this exposure can limit the health hazard. Most safety features related to slitters involve reducing the times the operator has to be near them. This reduction in exposure can be done by:

Blade guards: Blade guards should be provided that surround the blade, exposing only the area of slitting interface.

Removable blade cartridges: This feature allows the operator to remove the blade assembly as a cartridge. They would then perform the blade change at their workbench as opposed to at the machine site, which is a much safer practice.

Slit repositioning measuring devices: Using a measuring tape to determine correct slitter positioning requires the operator to be directly exposed to the blade. A machine-mounted measuring system eliminates this potential safety hazard.

Automatic slitter movement: Further operator exposure can be eliminated by automatically moving the slitters into position.

Linked slitters: Linking the bottom and top blades of a shear system will cut in half the operator's exposure when a mechanical slitting movement system is not employed. This linking is accomplished by a bar sliding down from the top blade cartridge into the bottom blade housing at the time of slitter movement. To prevent creating a pinch point, this bar must automatically retract if any objects are in line with the bar during engagement.

Trim support rings: A safety problem encountered during an edge trim break is that an operator must grasp the fluttering strip and re-insert it into the duct nozzle, risking coming near the slitter while doing so. The addition of a rotating or stationary trim support ring directs the strip around 90 degrees or so, absorbing both flutter and uncontrolled stresses in the ring rather than transferring it directly to the cut point. This reduces incidents of edge trim break and provides operators with a controlled edge trim path away from the cut point in the event of a breakage. (See Fig. 9.)

Fig. 9. Slitter with trim support

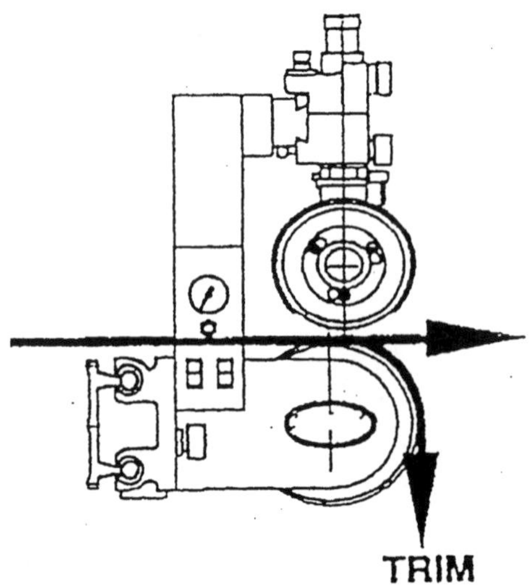

Treaters

In the world of treaters, ozone and electricity are the two safety culprits. The two main treater designs, covered and bare roll, handle these problems differently.

Covered roll treaters create a corona discharge between a dialectic covered roll and a bare electrode. A cabinet is placed around the entire area to both remove ozone and prevent access to the high voltage area. Safety interlocks are included that turn off the system if the door is open.

Bare roll treaters create the drop in voltage necessary to create corona within the covered electrode. This lessens the concern of electrical shock since the bare roll is at earth ground, creating less electrical resistance than the operator. For conductive substrates, precautions must be taken to make sure grounding does not occur through an operator in contact with the web. This is accomplished by running the substrate over a bare idler before or after the treating station. The roll should be stainless, plated aluminum, or steel to prevent oxidation. This will ensure proper grounding and eliminate "pitting" of metalized surfaces. Ozone is eliminated by a blower that both cools the electrode and also removes the ozone into the back of the unit where it is safely exhausted.

Corona treaters in general have a high voltage transformer. This transformer should be placed as near as possible to the electrode to avoid any necessary personnel exposure. If the transformer cannot be placed nearby, the high voltage wire should be run through a grounded metal pipe via plastic spacers providing clearance between the high voltage wiring and the inside wall of the pipe.

In either system, the electrodes should be interlocked so that, when they rotate for web thread-up, the power supply cannot operate. If these interlocks are defeated and the electrodes are rotated, high voltage will be present although the system will appear to be safe because corona is not being generated.

Drives

The safety threat in any extrusion drive system is the potential for electrocution during servicing and possible worker injury caused by an operator getting caught in the driven equipment.

Electrocution threats usually occur during troubleshooting and servicing. Increasing the reliability and diagnostic capability of any system will reduce service incidents and related safety problems. Modern drive systems addressed these issues by providing:

- Increased reliability is evident in all components from more advanced circuit board testing, reduced component count due to the increased use of microelectronics, and the insurgence of better manufacturing techniques in surface-mounted technology
- The use of automatic protection circuits offers increased protection. These circuits provide an orderly shutdown of the drive upon detection of a fault such as field loss, over-under voltage, and excess motor temperature.
- Increased diagnostics reduce the need for hands on servicing. Faults can be detected and displayed without requiring an operator to hunt around looking for the problem. More advanced distributed control systems allow global diagnostics from any control screen. This feature allows the serviceman to set up and troubleshoot a drive in full view of the drive in question.
- Isolation transformers reduce the potential for shock by creating a primary and secondary side of the circuit. Grounding the primary side will prevent operator shocks even if a second ground is created.

During the operation of an extrusion coating line, the possibility exists of the operator being caught in the driven equipment. In this event, a drive system must be able to stop the movement of the machinery during an emergency. There are two schools of thought as to how to accomplish it: One is stopping the drive through the use of the regenerative nature of the drive. The second method is using dynamic braking resistors. They are described as follows:

Regenerative braking: This is the quickest way to stop a drive. The full 150% current limit is used in regenerative mode to stop the drive. There is concern that if the control is lost, there will be no stopping force.

Dynamic braking: In this mode, a resistor is placed across the motor armature. During braking, a contactor switches the motor from the power source to the resistor. The motor then acts as a generator turning the rotational energy of the motor into electrical energy that is dissipated through the resistor. The time it takes to dissipate the energy is proportional to the power level generated from the motor/generator. Because the rotational energy reduces with time, the stopping energy does also.

GENERAL SAFETY GUIDELINES

When operating the described equipment together in an extrusion coating line, there are certain safety guidelines that should be adhered. Some of these guidelines are tabulated here with additional points described in Appendix II. This list is not all-inclusive, but it will give some indication of the type of questions that must be asked when devising a safety program.

- Have proper interlocks or timer lockouts been provided? For critical functions, have torque-type controls been used instead of pushbuttons to help prevent inadvertent activation?
- Are the most critical instruments in the most prominent location and easily read? Does the operator have to turn his head unduly or strain to read important instruments?
- Have operator consoles been designed so that readouts are above controls, preventing hands and arms that are making adjustments from blocking out readout instrumentation?
- Have programmed contingency measures been thought out to minimize requirements for hurried judgments at critical times? Have references been provided for making judgments such as marking meters with safe and unsafe limits?
- Is there similarity between controls located near each other so that inadvertent activation could cause an accident?
- Has a regular maintenance program been implemented? Many injuries are caused by an operator trying to compensate for some deficiency in the machinery caused by a lack of maintenance.

CONCLUSION

In view of increasing health and safety concerns and skyrocketing health costs, our own industry must be more responsive and ever more demanding of ourselves to increase our safety awareness.

REFERENCES

1. Accident, Facts National Safety Council, 1987-1990.
2. Smith, D. and Bayless, R., "Safety Considerations in the Design and Operation of Polymer Transfer Piping," TAPPI 1983 PLC Conference Proceedings, TAPPI PRESS, Atlanta.
3. Loss Prevention Data, Data Sheet 6-9S, Factory Mutual Engineering Corporation, May 1990.
4. ABC's of Revised Experience Ratings, National Council on Compensation Insurance, 1992.
5. Hammer, W., Product Safety Management and Engineering, Prentice Hall, 1980.

APPENDIX I-SAFETY ORGANIZATIONS

Occupational Safety and Health Act (OSHA)
Federal Office
200 Constitution Avenue
Room N3637
Washington, DC 20210

OSHA legislates and provides guidelines for the safe operation of equipment in a manufacturing organization.

Converting Equipment Manufacturers Association (CEMA)
66 Morris Avenue
P.O. Box 359
Springfield, NJ 07081

CEMA gathers a wide variety of information for converting equipment including safety.

Factory Mutual Insurance
10850 West Park Place, Suite 400
Milwaukee, WI 53224-3634

FM Insurance provides detailed safety information for machinery used in extrusion coating.

National Fire Protection Agency/National Electric Code (NFPA/NEC)
Battery March Park
Quincy, MA 02269

NFPA/NEC publish guidelines for the safe operation of equipment as it pertains to their areas of expertise. Unlike OSHA, they do not have the power to legislate, but many of their guidelines are used as standards for insurance companies.

National Safety Council (NSC)

The NSC furnishes all industries with safety-related information and services.

National Council of Compensation Insurance (NCCI)

The NCCI gathers information and performs statistical calculations on workers' compensation data.

APPENDIX II-SAFETY GUIDELINES

1. Have critical components been installed properly using asymmetrical configurations or polarization on electrical connectors or mechanical assemblies? Have male and female pairs that can only be assembled in one way been used on valves or other components where flow direction is vital?
2. Are openings in guards or protective covers small enough to prevent fingers from being inserted into hazardous locations?
3. To prevent guards from being removed by operators, could the guards be removed only by using a special tool that has limited accessibility?
4. Is there any point at which a person could touch a live conductor when the product is activated?
5. Are there any places that allow the accumulation of lint, grease, or other flammable material that can be ignited?
6. What is the possibility of inadvertently activating the product while a person is in a position to be injured?
7. Can the malfunction of a voltage regulator or other device cause damage to other components or equipment?
8. Are wires and cables protected against chafing, pinching, cutting, or other hazards that can damage the insulation so that a person could get a shock or weaken or cut the metal conductor?
9. Is the insulation on wires and cables nonflammable or self-extinguishing except where there is no possibility of a fire occurring?
10. Is there any source of ignition in the product that might be present during normal operation, for example, a source of high heat or of electric sparks or arcs?
11. If the occurrence of a fire is possible, has a means of fire extinguishing (either manual or automatic) been provided?
12. If the product must be handled, will it be of a temperature that will permit continuous contact?
13. Is a relief valve necessary on a pressure system to permit venting of fluid if the temperature increases its pressure?
14. Are containers of pressurized fluids that have high vapor pressures kept away from sources of heat?
15. Has piping for fluids been fastened securely to supports so that it will not vibrate when in operation?
16. Have safety-critical bolts and other fasteners been secured to ensure they will not loosen under vibration?

17. Have guards been provided to keep personnel from being injured by moving mechanical parts such as gears, fans, belts, or other components in motion?
18. Does the product have any pinch points, rotating components, or other moving parts that must be guarded?
19. Are ventilation openings small enough to keep fingers away from dangerous places?
20. Have all the lines and fittings been rated for the pressures they must withstand? Have they been pressure tested?
21. Have all flex hoses, their connections, and fittings been secured in order to prevent whipping if there is a failure?
22. Is the pressure system located near or can it be subjected to an unintended high heat input that would raise the pressure excessively?
23. Is it possible to accidentally connect the system to a source of pressure higher that that designed for the system or any of its components?
24. Is the relief valve, vent, or diaphragm set to operate at a pressure less than the pressure that can cause damage?

CHAPTER 18

Extrusion Coating: Some Basic Observations

Ray Edwards, Eastman Chemical Company

ABSTRACT

This chapter provides some helpful techniques and general information on extrusion coating for the novice in that field.

COAT WEIGHT VERSUS EXTRUDER OUTPUT, DIE WIDTH, AND LINE SPEED

A simple empirical equation permits an extrusion coater to predict a coating's basis weight (lb/ream) based on three variables: die throughput rate (lb/h), die width (in.), and line speed (ft/min):

$$\text{lb/ream} = \frac{600\ (\text{lb/h})}{(\text{ft/min})\ (\text{in.})}$$

(1 ream = 3000 ft^2).

(The equation can easily be manipulated to solve for any of the other three variables.)

Note that the equation gives coating weight, not coating thickness. To calculate thickness, one must know coating density. And coating density is dependent upon many variables.

COATING DENSITY (CRYSTALLINITY)

Coating density is always lower than the unextruded pellet density. Coating density is basically determined by the rate at which the molten extrudate is ultimately quenched, or annealed. The quenching does not occur fully at the chill roll; quenching continues beyond the chill roll because the coated structure continues to retain residual heat. This residual heat is usually sufficient to permit mobility of the polyolefin molecules to form more compact crystal sites.

Coating density, therefore, is dependent upon:

- Extrudate thickness
- Extrudate temperature
- Substrate mass and composition
- Substrate temperature
- Chill roll temperature
- Chill roll heat transfer efficiency
- Coating speed.

Figure 1 shows the three-dimensional relationship of coating density to extrudate/chill roll temperatures. (Not shown are the other influential variables, including substrate mass/temperature and coating speed.) Note that, as more energy is applied to the extrudate and/or the chill roll, the higher the resultant coating density is. And, conversely, the lower the temperatures of the melt and chill roll are, the lower the final coating density is.

Fig. 1. Coating density vs. extrudate and chill roll temperature

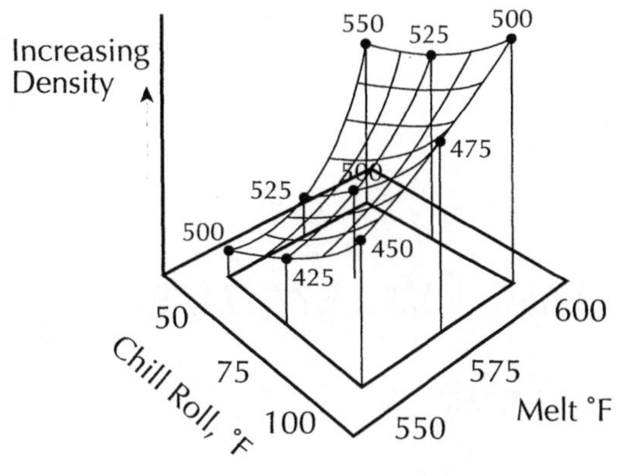

Polypropylene
135 = ft/min Constant Speed
3 - mil Coating

Note: Numbers by Coordinate Represent ΔT's

Density can usually be changed to a higher or lower desired value. If a lower density is desired, the extrusion coating operator should lower the temperatures of the substrate, extrudate, and chill roll, and increase the contact time on the chill roll as much as can be permitted. (Chill roll contact time can be increased by increasing line speed. If that is not practical, a variable stripper roll can be installed at the chill roll. The stripper roll can be positioned to provide the coating with a longer contact time against the chill roll.)

Conversely, if a higher density is desired, one should maximize all temperatures and decrease chill roll contact time.

What role does coating density play in the finished coating? Here are some examples:

- Higher density coatings give better barrier properties. Unfortunately, they also cause problems, e.g., excessive curl.
- Lower density coatings give less curl. They also give lower heat seal initiation temperatures, but they can cause coefficient of friction problems.

As mentioned, lower density coatings are desired for lower heat-seal initiation temperatures.

HEAT SEALING (POLYETHYLENE)

Heat seal properties, i.e., seal initiation temperature and hot tack range, are determined primarily by coating density and coating viscosity (melt index), respectively. The relationship of heat-seal initiation temperature (i.e., minimum sealing temperature) to density are obvious in Fig. 2. As polyethylene density increases, the temperature required to initiate its coating's heat seals likewise increases.

Fig. 2. Effect of density on hot tack (1-mil coatings on 40-lb kraft)

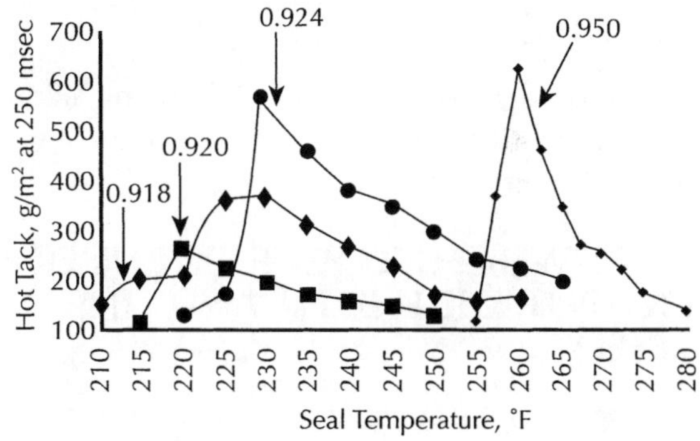

Figure 3 shows polyethylene heat seal "hot tack" as a function of coating viscosity (melt index). Note that, as melt index decreases, hot tack seal strength increases and the hot tack range is extended to hotter temperatures.

Fig. 3. Effect of melt index on hot tack (1-mil coatings n 40-lb kraft)

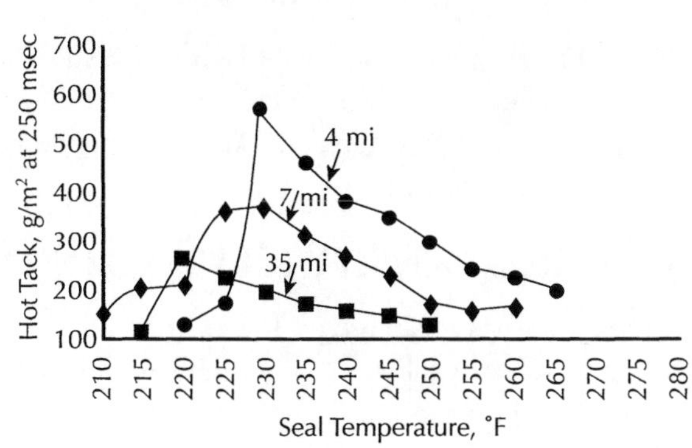

There is evidence that molecular weight distribution (MWD) and chain branching also influence heat seal properties. Current studies show that narrow MWDs, along with the proper branch distribution, can exhibit lower seal initiation temperatures.

BARRIER PROPERTIES AND THICKNESS

Density, besides influencing polyethylene heat seals, also determines a polyolefin coating's barrier properties. It is known that, at a given coating thickness, the higher the coating density is, the better the resistance is to gaseous diffusion through the coating.

In food packaging, it is desirable to exclude oxygen and water. For a given coating density, the diffusion of O_2 and H_2O through a polyolefin can be controlled by varying coating thickness. When plotted linearly, O_2 and H_2O transmission rates vary hyperbolically with thickness. (See Figs. 4 and 5.)

Fig. 4. O_2 transmission rate

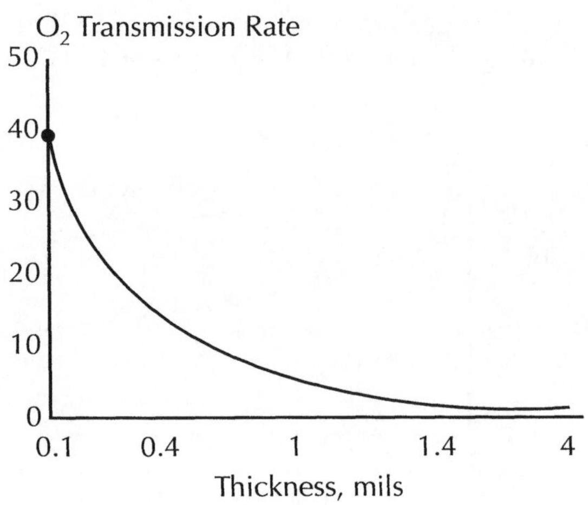

Fig. 5. WVTR, g/24 h/100 in.^{2}

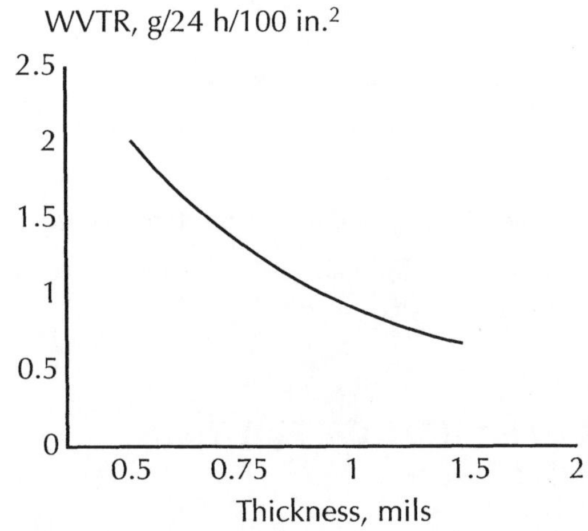

Such linear plots offer some disadvantages: (a) They are difficult to use in determining an exact thickness needed for a desired gas transmission rate, and (b) such plots require several data points to define the curve.

Experience has shown that a fairly accurate straight line relationship of transmission rate to thickness can be predicted using only one data point! This is done by plotting a single data point on logarithmic (log-log) graph paper, followed by sketching a straight line having a slope of -1 through that single data point. (See Fig. 6 as an example.) Advantage: A straight line permits better prediction of the coating thickness required for a given gas transmission rate.

Fig. 6. Log WVTR, g/24 h/100 in.^{2}

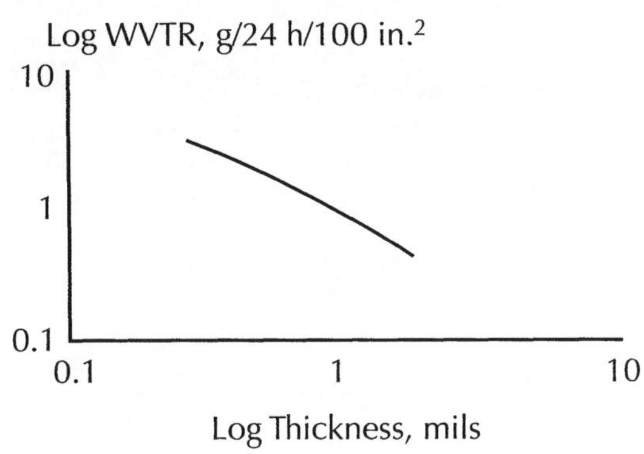

Remember this simple shortcut for hyperbolic curves involving gaseous diffusion versus coating thickness:

1. Use logarithmic (log-log) graphs.
2. Plot one data point (transmission rate, thickness).
3. Draw a straight line through that point.
4. The straight line should have a "negative one" slope, i.e., -1.

COEFFICIENT OF FRICTION CONTROL

Polyolefin density also influences a coating's coefficient of friction (COF) properties. The higher the coating density is, the lower the coating COF is. Most coating COFs are controlled by (a) adding slip agents such as amides, (b) varying coating thickness, or (c) a combination of the first two methods.

Linear graphs are usually used to predict the level of slip agenda and/or the coating thickness required to affect a given COF (Figs. 7 and 8). And, as in the case of gaseous diffusion through a polyolefin, such linear graphs result in hyperbolic relationships between COF vs. slip level and between COF vs. thickness. Such graphs do not lend themselves to accurate prediction of slip levels and/or coating thicknesses required to achieve a given COF.

Fig. 7. Minimum COF vs. erucamide content (1-mil thickness)

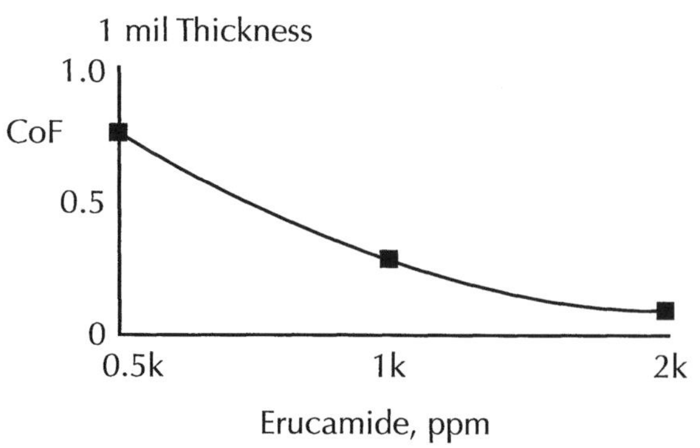

Fig. 8. Minimum COF vs. coating thickness (constant 0.1% erucamide)

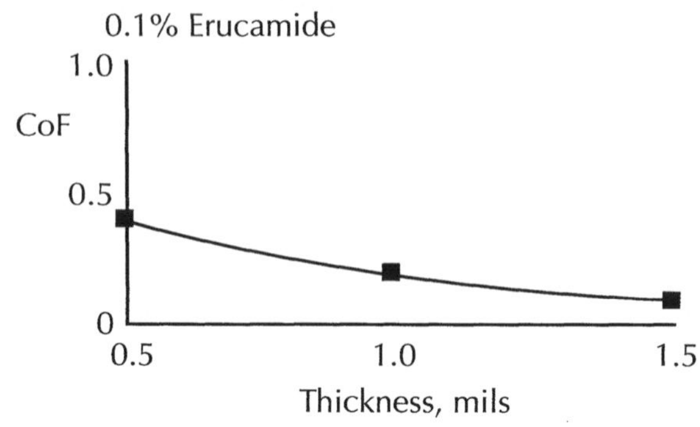

Straight line plots involving COF can be achieved, again with the use of logarithmic graphs. Experience has shown the following relationships of COF to slip level and COF to coating thickness. And, as with gaseous diffusion plots, only a single data point is needed:

COF vs. thickness. Figure 9 is a logarithmic graph of COF vs. thickness at a given level of slip agent. This relationship can be predicted fairly accurately with a straight line having a slope of -1.

Fig. 9. Logarithmic relationship of COF to coating thickness (0.1% erucamide)

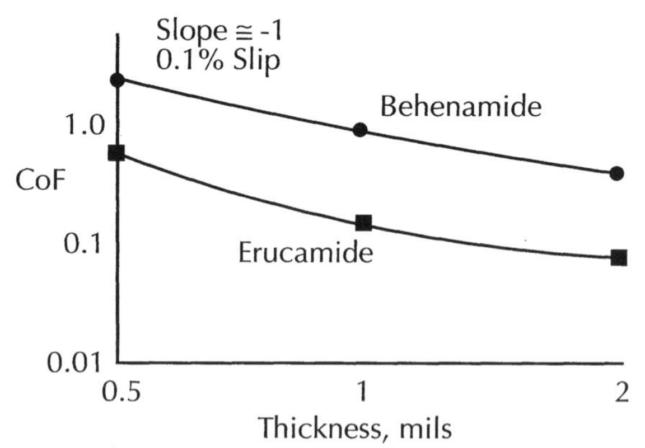

COF vs. level of slip agent. See Fig. 10, a logarithmic plot. Note that, for a given thickness, a straight line of relationship exists. But these straight line plots of COF vs. slip level require a slope of -2.

Fig. 10. Logarithmic relationship of COF to slip content (1-mil thickness)

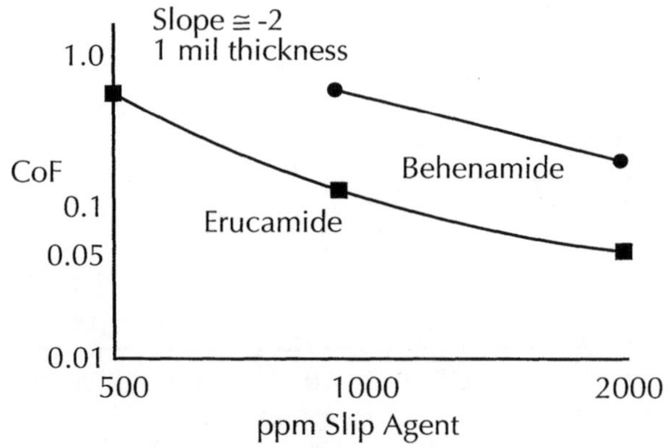

It is important to remember that the following logarithmic slopes apply to the use of erucyl amide only and in extrusion coating applications only. (They do not apply to other applications such as blown film or cast film.)

- COF vs. thickness: -1 slope
- COF vs. slip level: -2 slope

SURFACE TREATMENT

Polyolefin coating surfaces usually require some form of oxidative treatment to accept printing inks or glues. Two major forms of treatment are:

- Corona discharge
- Flame impingement.

A major problem with either method is decay of treatment with time. For example, in flexoprinting applications, a polyethylene surface is normally treated to about 40-44 dyne/cm. But if there is too long a wait after treatment, surface tension diminishes to a value too low to accept ink, i.e., about 35 dyne/cm.

Some extrusion coaters bombard their coatings with excessive treatment. This is with the belief that, if the surface tension starts at a high value, it will finally equilibrate at or above the required 40-44 dyne/cm after decaying with time. Such is not the case, however, as Fig. 11 shows.

Fig. 11. "Regular" treatment

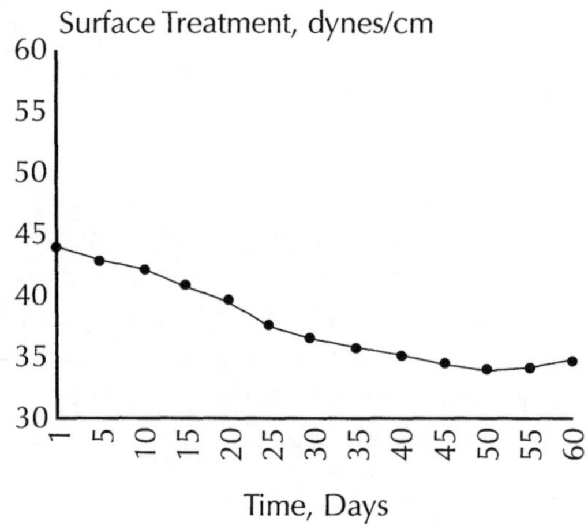

Figure 12 compares a "normal" initial treatment level of 44 dyne/cm to a typically "excessive" treatment of 56 -dyne/cm. The excessive treatment initially decays faster than normal treatment so that both treatments equilibrate to the same final surface tension.

Fig. 12. Excessive vs. "regular" treatment

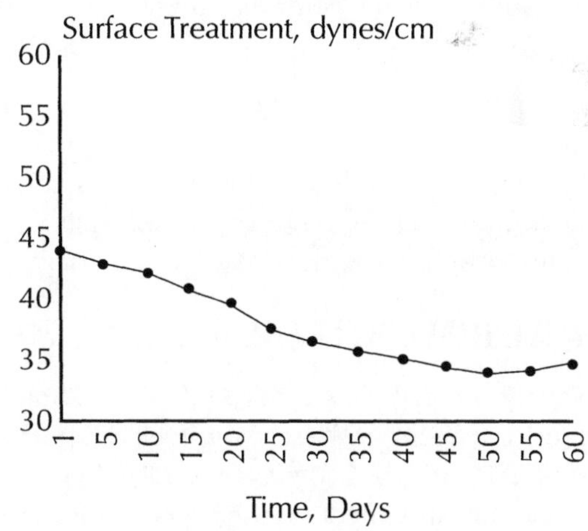

Excessive treatment usually results in major coating problems including:

- Pinholes from bum-through
- Heat-seal failures from excessive surface oxidation
- Ink scuffing, also from excessive oxidation.

On the subject of decay of surface tension with time, flame printing reportedly decays at a slower rate than corona treatment, and flame treatment reportedly equilibrates to a slightly higher surface tension.

When using flame treatment, it is imperative that the gas/air ratio be maintained at a "lean gas" mixture. This gas/air ratio provides optimum oxidation and complete combustion of the gas. When combustion is incomplete, hydrocarbon residues are deposited on the polyolefin surface. These hydrocarbons have been known to interfere with heat sealing and printing.

SURFACE GLOSS

Surface gloss of a polyolefin coating is inversely related to that polyolefin's molecular weight distribution (MWD). At one time, polypropylenes with their inherently narrow MWDs were considered to give the glossiest coatings of all polyolefins. Because of this, they were used primarily in packages requiring 11 aesthetic coatings, i.e., those high gloss coatings that attract buyers to the product (for example, frozen food packages). Now certain extrusion coating polyethylenes are available with coating surfaces more glossy than polypropylene.

This family of "glossy polyethylenes" has the most narrow MWD of all extrusion coating grade polyethylenes. They have lower-than-usual melt indexes to offset the tendency for excessive neck-in caused by their narrow MWDs.

CHILL ROLL RELEASE

Recently, polyethylenes have been developed for extrusion coating relatively low-melt temperatures (less than 480°F, or 250°C). Such polyethylenes are functionally modified to effect good adhesion to substrates. This functionality also lends itself to chill roll sticking. The resultant poor release from the chill roll usually manifests itself in either of two extremes: poor adhesion on the one hand, and substrate delamination on the other.

Experience has shown that an amide, specifically erucyl amide, can be added to such functionally-modified polyethylenes to eliminate or reduce the chill roll sticking. However, an optimum level of erucyl amide is needed (see Fig. 13, which depicts

Fig. 13. Chill roll sticking vs. % Additive A

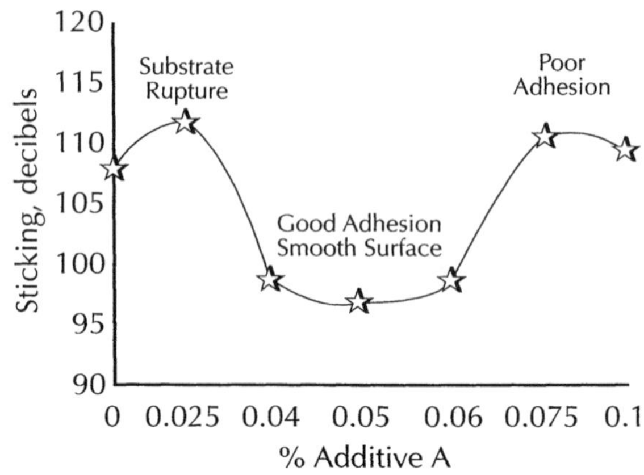

chill roll release by sound level, expressed in decibels):

Too little amide does not adequately permit good chill roll release. Result: substrate delamination. Excess amide in extrudate comes to the surface to prevent adhesion at the polyethylene/substrate interface. Result: a coated "product" with no structural integrity and no commercial value.

The optimum level of erucyl amide in the functionally modified low-temperature polyethylene was found to about 0.05%. Please note, however, that this level of erucyl amide is ineffective at extrusion coating temperatures above 500°F (260°C) in such functionally modified polyethylenes.

ALUMINUM FOIL CORROSION (ELECTROLYSIS)

Some multi-layered extrusion coated packages contain a metalized film layer. One major problem that erupted in recent years was corrosion of this metalized layer. With the metalized layer being essential to the copyrighted logo on the packaged, this corrosion detracted from the aesthetics of the packages.

The corrosion was traced to a carbon black-based ink on the surface of another layer. With the carbon black being a metal and the metalized surface being aluminum, a dissimilar metal, electrolysis was occurring. Result: electrolytic corrosion of the metalized surface.

Solution: Replace the metallic carbon black in the ink with an acceptable organic-based ink. This totally eliminated the electrolysis problem.

TITANIUM DIOXIDE

Titanium dioxide (TiO_2) is basically hygroscopic. It can contain water in three forms: absorbed water, adsorbed water, and water of hydration. In a TiO_2--modified polyolefin, the absorbed and adsorbed waters are removed by "normal drying," i.e., in a forced air dryer oven at 180°F (82°C).

Unfortunately, the water of hydration, being chemically bonded to the TiO_2, remains with the TiO_2 until driven off by the same high temperatures used in extrusion coating (about 500°F, 260°C, to well above 650°F, 342°C). See Fig. 14.

Fig. 14. 10% TiO_2 in LDPE ("dried" at 82°C)

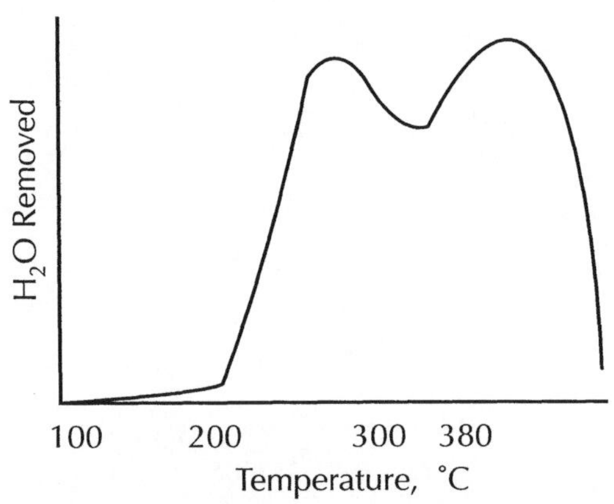

Excessive water of hydration can cause major imperfections in a coating if the TiO_2 is present at much above 15-20% by weight.

Corona "burn-through" in white-TiO_2--pigmented coatings is also attributed to the conductivity of residual water of hydration.

POLYPROPYLENE AND HIGH-DENSITY POLYETHYLENE EXTRUSION

Polypropylenes (PPs) and high-density polyethylenes (HDPEs) have relatively narrow molecular weight distributions (MWDs). Because of this molecular property, they tend to exhibit "draw resonance," or surging, during extrusion coating. These two polyolefins require a unique set of extrusion coating conditions to reduce or prevent draw resonance:

During extrusion (for any application, including the extrusion coating process), PPs and HDPEs must be permitted a maximum pressure drop across the die lands! This is a basic requirement.

To accomplish the required pressure drop across die lands, and to reduce or prevent surging, the extruder operator must reduce or, more ideally, eliminate all pressure-drop points in the extruder system-from the feed throat of the extruder to the die exit. The following parameters should be considered:

- Use a standard extrusion screw, i.e., a screw without mixing sections. Each mixing section creates a pressure-drop point. (And each pressure drop detracts from the all-important pressure drop across the die lands.) With mixing sections, the extruder operator has no control over the pressure (and shear) that a polymer is subjected to behind each mixing section.
- Use as high a screw speed (and as fast a line speed) as the process and/or quality product permits. The higher the volumetric flow is across the die lands, the higher the pressure drop is across those lands.
- Use a coarse screen pack, again, reducing pressure drop at that site.
- Open the pressure valve fully. This permits maximum "hydraulic" pressure to the die lands.
- Use no static mixers, for the same reason as explained for screen packs and mixing sections.
- Use short die lands and narrow die gaps. Ideally, die land length should be 0.25 in. (6.4 mm) or less.
- Die gap (land separation) should be adjusted to 0.020 in. (0.5 mm) or less.

The operator should combine as many of these parameters as possible when extruding PPs and HDPEs.

CONCLUSIONS

Many observations in extrusion coating applications have been made over the years. Those cited in this paper are designed to educate and assist the novice in the field of extrusion coating.

CHAPTER 19

Polyolefin Properties in Extrusion Coating

Ray Edwards, Eastman Chemical Company

ABSTRACT

Ethylene and propylene are the monomers used to make polyolefins. There are three important physical properties of polyolefins used in extrusion coating: molecular weight (MW), molecular weight distribution (MWD), and density. Their initial values determine how a polyolefin will perform during extrusion coating. The properties are significantly altered during extrusion, and these altered polymer properties determine the end-use performance of the resultant coating. This chapter follows each polyolefin property through the extrusion coating process and discusses how each physical property is changed and how these changes affect coating end-use performance.

POLYOLEFIN MOLECULAR STRUCTURES

Polyethylenes and polypropylenes are polyolefins. A polyolefin is a hydrocarbon made up of many long chain molecules of carbon and hydrogen atoms. The length of each chain determines the molecular weight of that chain. Each polyolefin contains molecules of various lengths or various molecular weights. This assortment of molecular weights is termed as the molecular weight distribution of that polyolefin.

Polyethylenes are made by chemically linking ethylene monomers (CH_2=CH_2) to form these long chain structures. These polyethylene molecules are not straight chains-as they also have side branches. These side branches determine polymer density, with more side branches contributing to a lower density.

Polypropylenes are made by chemically tying propylene monomers (CH_2=CH-CH_3) together to form long chain molecules. Every third, or "tertiary," carbon in the polypropylene chain is bonded to a single side chain carbon atom. These tertiary carbons are weakly bonded to adjacent atoms. These tertiary bonds, being fairly weak, are easily broken if exposed to chemical, mechanical, or thermal activity. Such activity severs, or degrades, the polypropylene molecules into shorter lengths, with lower molecular weights. The presence of these tertiary carbons therefore requires all polypropylenes to be highly stabilized to prevent this molecular degradation.

THE EXTRUSION COATING PROCESS FOR POLYOLEFINS

Polyethylenes with melt indexes in the 2-15 dg/min range are normally extruded at a melt temperature of about 310-332°C (590-630°F). There are now "low-temperature" functionally modified polyethylenes available with melt indexes >20 dg/min, and these are coated in the 177-260°C (350-500°F) range. At these colder melt temperatures, some form of substrate priming is needed to assure adhesion. Various primers are available such as chemicals, oxidizing flames, and high voltage corona discharge. These priming methods are sometimes used in combination.

Polyethylenes are fairly forgiving of tight screen packs and high back pressures. Die land lengths of 0.25-1.0 in. will easily permit uniform extrusion coatings of polyethylenes. A nominal die opening of 0.020 in. is apparently the optimum for polyethylenes, permitting the fastest line speeds and thinnest coatings.

Polypropylenes need as much pressure to the die lands as can be generated during extrusion. Polypropylenes cannot tolerate excessive pressure drops in the extrusion process. Pressure drop sites in an extrusion system include mixing sections on the extruder screw, screen packs, pressure valves, and static mixers. Polypropylenes perform best with a maximum pressure drop across the die lands. To achieve this, a high die throughput rate is needed (a minimum of 10 lb/h/in. is suggested). To gain the maximum pressure to the die lands, the extrusion system should contain a coarse (open) screen pack, an open pressure valve, the shortest die lands available, and as narrow a die gap (land-to-land) as can be adjusted and still maintain a uniform coating. (A 0.010-in. die gap was found to be the optimum in laboratory studies).

Polypropylenes normally coat well in the 282-316°C (540-600°F) melt temperature range. There are some functionally modified polypropylenes capable of coating as cold as 204°C (400°F). The colder melt temperatures provide a more viscous, tougher coating with improved barrier properties. The improved barrier properties come from reduced pinholing.

MOLECULAR WEIGHT IN THE EXTRUSION COATING PROCESS

Polyethylene molecular weight is expressed by melt index (MI). MI is actually a measurement of a polyethylene's viscosity under a given set of test conditions. MI, expressed in grams/10 minutes, tells how much of a molten polyethylene, at 190°C, will flow through a 2.1-mm orifice. A 9.5-mm diameter piston, weighted with 2160 grams, is used to drive the molten polyethylene through the orifice.

Polypropylene molecular weight is expressed by flow rate (FR). FR is also expressed in grams/10 minutes (dg/min). FR is determined in the same test equipment as melt index, but at a hotter temperature-230°C. MI and FR vary inversely with molecular weight. The higher a polymer's molecular weight is, the lower its MI or FR is. Both melt index and flow rate tests are contained in ASTM Procedure D1238-79.

For extrusion coating uses, polyethylenes and polypropylenes, in general, show the following melt index and flow rate ranges:

- Polyethylenes: 2-70 dg/min
- Polypropylenes: 5-70 dg/min.

EXTRUSION'S EFFECT ON MOLECULAR WEIGHT-FROM PELLET TO COATING

A given extrusion coating process exposes polyolefin to relatively hot temperatures and to mechanical shear. Such extrusion conditions always alter a polyolefin's molecular structure to change its physical properties in going from pellet state to coating state.

In a low shear environment, a polyethylene's molecules can chemically crosslink or mechanically entangle. Such cross-linking or molecular entanglement increases the molecular weight of the polyethylene extrudate/coating. This increased molecular weight produces a coating having a lower MI than the initial pellet MI. The coating will then exhibit end-use properties associated with low MIs, and not the MI of the original pellet state.

High shear mechanically severs, or degrades, a polyethylene's molecules, particularly its long chain molecules. This high shear lowers overall molecular weight, resulting in a higher MI coating. The coating will now exhibit different end-use properties than those subjected to low shear.

Regardless of low- or high-shear conditions, a polypropylene coating always shows a higher FR (lower molecular weight) than its original pellet state. This is because of the weak bonds of its tertiary carbons as previously discussed. End-use properties, then, will always be associated with some FR higher than that predicted by the FR of the pellets.

A coating's molecular weight will vary from extrusion process to extrusion process!

END-USE PROPERTIES OF A COATING ARE INFLUENCED BY ITS MOLECULAR WEIGHT

Polyethylenes contribute certain important properties to various coating applications, and polypropylenes offer their own unique properties. The melt index of a polyethylene coating influences the following end-use properties:

Low melt index coating:

- Broad heat seal range (hot tack)
- Pinhole resistance
- Stress crack resistance
- Toughness

High melt index coating:

- Fast heat seals
- Reduced curl.

The flow rate of a polypropylene coating affects these properties:

Low flow rate coating:

- Strong heat seals
- Pinhole resistance
- Abrasion resistance
- Stress crack resistance
- Grease resistance

High flow rate coating:

- Substrate wetting (good adhesion)
- Faster heat sealing.

MOLECULAR WEIGHT DISTRIBUTION IN THE EXTRUSION COATING PROCESS

The many molecules comprising a polyolefin vary in chain length and, therefore, their molecular weights vary. Each polymer contains an assortment of molecules ranging from low molecular weight oils to high molecular weight, long chain hydrocarbons. These different molecules can be separated by gel permeation chromatography to determine a polyolefin's molecular weight distribution (MWD). For a polyolefin with a given melt index (or flow rate) and at a given set of extrusion conditions, MWD plays the most important role in that polymer's ability to extrusion coat. This is because MWD directly determines the neck-in of the polyolefin during the extrusion coating operation. And neck-in, in turn, determines whether the polymer will coat satisfactorily or not coat at all. Neck-in, therefore, is a measure of MWD in the rheometer that we refer to as an extruder/die/laminator system.

In the extrusion coating operation, the chill roll/nip roll rotation draws, or pulls, the molten extrudate from the die. These draw forces are vertically perpendicular to the die opening, and they act only on that width of the coating that falls between the chill roll/nip roll. If a polymer's MWD is too narrow, it will exhibit too much neck--in. With too much neck-in, there are no draw forces acting on the outer edges of the extrudate and edge weave (and surging) results. Thus, too much neck-in can limit coating speeds.

A certain amount of edge bead at the outer edges of an extrudate is essential to extrusion coating. The edge bead gives strength to the edges of the extrudate to prevent edge tear. Because edge bead is a function of neck-in, the less neck-in there is, the less edge bead there is.

If a polymer's MWD is too broad, it will exhibit too little neck-in. Too little neck--in can limit coating speeds just as too much neck-in limits coating speeds. Too little neck-in can cause extrudate edge tear and heavy surging, and sometimes large voids in the coating.

Another interesting observation: Each melt index (or flow rate) requires a given MWD to assure an optimum neck-in. The higher the melt index is, the broader is the MWD needed for satisfactory coatability. This proper balance of MW and MWD is the responsibility of the resin supplier, but the extrusion coater has the responsibility of optimizing his extrusion operation to get the best performance from a polyolefin. For example, excessive temperatures can cause too much neck-in (and edge weave). Pressure drops (tight screen packs, high back pressures) can severely detract from the coating characteristics of polypropylenes. Die openings and extruder output rates will affect neck-in.

EXTRUSION'S EFFECT ON MOLECULAR WEIGHT DISTRIBUTION--FROM PELLET TO COATING

A polymer's molecular weight distribution is always narrowed by extrusion. Mechanical shearing, by the screw and mixing sections on the screw, breaks the polymer's molecules into shorter chains. Increasing back pressure or installing tighter screen packs will increase mechanical shear, as will increasing screw speed. High temperatures, in combination with long residence times in an extruder, also thermally breaks the molecules.

Note: Coating molecular weight distribution will vary from extrusion process to extrusion process!

MWD INFLUENCES END-USE PROPERTIES OF A COATING

As mentioned earlier, lower NM (i.e., higher MI, FR) polyolefins require a broad MWD to be extrusion coatable. Likewise, the high MW (i.e., low MI, FR) polyolefins require a narrow NMD. Here are some known end-use properties associated with the MWD of a coating:

Coating with narrow MWD:

- Higher gloss

Coating with broad MWD:

- Prevents leaks in bottoms of polythylene--coated milk cartons.

DENSITY IN THE EXTRUSION COATING PROCESS

As previously mentioned, density of a polyolefin is determined by its side chain branching. The more side chain branches there are, the lower the density is. Polyethylene pellet density is classified into three categories by ASTM:

- Low density: <0.926 g/cc
- Medium density: 0.926-0.940 g/cc
- High density: >0.940 g/cc.

Polypropylene pellet density is naturally lower than polyethylene. This is because of a single side chain carbon accompanying every other carbon atom in the polypropylene chain:

Polypropylene density: 0.90-0.91 g/cc.

DENSITY THROUGH THE EXTRUSION COATING PROCESS

A polyolefin's coating density is always lower than its original pellet density. This lower density is caused by the rapid quenching of the hot molten extrudate by a cold chill roll. This rapid quenching solidifies the polymer to stop any annealing process that would result in a maximum coating density. Variations in extrusion temperatures and the chill roll temperature, in combination with different line speeds (i.e., chill roll contact times), actually determine the final coating density.

Coating density will vary from extrusion process to extrusion process!

END-USE PROPERTIES OF A COATING INFLUENCED BY ITS DENSITY

Polyethylene coating density influences certain end-use properties:

Low-density coating

- Initial lower seal temperature
- Faster heat seals
- Reduced curl
- Better barrier properties

High-density coating

- Abrasion resistance
- Increased curl
- Improved barrier properties
- Greater stiffness.

CONCLUSIONS

There are three major physical properties that influence how a polyolefin will extrusion coat. These properties are always altered by the extrusion process. These altered properties determine the end-use performance of the final coating.

CHAPTER 20

Chemical, Physical, and Mechanical Properties of Common Polymers

Thomas Bezigian, PLC Technologies Consulting Co.

ABSTRACT

Polymers are used in almost every imaginable application today such as simple bags and pouches, medical and food packaging, aerospace applications, and everyday items such as photographic paper. Polymers are processed readily by many methods, extrusion being the most common. Within the class of extrusion processing, blown film, cast film, and extrusion coating are of particular interest to manufacture. The development of the extrusion coating process in 1946 began a revolution that has allowed complex structures to be manufactured to meet ever-increasing societal needs. Within a particular class of polymers, all physical and processing properties are derived from four fundamental polymer properties: molecular weight, molecular weight distribution, density, and chain branching. This chapter reviews the effect of these parameters on the chemical, physical, and mechanical properties of polymers.

INTRODUCTION

Polymers are long chain-like structures made up of many smaller units (monomers). There are unbranched (so-called linear) polymers such as high-density polyethylene (HDPE) and branched polymers such as low-density polyethylene (LDPE). Figure 1 depicts the branching phenomena observed in polymers. The type and extent of this branching determine (a) density, which in turn affects physical properties and (b) chain entanglement, which in turn affects processability and physical properties.

Fig. 1. Representative LDPE molecule

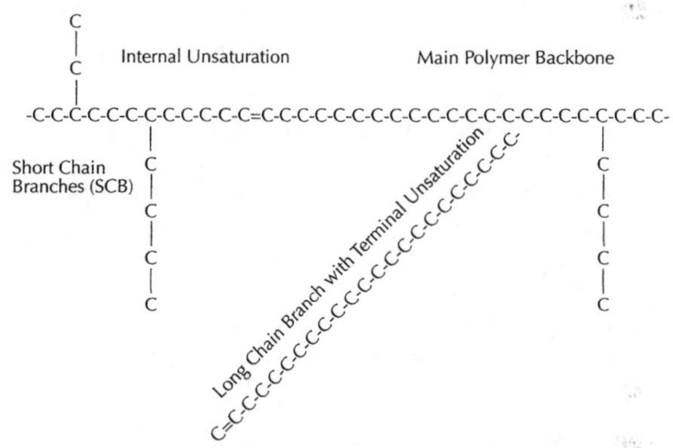

Internal unsaturation is prone to undergo **chain scission** due to shear and heat. Chain scission causes a localized decrease in **molecular weight**. A severe case can cause die smears or waviness in films.
Terminal unsaturation is prone to attack by oxygen or free radicals generated during chain scission.
Minor oxidation and/or **chain-linking** cause a localized increase in molecular weight, which can result in soft gels or flow lines in films.
Severe oxidation and/or **cross-linking** cause massive localized increases in molecular weight, resulting in resin discoloration and/or hard gels and holes in extrudates.

MASS FLOW OVERVIEW

In terms of unit operations, many things occur to the polymer we extrude that we either overlook or take for granted. There are multiple phase transformations that take place between the hopper and winder, consuming hundreds of horsepower and many tons of cooling. The electrical power required to turn the extruder motor is transformed into mechanical energy (heat) via the frictional loss between the polymer and extruder barrel. This is called viscous energy dissipation.

The rate of temperature rise while the polymer is in the extruder barrel, which ranges between 2 and 10 minutes, is slow when compared to the temperature drop experienced while on the chill roll. The typical time available for cooling on the chill roll is about 0.5 seconds, or 2-3 orders of magnitude (100-1000 times) less than that available for heating. This meaning that the rate of cooling of the extrudate is much greater than the rate of heating.

Special design considerations must be made when fabricating equipment, as well as in processing so that quality finished product is wound off the end of the machine. Excessive heating for example can decompose the polymer molecule, thus altering physical properties such as heat seal and hot tack strength, as well as odor, barrier properties, and more. Inadequate cooling can yield the same results by altering final crystallinity.

An overview of the processes that a plastic pellet sees is show below.

Spherical Pellets
Extruder Screw
Compacted, melted, pumped
Filtered
Homogenized, Coextruded
Formed into shape through the die
Exits die, Extensiona lFlow, Oxidation
Chemical and Physically married to substrate
Solidified

SIMPLIFIED POLYMER CHEMISTRY

This section will review simple polymer chemistry to establish a baseline from which the remainder of the paper will be based.

Chain branching and unsaturation

Figure 1 shows a simplified structure for branched polyethylene, short chain branches, long chain branches, and unsaturation. Unsaturation is important in a polymer as it forms sites for polymer degradation and oxidation to take place during processing.

Critical polymer properties

There are four *key* parameters that determine *all* properties within a polymer class:

- Molecular weight (MW)
- Molecular weight distribution (MWD)
- Density
- Chain branching.

MW and **MWD** heavily influence processability. **MW** and **density** heavily influence finished physical properties. **Chain branching** affects processing and physical properties.

With the exception of density, the tests available to determine these key parameters are time consuming at best and difficult and not repeatable at worst. Because of this, a quick and easy test has been developed to determine the quality of polymers before extrusion. This test is called ***melt index***.

Melt index (MI) measures the amount of plastic that flows through an orifice under a certain load and temperature over a period of time. It is inversely proportional to molecular weight and viscosity. Its units are g/10 min or dg/min. It is a relatively crude test, but nonetheless popular because of its simplicity and ease of use.

FINISHED PROPERTY CONSIDERATIONS

These four critical parameters determine all the properties within a class of polymer. Though there are many, many types of products made via extrusion coating and laminating, several generalizations can be made. For example, the layers within the extrusion lamination generally must have good adhesion to each other, and must possess some combination of chemical, physical, and optical properties as required by the end-use application. A brief listing of some of the properties required in the many finished products made by the blown and cast film processes, as well as extrusion coating and laminating, are shown below.

Some physical property requirements

Physical barrier	Surface appearance
Moisture vapor barrier	Gloss
Chemical resistance	Haze
Heat sealability (strength, speed)	Temperature resistance
Hot tack strength	Coefficient of friction
Peelable seal	Physical strength
Tensile strength	Stress crack resistance
Tear strength	Electrical or RF properties
Stiffness	Odor free
Elongation at break	Curl free
Flex crack resistance	Printable

CRITICAL FACTOR RESPONSES

This section of the chapter reviews the effect of the four critical polymer attributes on the physical and processing properties of polymers.

Molecular weight

As $\mathbf{M_w}$ increases, these physical properties:

Increase	Decrease
Impact strength	Melt index
Tensile strength	Cold flow
Film tear strength	Brittleness
Stress crack resistance	Brittleness temperature
Chemical resistance	
Fusion temperature	
Heat seal range	
Hot tack strength	

As $\mathbf{M_w}$ increases, these processing properties:

Increase	Decrease
Melt viscosity	Melt index (ease of flow)
Extruder horsepower	Drawdown ability
Extruder head pressure	Extruder output
Melt temperature override	
Melt strength	

Critical factor responses

As **MWD** increases, these physical properties:

Increase	Decrease
Heat seal range	Melt index
Brittleness temp. (BAD)	Gloss
Haze (BAD)	Stress crack resistance
	Overall strength properties

As **MWD** increases, these processing properties:

Increase	Decrease
Melt strength	Viscosity at high shear rates
Overall processability	Critical point
Shear heating	Neck-in
Melt temperature override	

Critical factor responses

As (solid) **density** increases, these physical properties:

Increase	Decrease
Crystallinity	Tear strength
Tensile strength	Heat seal range
Stiffness	Permeability (GOOD)
Chemical resistance	Deadfold
Fusion temperature	
Initial heat seal temperature	

As **density** increases, these processing properties:

Increase	Decrease
Extruder horsepower	Extruder output
Energy to melt	

Critical factor responses

Short chain branching (SCB) affects physical properties. **Long chain branching** (LCB) affects processing properties.

As **SCB** increases, these physical properties:

Increase	Decrease
Impact strength	Density (crystallinity)
Film tear strength	Stiffness
Stress crack resistance	Tensile strength
Heat seal range	Brittleness temperature
Permeability (BAD)	Chemical resistance
Tackiness (blocking)	Fusion temperature
	Shrinkage (curl)

As **LCB** increases, these processing properties:

Increase	Decrease
Melt strength	Surging (draw resonance)
Edge tearing	Viscosity at high outputs
Die swell	Extruder horsepower
Surface roughness	Head pressure
Haze	Melt temperature override
	Gloss

LOW-DENSITY POLYETHYLENE

This simplest of olefin polymers accounts for nearly 75% of the volume of plastics used in packaging applications. For many applications, it possesses an ideal combination of physical, chemical, and rheological properties. These properties, combined with its low price, account for its extensive market penetration. It has been adapted well to screw extrusion for film, sheet, coating, as well as compounding.

Simplest fundamental unit [CH_2CH_2]n: For every 1000 chain carbon atoms, LDPE's usually have 20-30 short chain branches, with some exceeding 10 carbon atoms in length. This is the factor that controls density. LDPE has many (up to 8 per 1000 atoms) long chain branches. This is the factor that controls processability.

Molecular weight (M_w): Up to 150,000.

Molecular weight distribution: Typically 8-10.

Degree of crystallinity: About 55%.

The growth of crystallites to form three-dimensional ordered structures *(spherulites)* is promoted by annealing. Shock-cooling (quenching) produces a more random arrangement with shorter crystalline segments and smaller spherulites, resulting in a high degree of optical clarity.

Density (g/cc):
0.910 to 0.925 (LDPE)
0.926 to 0.940 (MDPE)
0.941 to 0.965 (HDPE)
Approximately 0.750 (molten).

Density is normally related linearly to *crystallinity,* but crosslinking produces an increase in density with a decrease in crystallinity.

Permeability: Transmission of gases and vapors through polyethylene decreases with increasing crystallinity and, therefore, is least in HDPE. Most organic solvents readily permeate through LDPE. A typical MVTR (moisture vapor transmission rate) is approximately 4 g/m^2/24 h/25μ.

Physical effects of temperature: Low- and high-density types become brittle only below -70°C; low molecular weight polymers (waxes) are brittle in the range of -30°C to room temperature.

Softening points: 60-80°C (LDPE); (waxes, 20-60°C).

Melting point: (loss of crystallinity), 109-125°C.

Mechanical properties: Properties involving small deformations (e.g., modulus, creep) depend upon crystallinity and, therefore, upon density. For large deformations (e.g., tensile strength, tear), molecular weight and branching (as indicated by melt index) are determining factors. Typical tensile strength is 3000 psi.

Melt index: A viscosity measurement inversely proportional to molecular weight. Typical values range from 4 to 15 (g/10 min).

Processability: Output: 6 lb/h/hp (3.7 kg/h/kW); melt temperature: 580-620°F (304-325°C); head pressure: 800-2500 psi (5.5-17.2 MPa); neck-in: 2-6 in. (50-150 mm), depending on melt index, melt temperature, line speed, and air gap; and edge weave: zero.

HIGH-DENSITY POLYETHYLENE

Because it is made from the same *monomer* (or building block) as LDPE, HDPE is essentially void of the side branches indicative of LDPE. The lack of these branches increases crystallinity of the polymer, which improves most of the physical properties of the end product. However, the absence of *chain branching* reduces *melt strength,* and thus degrades the processability of HDPE. Extrusion coating grades are blends of HDPE with LDPE to improve processability.

Simplest fundamental unit [CH_2CH_2]n: For every 1000 chain C-atoms, HDPEs can have fewer than 5 short branches and are, therefore, nearly linear in structure. There is usually some degree of unsaturation present: from 0.1 to 2.5 double bonds per 1000 C-atoms.

Molecular weight (M_w): 20,000-40,000 (Phillips Process) and up to 200,000 (Ziegler Process).

MWD is 4-6 and is considered narrow.

Degree of Crystallinity: Up to 95%. The growth of crystallites to form three-dimensional ordered structures *(spherulites)* is promoted by *annealing.* Shock-cooling *(quenching)* produces a more random arrangement with shorter crystalline segments and smaller spherulites, resulting in a high degree of optical clarity.

Density: 0.94-0.95 g/cc (Ziegler), 0.96-0.97 g/cc (Phillips).

Permeability: Transmission of gases and vapors through polyethylene decreases with increasing crystallinity and, therefore, is less in HDPE.

Softening point: 110-125°C.

Melting point: (loss of crystallinity), 130-135°C.

Mechanical properties: Properties involving small deformations (e.g., modulus, creep) depend upon crystallinity and, therefore, upon density. For large deformations (e.g., tensile strength, tear), molecular weight (as indicated by melt flow index) appears to be the determining factor. Tensile strength is about the same as LDPE, but elongation at break is less, yielding a higher modulus (stiffer) material.

Melt index: Typical values are 2-10 g/10 min at 190°C.

Processability: Output: 5 lb/h/hp (3.1 kg/h/kW); melt temperature: 610°F (320°C); head pressure: 1000-2500 psi (6.9-17.2 MPa); neck-in: 3-6 in. (75-150 mm), depending on melt index, melt temperature, line speed, and air gap; and edge weave: 1/4-1 in. (6-25 mm).

ETHYLENE/VINYL ACETATE COPOLYMERS (EVA)

Combining vinyl acetate monomer with ethylene during polymerization yields EVA. The relatively large pendant vinyl acetate group serves to disrupt crystallinity, and thus reduces crystalline dependent properties such as melting point and stiffness. These properties, however, are advantageous in low-temperature heat seal and low-temperature freezer applications.

Simplest repeat unit:

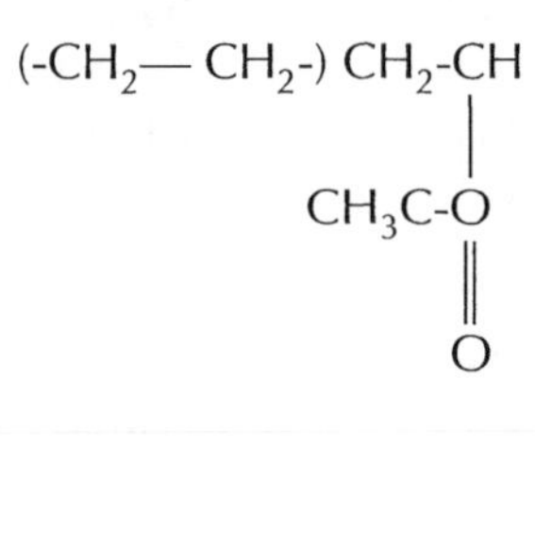

Crystallinity: The presence of vinyl acetate reduces crystallinity and at 20-30% concentration produces rubbery copolymers.

Density: 0.926-0.95 g/cc.

Processability: Output: 6 lb/h/hp (3.7 kg/h/kW); melt temperature: 450°F (230°C); head pressure: 800-2500 psi (5.5-17.2 MPa); neck-in: 2-4 in. (50-100 mm); edge weave: zero.

IONOMERS

These are ethylene copolymers having cross-links based on ionic intermolecular bonds between oppositely charged groups. The "free swinging" anionic carboxyl groups are attached to the ethylene main chain by copolymerization and metal cations (Mg^{2+}, Zn^{2+}, Na^{+}, and K^{+}) are introduced. The cross-linked structure does not impair flow properties at processing temperatures, but provides extraordinary strength at temperatures near or below the melting point. The chemical nature of ionomers make them very difficult to flame retard.

Simplest fundamental unit:

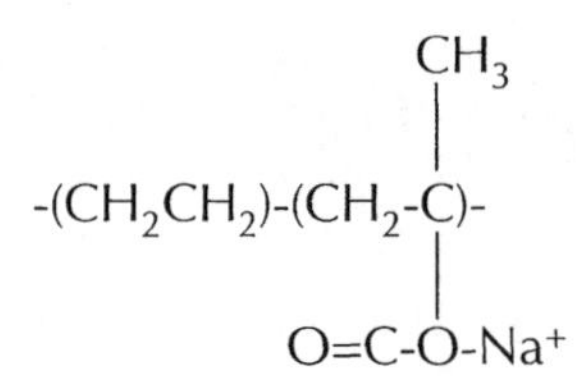

Density: 0.93-0.97 g/cc.

Physical effects of temperature: Brittle point: below -100°C; softening point: 70-90°C.

Melt index: 1-14 g/10 min at 190°C.

Processability: Output: 4 lb/h/hp (2.5 kg/h/kW); melt temperature: 550-620°F (290-325°C); head pressure: 1000-2500 psi (6.9-17.2 MPa); neck-in: 1-3 in. (25-75 mm); and edge weave: zero.

POLYPROPYLENE

Due to its structured crystalline nature, the polypropylene backbone is rigid and strong. It is chemically more resistant, has a higher melting point, is stiffer, and is clearer than polyethylene. The oriented form of PP possesses excellent water vapor transmission rates. Metallized OPP is even more superior. It is a staple in the snack food industry.

Simplest fundamental unit:

The crystallinity and rigidity of the commercial *isotactic* (same side) polymer arise from the regularity of the structure. *Syndiotactic* (alternating sides) polypropylene is more readily soluble and fusible and is not commercially used. *Atactic* (random placement of pendant methyl group) polypropylene is amorphous and is used as an industrial adhesive.

Molecular weight: Above 80,000-200,000. The melt flow index is indirectly related to molecular weight. Melt flow index (g/10 min at 230°C), about 8.0-60. MWD values of 4-6 (narrow) are typical and are responsible for neck-in/edge weave problems seen with PP.

Degree of crystallinity: -60 to 70% for commercial polymers, 95% isotactic in structure.

Density: Cast films (chilled from melt) = 0.89; oriented = 0.90-0.91.

Softening point: 85-105°C; melting point: 165-175°C.

Physical effects of temperature: Polypropylene embrittles more readily than polyethylene (which is brittle only below -70°C). Films can be heat sealed at 170°C. The polymer is subject to oxidation at 100°C, unless suitably protected.

Mechanical properties: The $-CH_3$ groups, attached isotactically to alternate C-atoms in the molecular chain, modify the properties in certain ways: (a) causing stiffness, which raises the melting point; (b) rendering it less stable than polyethylene with regard to oxidation. An increase in density (i.e., crystallinity) raises the softening temperature, stiffness, and yield strength, but decreases permeability, low temperature brittleness, and tensile strength.

Processability: Output: 4 lb/h/hp (2.5 kg/h/kW); melt temperature: 560°F (293°C); head pressure: 600-1500 psi (4.1-10.3 MPa); neck-in: 3-6 in. (75-150 mm); and edge weave: 1/2-1 in. (15-25 mm).

POLYESTERS

As perhaps the most common condensation polymer, polyesters find use where high-temperature resistance is needed such as in dual ovenable food trays or where high strength and stability are needed such as in magnetic tape media. *Metallized OPET* possesses low moisture and oxygen vapor transmission rates, desirable properties for packaging materials, and novelties such as balloons.

Polyesters (PET) are designated *condensation* polymers because water is a by-product of the polymerization reaction. Because of this, PET, and all condensation polymers are sensitive to moisture and *must be dried* before processing. Condensation polymers are fundamentally different from addition or *vinyl-type* polymers. Their structures are generally much more complex, their properties being more like engineering resins and their processability being much less desirable than olefin resins such as LDPE.

Their narrow molecular weight distribution, typically 2-4, accounts for this troublesome processability. The bulky side groups generally associated with condensation polymers limits chain mobility, which translates to increased motor torque requirements, melting point, and stiffness.

SIMPLEST FUNDAMENTAL UNIT

Polyethylene terephthalate (PET) has the following repeat unit:

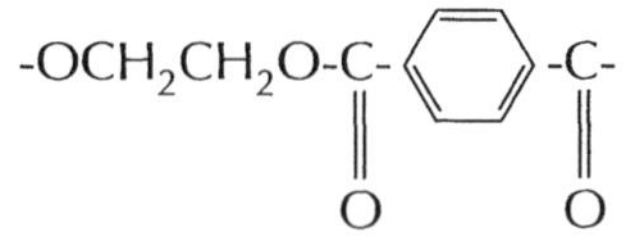

Molecular weight: 20,000-30,000.

MWD: 2-4.

Density: Amorphous, 1.38 g/cc; crystalline, 1.455 g/cc. The polymer is substantially amorphous when quenched from a melt but crystallizes after heating above 80°C.

Permeability: Films have relatively low permeability to most gasses and vapors.

Physical effects of temperature: Flexible to below -70°C., with increase in strength but decrease in extensibility. Untreated films shrink when heated to about 100°C. Heat set films are serviceable up to 150°C. It softens (sticky, can be heat sealed) at 230-240°C.

Melting point: 256-260°C (pure polymer, 265°C).

Processability: Output: 4 lb/h/hp (2.5 kg/h/kW); melt temperature: 555°F (290°C); head presure: 600-1500 psi (4.1-10.3 MPa); neck-in: 3-6 in. (75-150 mm); and edge weave: ½-2 in. (12-50 mm).

POLYAMIDES (NYLON)

As another type of condensation polymer, polyamides are essentially made from joining an acid and an amine. When combined, water is extracted and the polymer formed. Its chemical, physical, and processing properties are similar to that of polyester. Polyamides are quite often used as barrier layers in multilayer packaging.

Simplest fundamental unit:

Repeat Unit	Name
$-NH(CH_2)_5CO-$	Nylon 6
$-NH(CH_2)_{10}CO-$	Nylon 11
$-NH(CH_2)_6NH-CO(CH_2)_4CO-$	Nylon 6-6
$-NH(CH_2)_6NH-CO(CH_2)_8CO-$	Nylon 6-10

Unstretched polymers are amorphous or randomly polycrystalline.

Molecular weight: 10,000-100,000 (Nylon 6).

MWD: 3-5.

Physical effects of temperature: Strength is retained to about -40°C; Tg: Nylon 6, 70°C (lowered by moisture); Nylon 11, 46°C; thermal decomposition, in the absence of oxygen, commences at about 300°C; the primary mechanism involves the scission of $-NH-CH_2-$ groups.

Melting points: Nylon 6, 215-220°C; Nylon 11, 185°C.

Processability: Output: 4lb/h/hp (2.5 kg/h/kW); melt temperature: 540°F (280°C); head pressure: 1000-2500 psi (6.9-17.2 MPa); neck-in: 3-6 in. (75-150 mm); edge weave: ½-2 in. (12-50 mm).

POLYCARBONATE

The amorphous nature of PC yields excellent optical properties. It requires special considerations such as drying, a special screw design, and a powerful motor/gearbox assembly.

Simplest fundamental unit:

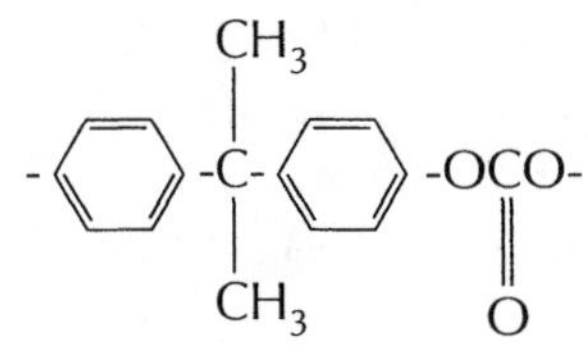

Molecular weight: Up to 30,000, beyond which the increasing viscosity limits processing.

Crystallinity: Does not crystallize.

Density: 1.20 g/cc.

Melt index: 3.0-6.0 (g/10min).

Physical effects of temperature: Brittle only below -135°C; Tg about 145-150°C; softens, 160°C; decomposes, 310°C; accelerated by moisture.

Melting point: = 215-230°C; crystalline melting point, 268°C.

Processability: Output: 4 lb/h/hp (2.5 kg/h/kW); melt temperature: 525°F (275°C); head pressure: 800-1500 psi (6.9-17.2 MPa); neck-in: 3-6 in. (25-150 mm); edge weave: 1-2 in. (25-50 mm).

POLYURETHANE

An interesting specialty resin is polyurethane. It is very tough and finds application in medical areas or where excellent abrasion resistance is needed. Due to ULDPE's linear nature, polyurethanes are tough to process because of the excessive torque requirements on the extruder motor.

Simplest fundamental unit:

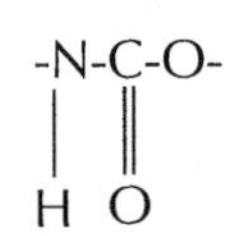

Molecule Weight: Linear polymers, about 10,000.

In complex structures, the molecular weight of the initial polyester or polyether can range from 500 to 6000, but the value rises several-fold by extension through urethane linkages. In the final network, molecular weight is "very large."

All polyurethanes undergo slow degradation at about 200°C (toxic isocyanates are released) and further degradation at 200-300°C to produce amines and olefins.

Processability: Output: 3.5 lb/h/hp (2.2 kg/h/kW); melt temperature: 620°F (325°C); head pressure: 1000-2500 psi (6.9-17.2 MPa); neck-in: 1-3 in. (25-75 mm); and edge weave: zero (edge tear possible).

CAST FILM AND EXTRUSION COATING RESIN BENEFIT ANALYSIS

This section briefly analyzes the benefits of the major resins utilized in the cast film and extrusion coating industries. Their properties are weighed against each other in a logical manner as a practicing polymer engineer does in his daily routine. The information discusses major features and benefits, and presents the objections and solutions to these objections.

Table 1. LDPE

FEATURE	BENEFIT	OBJECTION	OVERCOMING OBJECTION
Moisture barrier	Longer shelf life	HDPE has better moisture barrier and stiffness	HDPE has poor puncture strength and heat sealability
		Poor oxygen barrier	LDPE must be combined with materials like EVOH, Nylon, and PET when good oxygen barrier is required
Impact, tear, and puncture strength	Higher profits because fewer product returns from leakers in shipment and handling	LLDPE is stronger and has better hot tack	LLDPE is more expensive and has a higher heat seal temperature
			LDPE and LLDPE can be coextruded or blended to optimize properties
Heat sealable	Good machinability at low cost	EVA heat seals at lower temperatures	EVA is more expensive and has higher risk of odor
		EAA and Surlyn have better hot tack to make more reliable seals	LDPE is lower cost, but better hot tack might be needed in some applications, such as vertical FF&S of heavy packages

Table 1. LDPE (continued)

FEATURE	BENEFIT	OBJECTION	OVERCOMING OBJECTION
		Seals of LDPE to itself are not easy-open	Easy open seals require more expensive resins or additives such as Surlyn or polybutylene
Low cost	Higher profits strength	Inadequate barrier other resins to optimize	Coextrude or blend with cost/performance
Many grades and suppliers	Material availability	Older technology with fewer new products	LDPE is still in use because of good Properties at low cost
Strong bond to paper and varying bond to other substrates in extrusion coating and laminating		Weak bonds to substrates	Coextrude with other resins such as EAA to obtain better bonds
		Extrusion coatings have odor problems because of high extrusion temperatures	Coextrude at lower temperature to obtain strong bonds without odor
			Adhesives also have odor problems because of incomplete solvent drying
Clear and printable films and coatings	Increased sales due to improved graphics	MDPE has much better clarity and gloss	LDPE is less expensive
			Process printing provides Superior graphics
Coextrudable	Lower packaging Costs by combining with other resins to optimize cost/properties	Not all converters can coextrude	Upgrade equipment

LLDPE

Linear low-density polyethylene has the same features as LDPE but is stronger and has better hot tack strength. LLDPE film grade resins are more expensive than LDPE, and extrusion coating grades are even more expensive. Butene-, hexene-, and octene-based LLDPE are all available. Hexene and octene have better tear and puncture strengths. Octene is the strongest and most expensive.

Table 2. LLDPE

FEATURE	BENEFIT	OBJECTION	OVERCOMING OBJECTION
Tear, impact and puncture strength	Higher profits from improved quality or source reduction, resulting in fewer leakers	Octene is the strongest type, but price is high and supply limited	LLDPE can be coextruded or blended with LDPE to yield optimum cost/properties Fewer returns might justify cost
Hot tack strength	Better packaging machine efficiency because of fewer rejects and re-wraps	EAA and Surlyn have better hot tack and lower heat seal temperatures	EAA and Surlyn are more expensive and used only when absolutely needed
Seal strength	Fewer product returns from seal failures in shipment and handling	LDPE seals at a lower temperature	LLDPE seals are stronger and provide an extra margin of safety LLDPE must be use in hot-fill applications because LDPE has insufficient heat resistance
Flex-crack resistance	Fewer product returns because of reduced pinholes	18% VA EVA has good flex-crack resistance and a lower heat seal temperature	LLDPE has better hot tack strength, so fewer leakers on VFF&S machines LLDPE is less expensive than 18% VA EVA

NOTE: LMDPE is available for improved stiffness, gloss, and reduced flavor adsorption

EVA

Ethylene vinyl acetate is available in many grades for a variety of uses. Resins containing 18% vinyl acetate are common because of their low seal initiation temperature and excellent flex-crack resistance. Other grades are used for applications such as frozen food.

Table 3. EVA

FEATURE	BENEFIT	OBJECTION	OVERCOMING OBJECTION
Low seal temperature	Higher packaging speeds with fewer rejects and re-wraps off packaging machine	Poor hot tack strength applications	Surlyn or EAA might be needed for some
		Seals of EVA to itself aren't easy-open additives	Easy open seals require more expensive resins or
Flex-crack resistance pinholes from flexing	Fewer product returns because of less and is less expensive during shipping	LLDPE has good flex-crack resistance easily seals off channel leakers and makes a hermetic package	EVA has much lower seal temperature so more
		EVA has potential odor problems odor; it is extensively used in food packaging without problems	EVA is extruded at low temperatures to minimize
Co-extrudable	Lower package costs by combining with other resins to optimize cost/properties		
Impact, tear, and puncture strengths	Higher profits because fewer leakers in shipping and handling	LDPE costs less	EVA has a much lower heat seal temperature, and fewer returns may justify cost.

EAA AND EMAA

Ethylene Acrylic Acid and Ethylene Methyl Acrylic Acid are acid copolymers. They have excellent adhesion to foil in extrusion coating and laminating, excellent stress crack resistance, and very good hot tack strength. EAA is Dow's PRIMACOR and EMAA is DuPont's NUCREL. They are similar, but many grades are available, so selection for a particular application depends on the cost/performance ratio required.

Table 4. EAA and EMAA

FEATURE	BENEFIT	OBJECTION	OVERCOMING OBJECTION
Excellent foil bond	Longer shelf life because bonds won't delaminate even when packaging aggressive liquids	Lower cost LDPE can be used if foil is primed	Incomplete drying of packages can cause odors in packages
			EAA and EMAA have better chemical resistance, preventing permeation of aggressive liquids that can delaminate and corrode the foil
Stress crack resistance	Improved quality with fewer product returns caused be leakers from stress cracking caused by aggressive liquids	EAA and EMAA are too expensive	Fewer product returns might justify the cost
			EAA and EMAA coextrusions with LDPE are less costly and give adequate product resistance in many applications
			Coextrude EAA and EMAA with LDPE to give optimum cost properties
			EAA and EMAA can be coextruded with Surlyn to optimize foil bond, hot tack and cost
Hot tack strength	Better quality with fewer rejects and re-wraps on packaging machine	Surlyn has better hot tack	EAA and EMAA are less costly than Surlyn and their hot tack is as good as some Surlyn grades

Grease resistance	Improved product because no grease stains on package	LDPE is adequate and less expensive	EAA and EMAA can increase shelf life
Co-extrudable	Lower packaging costs by combining with other resins to optimize cost/ properties	Coextrusion equipment unavailable	Buy new equipment

IONOMERS

Ionomer resins possess excellent hot tack and heat seal strengths, grease resistance, stress crack resistance, and adhesion to foil. It is the standard of the industry for these applications.

Table 5. Ionomers

FEATURE	BENEFIT	OBJECTION	OVERCOMING OBJECTION
Excellent heat seals and seals through contamination	Fewer product returns because fewer leakers	EVA seals at lower temperatures	Surlyn has much better hot tack and grease resistance
			Surlyn 1702 seals as low as 18% VA EVA
		Insufficient heat resistance for hot fill	LLDPE required in hot fill
Hot tack	Better packaging machinery efficiency faster, more reliable sealing	LDPE and EVA are less expensive than Surlyn	Improved machine efficiencies with because of Surlyn might reduce total packaging costs
		LDPE has better moisture barrier	Surlyn has better grease resistance and foil adhesion and can be coextruded with LDPE for optimum cost/performance
		EAA is less costly than Surlyn and has equivalent hot tack	Some EAAs are as good as some Surlyns, but Surlyn 1702 has better hot tack and lower seal temperatures than EAA

Table 5. Ionomers

FEATURE	BENEFIT	OBJECTION	OVERCOMING OBJECTION
Coextrudable	Lower packaging costs by combining with other resins to optimize cost/performance	Some converters can't coextrude	The excellent cost/ performance ratio of Surlyn coextrusions might allow end users to reduce total packaging costs
Abrasion resistance	Fewer leakers in shipment	Doubling the thickness of LDPE can increase abrasion resistance and it is still less costly than Surlyn	Surlyn has 5-10 times the abrasion resistance of LDPE
Chemical resistance	Longer shelf life with structures like paper/PE/foil/Surlyn because Surlyn resists permeation of many aggressive liquids	Surlyn absorbs flavors from product, leaving the product less tasty	LDPE, LLDPE, and EAA also absorb flavors; coatings available that do scalp flavors, but heat seal like Surlyn
Grease resistance	Reduces grease stains on package	Grease resistance with LDPE appears adequate	Surlyn can increase life because it takes a longer time for grease to penetrate
Strong bond to foil in extrusion coating	Longer shelf life because bonds won't delaminate even when packaging aggressive liquids	Lower cost LDPE can be used if foil is primed	Incomplete drying of primer can cause odors in package
			Surlyn has better chemical resistance so it prevents permeation of aggressive chemicals that can delaminate and corrode the foil
Good processability	Fewer gels, more uptime, better product quality	EAA is not moisture sensitive and requires less horsepower	Surlyn can be extruded 330°C without degradation, and coating weights not limited by low screw speeds

POLYPROPYLENE

Polypropylene is a higher order polyolefin with properties generally superior to LDPE. PP has superior temperature, chemical, and abrasion resistance, and has better optical and moisture vapor barrier properties than LDPE or its copolymers. While slightly more difficult to process, coextrusion with more stable resins eliminates neck-in and draw resonance problems.

Table 6. Polypropylene

FEATURE	BENEFIT	OBJECTION	OVERCOMING OBJECTION
Improved temperature resistance	Provides product design capability for rigid or flexible packaging hot fill applications	Not as temperature resistant as PET or nylon	Much easier to process PP than PET, as has sufficient resistance for microwave applications
Improved chemical resistance	Offers superior protection in demanding chemical environments	Surlyn offers better chemical resistance	Surlyn has inferior temperature, optical, and moisture barrier properties
Good optical properties	Provides clear packages for product display	Not as clear as polycarbonate	PP is easier to process and less expensive than polycarbonate, and generally has sufficient clarity and gloss
Good moisture barrier	Provides extended shelf life to a wide variety of food and chemical products	Foil laminates have superior MVTR	Foil laminates are more costly, require special handling during manufacture, and are prone to pinholes
Processability	Good manufacturing efficiencies	More neck-in, edge weave, and gels than LDPE, LLDPE, HDPE	Optimizing process conditions helps; better processability than condensation polymers

CHAPTER 21

Polyethylene for Extrusion Coating

Gary D. Jerdee, Chevron Chemical

INTRODUCTION

High-pressure low-density polyethylene (LDPE) originated in the early 1930s at Imperial Chemical Industries in England. Since its inception, a multitude of uses have been found for the material in many areas including food packaging, consumer goods, construction, agriculture and the electrical industry. The first major application was as an insulating material for electrical wire and cable. During the late 1940s, the H.P. Smith Company offered extrusion coated paper for freezer wrap applications. In the years to follow, other applications for extrusion coated products grew quickly. Multi-wall bags, milk cartons, folding carton board, flexible packaging, and release liners were markets penetrated early with extrusion coated products. Milk cartons alone were responsible for a significant increase in usage of polyethylene in the late '50s and early '60s.

Low-density production capacity in the United States for 1996 was 3,658,838 million metric tons (8,068,000,000 lb) The capacity is split at 54.8% autoclave and 45.2% tubular. Autoclave-produced polyethylenes are more suitable for extrusion coating due to their inherent high degree of long chain branching. LDPE (including copolymers) demand in the extrusion coating market grew from 339,072 metric tons (786 MM lb) in 1990 to 394,701 metric tons (894 MM lb) in 1995 (according to SPI data as shown in Table 1). It is projected to grow at an average 2.5-3% annually for the next three years. The coating and laminating market for LDPE, including EVA resins, is divided into the following market segments:

Table 1. Major end-use markets (SPI 1995 data)

Market	Share
Liquid packaging board	24%
Folding cartons	15%
Paper/aluminum foil & film/aluminum foil laminations	12%
Snack food laminations	11%
Carpet backing	9%
Cup stock	7.5%
Laminating film	6%
Aseptic brick package	5.5%
Photographic paper	3%
Release paper	3%
Industrial wrap	2%
Multi-wall sacks	1.5%
Laminated squeeze tubes	0.5%

PROPERTIES REVIEW

General

The four main categories of polyethylene homopolymer resins for extrusion coating are identified in Table 2. Although other properties such as molecular weight distribution influence the behavior of these resins, the two properties used most by both processors and suppliers are melt index and density. Melt index is inversely related to molecular weight. As the melt index of a resin increases, the molecular weight decreases. Molecular weight has an impact on physical strength properties such as tear, tensile, impact, and toughness.

Table 2. Polyethylene coating resins

Density Range g/cc	Melt Index g/10 min	Process Temperature	End Use
0.910-9.919	4-16	288-343°C 550-650°F	General purpose flexible; packaging/laminations; snack foods
0.920-0.925	3-10	288-343°C 550-650°C	General purpose board coating; liquid packaging, cupstock; folding carton
0.925-0.940	5-12	288-343°C 550-650°C	Multi-wall bags; boil-in-bags; improved barrier
0.941-0.955	8-15	288-343°C 550-650°C	Spiral wound cans; release liners; best barrier

Physical properties

Resin properties

Values are typical for extrusion grade resins:

Melt index (g/10min)	3.5-16.0
Density (g/cc)	0.910-0.925 (LDPE)
DSC melting point (°C)	100-114
Vicat softening point (°C)	85-102
Tensile at yield, MPa	8-12.5
Flexural stiffness, MPa	110-262

The main properties of the LDPE coating in the finished construction are adhesion, barrier, and heat seal.

Adhesion

Adhesion is one of the most important properties of the finished product and is obtained by mechanical or chemical bonding. Mechanical bonding occurs where molten polymer physically penetrates the fibers of a porous substrate such as paper or board. Chemical bonding to nonporous substrates is achieved by the oxidation of the extrudate and also by the treatment or priming of the substrate.

Higher melt temperatures and long draw distances increase adhesion. Both conditions create higher oxidation of the extrudate.

Coating weight and line speed also are factors affecting adhesion. Heavier coating weights carry more heat to the substrate and enhance adhesion. Likewise, slower line speeds allow more time for oxidation in the air gap, further enhancing adhesion.

Lastly, adhesion can also be improved by use of priming. The substrate can be corona treated, flame treated, or chemically primed. Additionally, the application of ozone onto the molten extrudate has proven to be beneficial to adhesion. The adhesion of LDPE to various substrates compared to ethylene copolymers is shown in Table 3.

Table 3. Adhesion to various substrates

	LDPE	EMA	EVA	EAA
Kraft paper	Excellent	Excellent	Good	Excellent
OPP (unprimed)	Fair	Excellent	Fair	Fair
OPET (unprimed)	Fair	Excellent	Fair	Fair
Cellophane (PVdC coated)	Fair	Excellent	Fair	Fair
Aluminum foil (unprimed)	Fair-Good	Fair	Fair	Excellent
LDPE film	Inseparable	Inseparable	Inseparable	Inseparable

Melt index is the property that indicates how easy a resin will flow. The higher the melt index is, the easier the resin flows, resulting in lower head pressures and lower power requirements to extrude. Higher melt index resins tend to heat seal easier due to their lower viscosities. Other properties directly affected by melt index are neck-in and drawability.

The density of a resin has a pronounced effect on barrier properties and heat sealability. Generally, the higher the density is, the better are the barrier properties such as grease resistance and moisture vapor transmission rate (MVTR). Conversely, the higher the density is, the higher is the temperature it requires to seal. Higher density resins also result in a lower coefficient of friction and usually result in higher neck-in in the extrusion coating process.

Barrier properties

Barrier properties of homopolymer coating resins are dependent on density and coating weight. The higher the density is, the better are the barrier properties such as moisture vapor transmission rate and pinholes.

Figure 1 shows that the finished coating density is much less than the original resin density. This is due to the quench cooling action of the chill roll which prevents crystalline growth. All density ranges are similarly affected so that the higher the original resin density is, the better are the barrier properties of the finished construction.

Fig. 1. Coating density vs. resin density

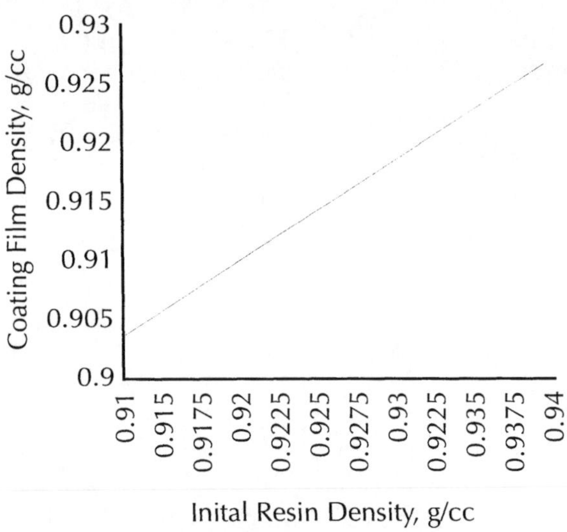

Figure 2 shows that increased thickness enhances MVTR of two density ranges. It is possible to achieve the same MVTR of the lower density range by using lower thickness of a higher density resin.

Fig. 2. Moisture vapor transmission rate vs. coating weight

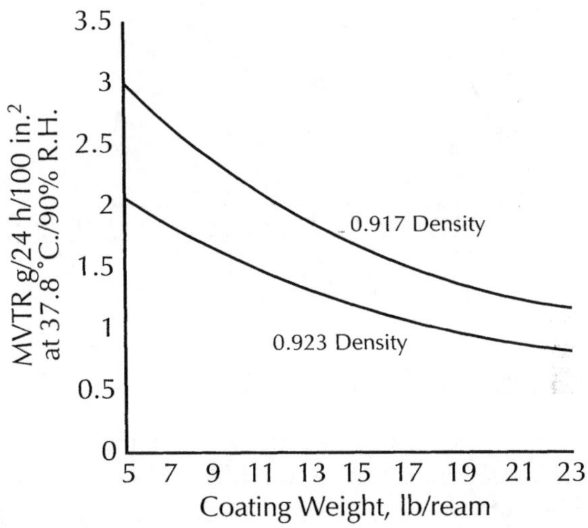

Heat seal

Heat seal is dependent on melt index and, more importantly, on density. The higher the density is, the more heat is required to melt the coating, thereby affecting the minimum seal temperature. The lower the melt index is, the more difficult it is for the two layers to flow together.

Figure 3 shows a typical heat seal curve of seal strength vs. sealing temperature for LDPE. As seal temperature is increased, seal strength increases up to a point where seal strength begins to decrease due to the coating becoming thinner in the seal area.

Fig. 3. Seal strength vs. sealing temperature

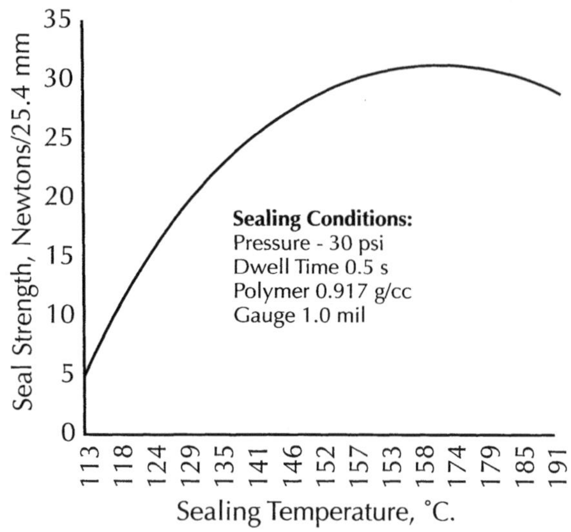

CHEMISTRY AND MANUFACTURE

Low-density polyethylene is manufactured by polymerizing ethylene in a high-pressure process. The syllable poly means "many" and mer means "unit" in Greek. Hence the word polymer means "many units." One molecule of ethylene has a molecular weight of 28. In the manufacturing process, many ethylene molecules attach to each other to create a long repeating unit called polyethylene. Typical polyethylene chains have molecular weights ranging from 50,000 up to 1,000,000. Molecular weight affects both the melt and physical properties of the polymer.

High-pressure low-density is unique by virtue of its high degree of long chain branching. On average, for every 1000 carbon atoms, there are usually 8-40 polymer chains extending as branches from the backbone of the main chain. These chains are the principal source of melt strength during the extrusion coating process.

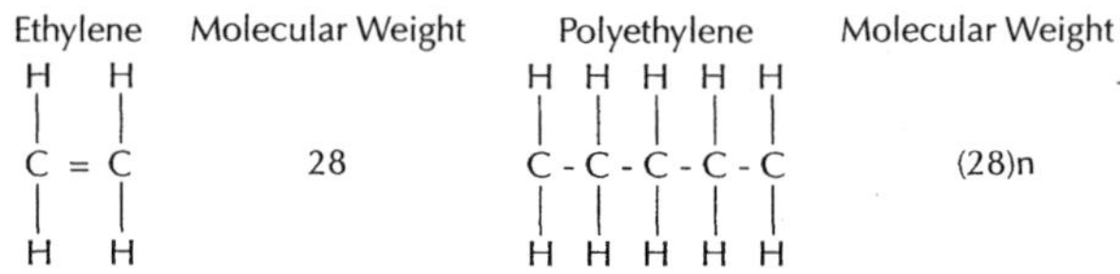

In the manufacturing process, ethylene gas is compressed to pressures ranging from 15,000 to 30,000 psi nominally. The compressed ethylene combines with a catalyst in a reactor vessel where the actual polymerization occurs. The two types of reactors used for manufacturing LDPE are autoclave and tubular reactors. In both reaction processes, molten polymer is separated from the unspent ethylene and pelletized in a downstream extrusion operation. The unspent ethylene is then purified and cooled and returned to the reactor to continue the process. If necessary, additives can be incorporated during the pelletizing step.

LDPE PROCESS SUMMARY

Operating Parameters	Autoclave	Tubular
Reactor temperature	176-260°C	176-260°C
Reactor pressure	15,000-30,000 psi	30,000-50,000 psi
Catalyst	Organic peroxides	Organic peroxides or O_2
Reactor diameter	685 mm	25 mm
Reactor length	6.0 meters	1600 meters
Residence time	1 minute	1 minute
Operating costs	High	High

PROCESSING

Low-density polyethylene is the easiest and most forgiving polymer to use in extrusion coating. It can sometimes be grossly mishandled in the extrusion coating process and still serve its original purpose adequately.

Although given this high degree of flexibility in regard to processing conditions, polyethylene is still sensitive to the high extrusion temperatures used in extrusion coating. Some polymers work better than others in certain applications, which is a measure of their melt stability.

The melt stability of resins can be measured by the structural change of the polymer after extrusion, namely by measuring the melt index of the coating. This can best be shown in a plot of extrusion temperature versus extrudate melt index (Fig. 4). This plot shows that, as extrusion temperatures are increased for a given output rate, structural changes are occurring in the polymer resulting in a lowering of the melt index. This is caused by a cross-linking of the polymer chains. As extrusion temperatures continue upward, we see a new change occurring, that of increasing melt index. This breaking up of the molecular chains into smaller chain lengths manifests itself in the coating as high neck-in, odor, smoke, and sometimes as surging. All LDPE resins react in this way; some more than others. This polymer change can also be induced by excessive residence time in the extruder (low output rates) as well as exposure in the air gap.

Fig. 4. Extrudate melt index vs. melt temperature

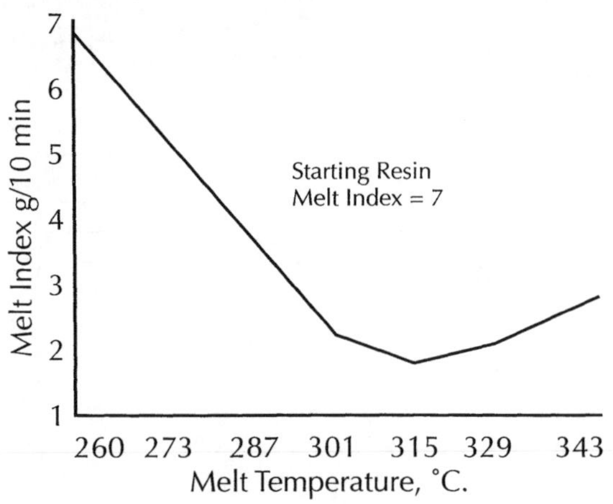

An extruder should always be purged out if a run has been stopped and the screw allowed to turn at minimum rpm or allowed to sit without the screw turning. Failure to purge out the overheated resin before resuming coating could result in coated product exhibiting sealing problems, high odor, and high pinhole frequency. When shutting down an extruder for an extended period, it is recommended the extruder be allowed to cool down to below 204°C (400°F) before stopping the screw completely. By doing so, extensive purging can be avoided when the extruder is restarted.

Most extrusion coating with low-density polyethylene homopolymer resins is done at temperatures of 315-326°C (600-620°F). The high temperatures are necessary for adhesion and drawability.

There are two areas that are very important in setting up a temperature profile on the extrusion coater. One is the feed zone, and the other is the metering zone of the screw. The temperature at the feed section should be low enough to prevent bridging or blockage at the throat section by melting polymer in the hopper feed port. Then, temperature along the barrel should be elevated to produce a uniform temperature at the desired melt temperature before reaching the last section of the metering section. Temperatures of the adapter, connecting pipe, and die should be set to maintain melt temperature only.

Factors such as screw diameter, length to diameter ratio, desired output, melting point of the resin, screw design, screw wear, and back pressure must be considered when setting a barrel profile. A typical profile range for LDPE on extruders ranging from 63.5 mm (2.5 in.) to 152 mm (6 in.), depending on these variables, is as follows:

Barrel Zone 1
176-232°C
350-450°F

Barrel Zone 2
204-287°C
400-550°F

Barrel Zone 3
260-315°C
500-600°F

Barrel Zone 4
287-332°C
550-630°F

Barrel Zone 5
287-332°C
550-630°F

An extruder should offer a steady-state output with a low melt temperature variation at the desired rate and temperature to be considered optimized. Extruder screws are typically designed to run one or more polymers. Generally, the more compromise there is in the design, the lower the rate is that can be obtained by the particular screw.

Effects of extrusion variables

Neck-in is primarily affected by temperature. As temperature increases, so does neck-in. Figure 5 shows the effect of melt temperature on neck-in. Another variable is draw distance or air gap. Figure 6 shows that as draw distance is increased, neck-in also increases since the polymer has more time to contract due to its elasticity effect. A proper draw distance is necessary for oxidation of the extrudate to promote adhesion. Typical draw distances for most extrusion coaters ranges from 127 mm (5 in.) to 304 mm (12 in.) On average, 177 mm (7 in.) to 228 mm (9 in.) is usually observed.

Fig. 5. Neck-in vs. melt temperature

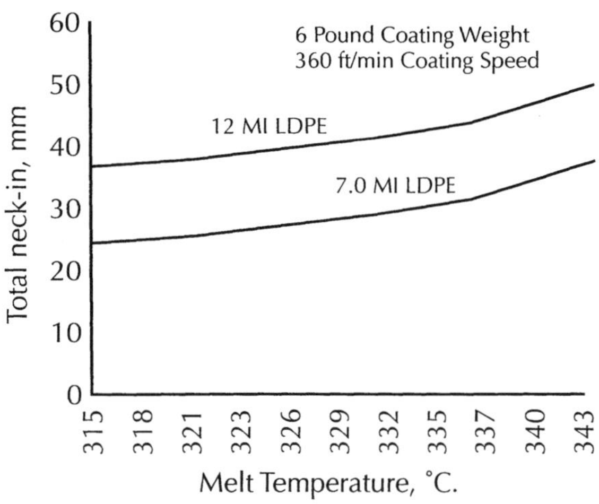

Fig. 6. Effect of draw distance on neck-in 6 lb/ream coating weight

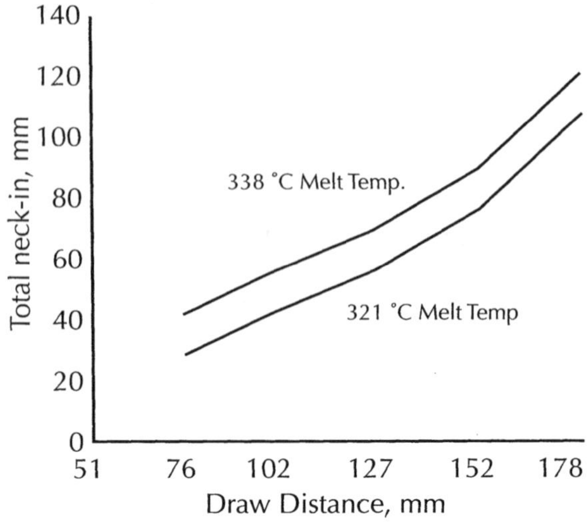

Line speed also has an effect on neck-in as shown in Fig. 7. As line speed increases, the amount of neck-in decreases. As speed is further increased, only smaller decreases in neck-in are gained. The amount of time that the extrudate is exposed in the air gap due to low line speeds allows time for contraction of the extrudate causing high neck-in.

Fig. 7. Effect of coating speed on neck-in (3.7 miles 0.923 LDPE)

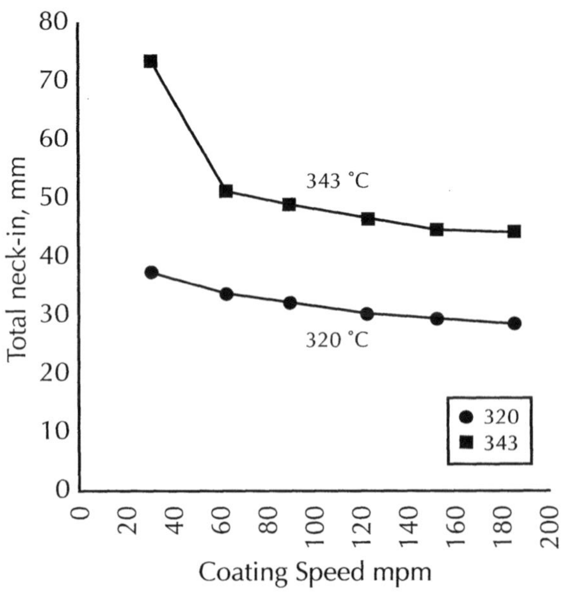

Safety

Due to the extreme temperatures and pressures involved in extrusion coating, utmost care should be taken when working around extruders and dies. Proper protective clothing and safety glasses should be worn at all times. When close work is required, a face shield is recommended.

Starting up a loaded extruder from a cold start must be performed with extreme care. The entire extrusion system should be pre-heated to a point where the plastic is fluid enough to be pumped. It is important to allow the extrusion system to soak for a sufficient time at the desired temperature to ensure the plastic is molten throughout the entire system. The soak time starts when all the zones have reached the desired settings. For most low-density polyethylenes, a minimum temperature of 204°C (400°F) should be achieved before starting a "cold" extruder.

When starting up a properly heated extruder, you should turn on the extruder drive and rotate the screw at a slow speed (5 rpm maximum). Keep a close watch on the extruder head pressure and drive amps. If either of these indicators appears to be excessive or abnormal, stop immediately and continue to pre-heat the system. If conditions appear normal and polymer is exiting the die, then proceed with setting the desired profiles. Normal operations can resume when the desired profile settings are reached.

NOTE: Never operate an extruder where the rupture disc has been removed! These devices allow pressure relief and are crucial to the safe operation of the equipment.

When shutting down an extruder with low-density polyethylene, it is recommended to cool the extrusion system down to less than 204°C (400°F) to avoid thermal degradation in the system.

Risk from physical hazards

Polyethylene is not flammable or reactive. There can be an explosion hazard from dust at concentrations above 20 mg/l, and bulk storage of pellets, from ethylene gas at concentrations above 2.7%. There can be a slipping hazard from spilled pellets. Risk from physical hazards is low.

Risk to human health

The potential human exposure is considered high, due to the ubiquitous use of polyethylene products. Dermal exposure to polyethylene by consumers or employees is not expected to cause any adverse health effects. In work locations where polyethylene dust is generated, lung effects can occur due to high-level exposure to dust [above the OSHA Permissible Exposure Levels (PELs) for dusts]. Emissions from the combustion or processing of polyethylene can cause health effects at high exposure levels; however, worker exposure is not expected to exceed permissible levels. Exposure to dust in the workplace is routinely below the OSHA PELs; respirators are worn if needed. Exposure following migration of ethylene monomer, comonomers, or additives from polyethylene in contact with food is very low. The risk of health effects due to exposure to polyethylene from occupational and non-occupational exposure is considered low due to (a) the low toxicity and low bioavailability of the polymer, (b) the low content of additives, and (c) the FDA-regulated status of all additives in food-use resins.

Risk to the environment

The potential environmental exposure is considered high due to the widespread use and disposal of polyethylene products. Polyethylene is not expected to cause adverse effects to aquatic organisms; however, fish and birds might eat plastic pellets which can obstruct their digestive tracts. Polyethylene does not readily degrade in the environment. The risk of adverse effects to the environment from polyethylene products is negligible. If polyethylene pellets are inadvertently released to the environment, there is some risk of adverse health effects to birds and fish.

ACKNOWLEDGMENTS

R. F. Buesinger, Chevron

M. J. Armand, Chevron

M. L. Machado, Chevron

CHAPTER 22

Extrusion Coating Linear Low-density Polyethylene

Michael W. Potts, The Dow Chemical Company

ABSTRACT

Linear low-density polyethylene (LLDPE) offers many performance advantages in a variety of packaging applications. As with any structure component, understanding the material allows full utilization of these benefits. A basic understanding of the resins also ensures optimum processing of the products on conventional extrusion coating/laminating equipment.

INTRODUCTION

Since the introduction of LLDPE, converters have been taking advantage of its unique properties by lamination of free film to a variety of substrates. Applications using LLDPE have generally been limited to only those with the most demanding property requirements due to the economic disadvantages of film lamination compared to extrusion coating. The development of extrusion-coating grades of LLDPE makes it economically attractive to use LLDPE in many additional applications.

Full utilization of the exciting attributes offered by LLDPE extrusion-coating resins requires converters to become better acquainted with LLDPE's characteristics.

What Is LLDPE?

Prior to the introduction of LLDPE, two types of polyethylene were available. The two types, low-density and high-density, are manufactured by different processes that provide the molecular structure variations responsible for the distinctive properties exhibited by each type of polyethylene.

Understanding LLDPE is easier if its similarities to the two more familiar types are considered (Fig. 1). The differences between these types of polyethylene are a result of differences in the number and length of branches in the molecular structure.

Fig. 1. Polyethylene comparison

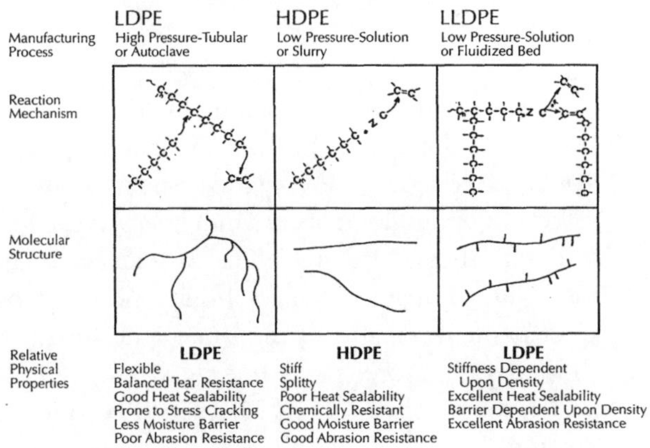

How do these differences (and similarities) affect properties?

These differences in branching are responsible for how each polymer behaves in both the molten phase during processing as well as in the solid phase in the finished structure.

MELT BEHAVIOR

Two factors which determine the melt behavior of a polyethylene are the molecular weight and the type and degree of branching present. For a given molecular weight, the response of polymer viscosity to varying shear rates is mainly dependent upon branching. Figure 2 shows the effect branching has on the shear viscosity of polyethylenes of like melt index throughout a range of typical shear rates.

Fig. 2. LLDPE displays higher viscosity at extrusion shear rates

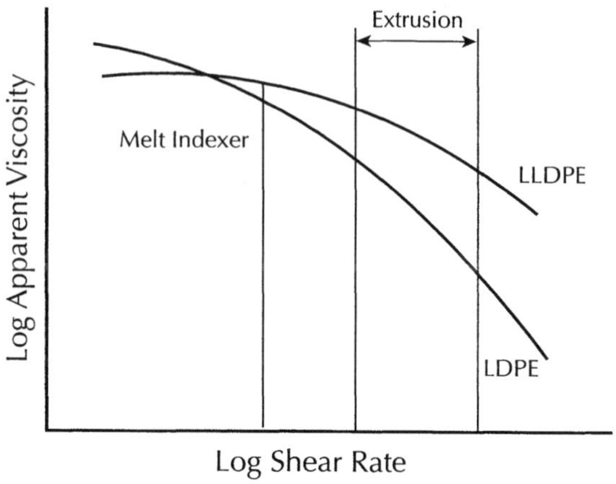

At low shear rates, the entanglements of the long chain branching of LDPE contribute strongly to the apparent viscosity. At higher shear rates, the effects of entanglements are much less, resulting in a lower "shear-thinning" effect. At extrusion shear rates, linear resins retain a higher melt viscosity than branched resins of the same melt index, and thus absorb more shear energy from the screw which requires higher drive horsepowered for processing.

The extensional viscosity of polyethylene is also strongly dependent upon branching (Figure 3).

Fig. 3. LLDPE has lower extensional viscosity at high deformations

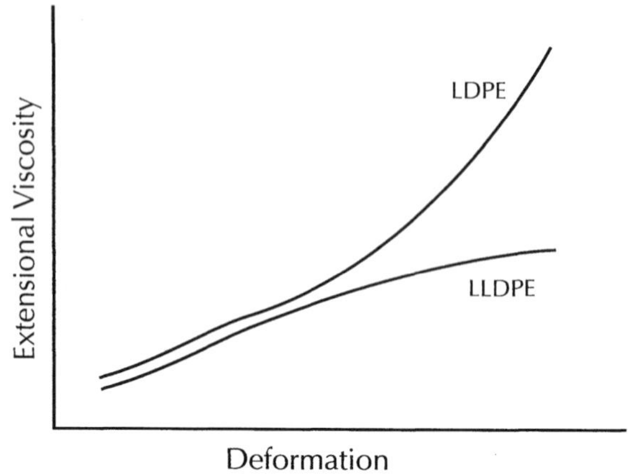

At normal draw ratios, disentanglement of the long-chain branches of LDPE does not really occur, causing the extensional viscosity to increase with the degree of deformation in an elastic fashion. The short branches of LLDPE, on the other hand, offer little resistance as the molecules slide by each other. This low extensional viscosity at high deformation allows high extrusion coating drawdown speeds with resulting low minimum coating weights.

SOLID PROPERTIES

The branching variations of polyethylenes also affect the crystalline structures of the solidified polymers and, therefore, are responsible for many of the solid-phase property differences between polyethylenes. Figure 4 shows a simplified representation of the crystalline lamellae of HDPE and LLDPE. The lamellae are held together by tie molecules whose formation are facilitated during crystallization by irregularities in the polymer chain such as branching.

Fig. 4. LLDPE forms more tie molecules than HDPE

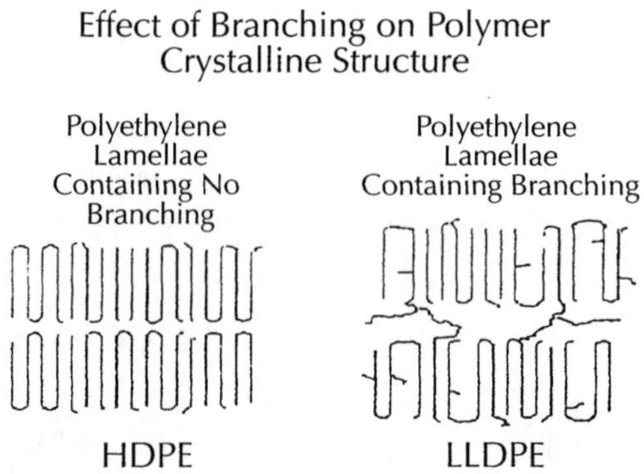

The number and degree of incorporation of the tie molecules in separate lamellae has a direct bearing upon tensile and elongation properties (Fig. 5) which, in turn, largely impacts commonly measured performance properties such as heat seal strength, tear resistance, environmental stress crack resistance, and abrasion resistance.

Fig. 5. Tensile elongation characteristics of polyethylene

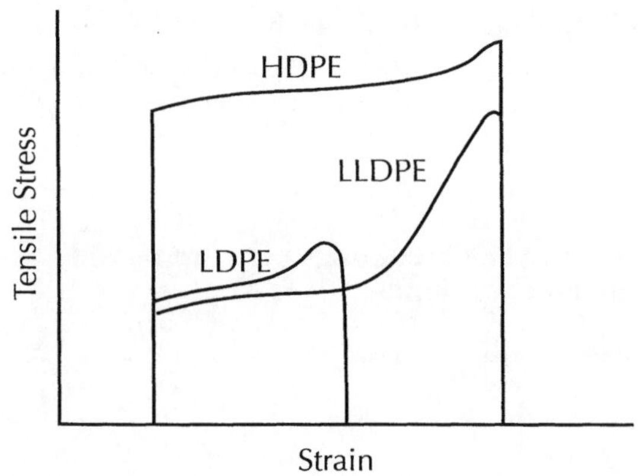

HDPE has few tie molecules which tends to result in cross direction splitting when the material is oriented and crystallized as occurs in the extrusion-coating process. However, this orientation results in excellent machine direction tensile. On the other hand, LLDPE has many tie molecules which occupy the amorphous region between the lamellae which disrupts its crystallinity compared to the HDPE. A similar effect is seen with LDPE since it is also branched.

What are the performance properties of LLDPE?

With an understanding of the basic polymer structures of LDPE and LLDPE extrusion-coating resins, it becomes easier to understand their performance properties.

HEAT SEAL STRENGTH

The heat seat strength of a material coated onto a substrate is a function of three factors:

1. Interfacial mingling of the layers during sealing
2. The tensile strength of the coating material
3. The adhesion of the heat seal layer to the substrate.

Any one of these three can limit the heat seal strength of a structure.

The excellent LLDPE heat seal strength illustrated in Fig. 6 is a result of the LLDPE's relatively low viscosity at low shear rates and superior tensile strength.

The lower viscosity at low shear rates increases LLDPE's adhesion to paper as well as improving interfacial mixing. The superior tensile strength allows heat seal strengths stronger than most substrates in structures which are not limited by adhesion.

Fig. 6. LLDPE offers excellent heat sealability

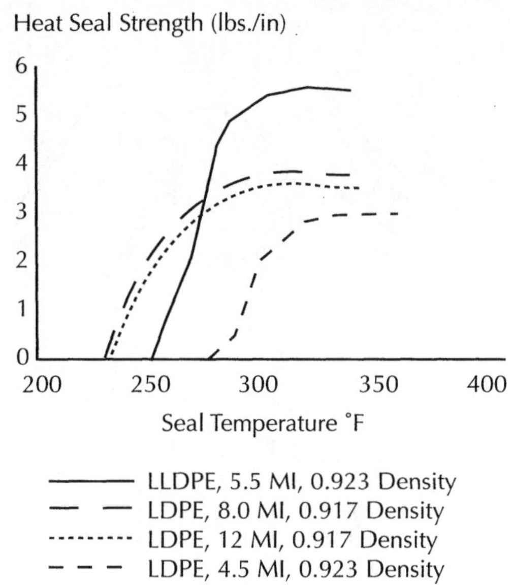

TEAR RESISTANCE

The tear resistance of a coating is largely determined by the tensile strength. Figure 7 shows the kinds of tear resistance improvements possible using LLDPE in a common structure, paper/poly. While LDPE typically contributes some to the overall tear resistance of a structure, LLDPE can make contributions allowing downgauging of coatings on substrates.

Fig. 7. Paper/poly structures

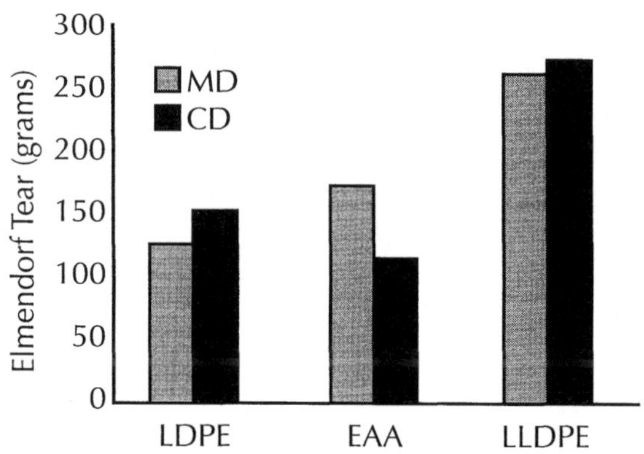

For improved package integrity, the elongation characteristics of LLDPE provide a bridging phenomenon, should the substrate tear (Fig. 8).

Fig. 8. LLDPE provides a bridging phenomenon

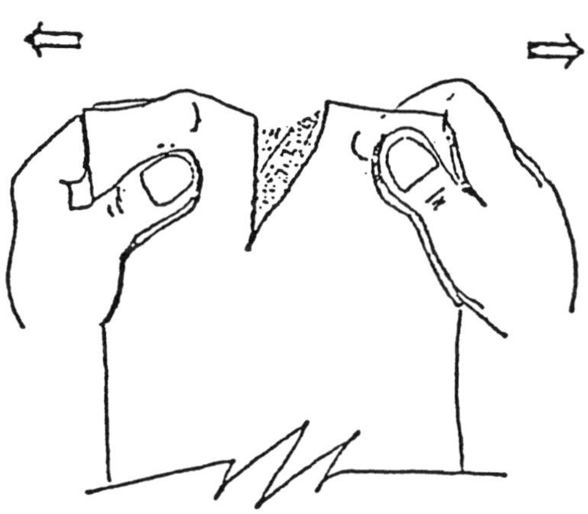

ENVIRONMENTAL STRESS CRACK RESISTANCE (ESCR)

LLDPE shows dramatic improvements in ESCR over LDPE and ionomers as shown in Fig. 9. This impressive ESCR performance is due to LLDPE's ability to relieve stresses between lamellae without breakage and its superior tensile strength.

Fig. 9. LLDPE offers improved ESCR

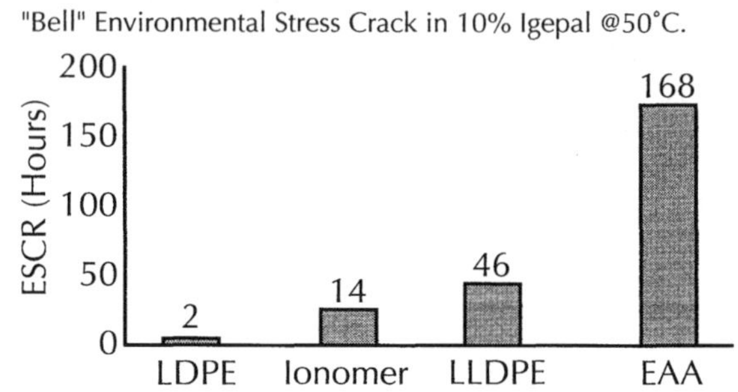

ABRASION RESISTANCE

The abrasion resistance of a material is determined by factors such as tensile strength, molecular weight, coefficient of friction, and degree of crystallinity. Figure 10 shows how LLDPE (0.921 density) compares to LDPE (0.923 density) and several other commonly used polymers, including a sodium ionomer.

Fig. 10. LLDPE coatings offer improved abrasion resistance.

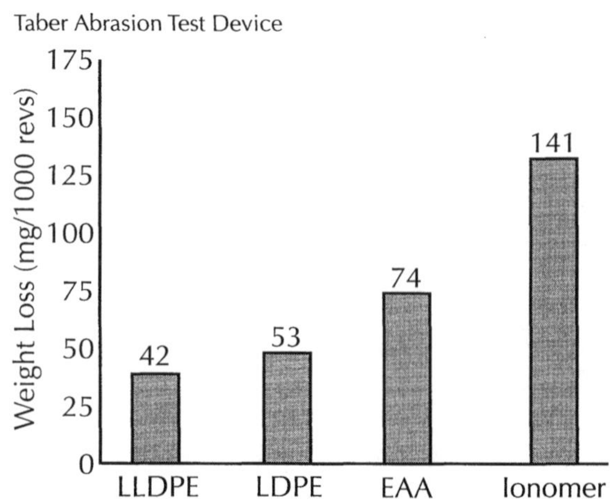

Fig. 11. Typical extruder conditions for LLDPE

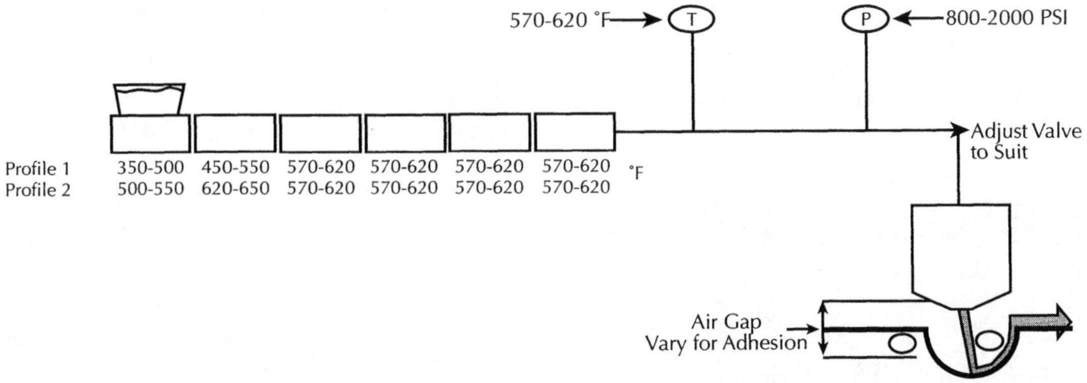

APPLICATIONS

The performance properties of LLDPE extrusion coating resins provide numerous applications with exciting economics such as:

1. Aseptic packaging—Increased heat seal strength can reduce leaker rates and improve package integrity.
2. Spiral-wound can liners—Increased heat seal strength helps contain hard-to-hold products such as frozen juice concentrates. Enhanced abrasion resistance also makes LLDPE perfect for the contact surface for abrasive snack foods (chips).
3. Fiber drum liners—Superior resistance to aggressive chemicals makes LLDPE an excellent choice for drum liners for a variety of products.
4. Paper/poly/aluminum foil/poly pouches—The excellent heat seal strength, tear resistance, and puncture resistance improve package integrity for the toughest food products.
5. Novelty balloons—The excellent heat seal strength provides the helium-filled balloons with excellent burst resistance and gives longer inflation life.

PROCESSING TIPS: EXTRUDER CONDITIONS

Remembering the rheology of LLDPE compared to LDPE as shown in Fig. 2, it is not surprising that special extruder conditions might be needed to run LLDPE on equipment designed for LDPE. Since the apparent viscosity of LLDPE is higher at extrusion shear rates, more energy is absorbed from the screw, which results in higher drive horsepower requirements. If excess horsepower is not available, it might be desirable to use a "reverse" temperature profile (Profile 2) to heat the polymer in the first two zones rapidly in order to reduce viscosity (Fig. 11).

This profile often requires the use of barrel cooling on the final zones to limit the resin temperature. Even when using a more conventional profile (Profile 1), the shear energy supplied by a screw designed for LDPE can cause resin temperatures to exceed barrel settings. Lowering the back pressure can improve this situation somewhat.

Also, it is important that thermocouples are properly placed and are not unduly influenced by the temperature of the metal in which they are mounted.

As a general rule, the temperature should not exceed 630°F and, if excessive smoking occurs, it is too hot!

Since the adhesion chemistry is identical to LDPE's, the same adjustments to air gap are appropriate to adjust bond levels.

SCREW DESIGN

While scheduling might present difficulties running LLDPE with a screw specifically designed for LLDPE, knowing desirable characteristics of an LLDPE screw can allow scheduling runs on the most appropriate equipment available.

Although sophisticated, computer-aided techniques are often used to design extruder screws, the general concepts are not difficult to understand.

A typical screw for extrusion coating (Fig. 12) consists of the following four elements:

Fig. 12. Typical screw for extrusion coating

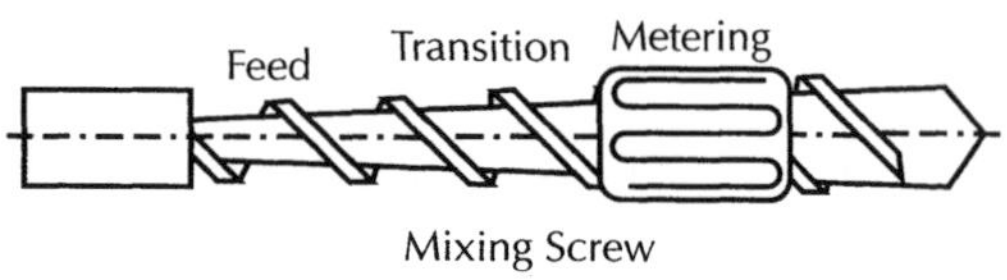

1. Feed zone. Pellets forwarded into the extruder. The pellet flow of LLDPE is very similar to LDPE.
2. Compression zone. Melting begins, and air is squeezed out. LLDPE melts at a higher temperature so a longer transition might be necessary.
3. Metering zone. Melting is completed; further heating, melt homogenization, and pressure build up. Increasing metering zone depth lowers the amount of energy transferred to the resin, which helps to limit temperature rise.
4. Mixing zone. High shear, blocks passage of unmelted polymer. It can cause high temperatures and high horsepower requirements if clearances are too tight.

PROCESSING AIDS

Numerous processing aids have been offered for LLDPE. Some offerings such as those containing fluoroelastomers address problems unique to the blown film process like melt fracture and high pressure drops. These products coat metal surfaces which smooths flow at the surface and results in a smoother film surface. Usually, these products do not reduce the power required for extrusion and, therefore, are of little use in extrusion coating.

Processing aids that do reduce horsepower requirements are of interest for extrusion coating. Low- molecular weight waxes are available which, when dry-blended by the converter, can reduce amperage. It appears these products work by melting sooner than the LLDPE and acting as a lubricant to limit the transfer of energy from the screw to the polymer. However, converters must be aware of the potential for die-lip buildup and excessive smoke, especially at very high melt temperatures, since the waxes are low molecular weight materials which increases their tendency to exude to the surface of the extrudate.

A more common process aid for extrusion coating LLDPE can be an extrusion-coating grade of LDPE. Typically, adding 5-10% of LDPE will reduce the amps and back pressure of the extruder. Essentially, the LDPE is again melting sooner than the LLDPE, but also provides long-chain branching which further increases its effectiveness. Another benefit is the significant reduction in the risk of die-lip buildup and excessive smoke.

CONCLUSION

A LLDPE offers many performance advantages to the packaging industry. Processing, though not identical to LDPE, is well within the capabilities of most extrusion coating equipment in use today.

Notice: The information in this article is presented in good faith, but no warranty is given nor is freedom from any patent to be inferred.

CHAPTER 23

Single-site Catalyst Produced Polyolefin Plastomers for Extrusion Coating and Laminating

James L. Cooper, David C. Kelley, and Sharon L. Baker, The Dow Chemical Company

Conventional Ziegler-Natta catalyzed semi-crystalline linear polymers have been available since the middle 1960s *(1)*. Single-site catalyzed polymers date from the early 1970s *(2)*. They range from semi-crystalline to elastomeric. Since then, industry research activity in single-site catalysts has been tremendous. One industry consultant stated in 1993, "There are more than 60 companies and institutions working in the area that have made patentable discoveries, and there are many more that have not yet reached the patenting stage *(3)*." This paper discusses the structure of Ziegler-Natta catalysts and single-site catalysts and describes how the structure of the catalyst can influence polymer properties. The discussion also covers specific polymers and their performance properties with emphasis on polymers produced using constrained-geometry catalysts. Constrained-geometry catalysts differ from other single-site catalysts and can produce polymers with unique physical properties.

CATALYST STRUCTURE

Ziegler-Natta catalysts

Conventional Ziegler-Natta catalyst systems typically have several catalytic sites attached to each support particle as Fig. 1 shows. The activity of each transition metal site for polymerization is different. This causes the formation of polymer molecules with a plurality of molecular weights and copolymer molecules with a variety of compositions. These attributes strongly influence other polymer properties.

Fig. 1. Conventional Ziegler-Natta catalyst structure

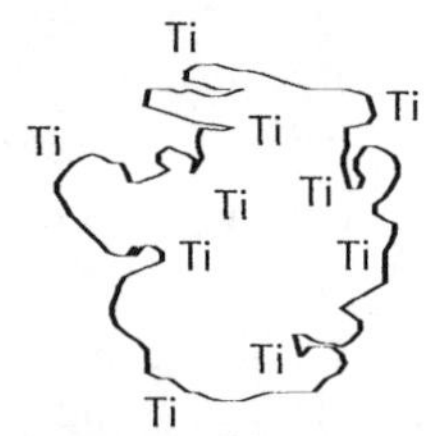

Single-site catalysts

Single-site catalysts have a singular polymerization site. One type of single-site catalyst that has had extensive study uses a sandwich structure containing two cyclopentadiene rings. Using the term originally applied to these sandwich structures in the 1950s, the catalysts have the name metallocene catalysts. Figure 2 is a diagram of such a structure. Polyethylene products produced with these metallocene catalyst systems were available commercially at least as early as 1991.

Fig. 2. Typical metallocene catalyst structure

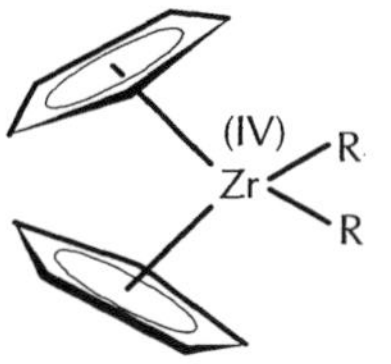

A new class of single-site catalysts has recently come into use. These catalysts use constrained-geometry complexes that have only one cyclopentadiene ring structure attached to the metal. Constrained-geometry catalysts are highly effective for olefin copolymerization because of their unique structural attributes. Under appropriate conditions, these allow the production of polymer products with unique physical properties that have commercial significance.

A new family of polyolefin plastomers (POP) uses constrained-geometry catalyst technology (CGCT). The patented CGCT *(4)* differs from other catalyst technologies. The catalytic activity of CGCT uses group IV transition metals covalently bonded to a monocyclopentadienyl group and bridged with a heteroatom. The three components connect to form a constrained cycle structure with a metal center. Figure 3 shows this geometric feature. The constrained angle at the ligand site results in an enlarged exposure of the single catalytic site that is unique to this class of molecules. Constrained-geometry catalysts produce polymers that have unique structural and performance properties *(5)*.

Fig. 3. Typical structure of a constrained-geometry catalyst

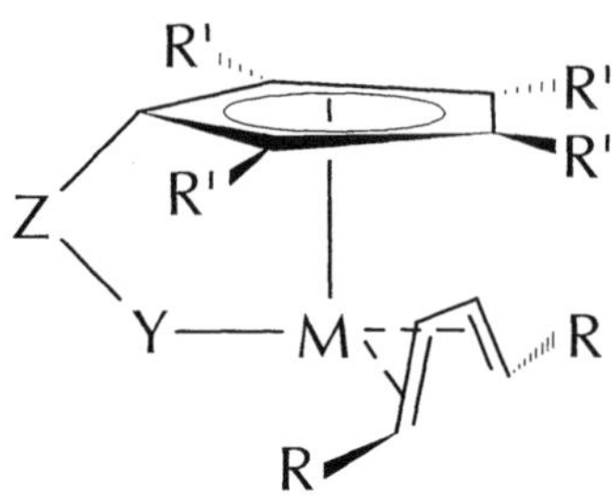

POLYMER PROPERTIES

Ziegler-Natta catalysts

Conventional Ziegler-Natta catalyst systems typically have several catalytic sites attached to each support particle. The activity toward monomer insertion at each site therefore differs. Molecules produced by Ziegler catalysts have an uneven distribution of the co-monomer side chains along the backbone of the main polymer chain as Fig. 4 shows and contain a high density fraction.

Fig. 4. Traditional Ziegler-catalyzed polymers

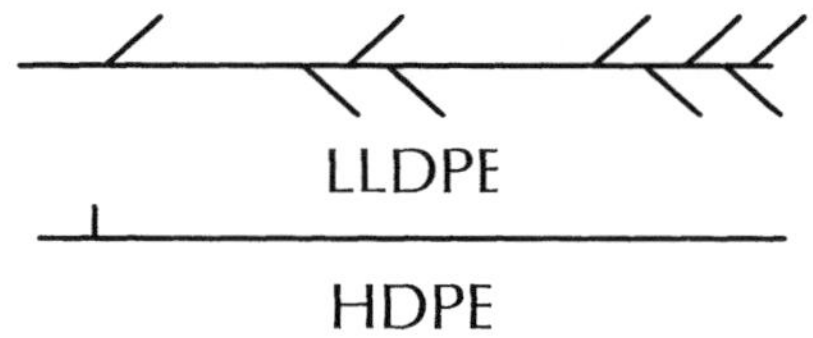

Some catalytic sites are deep within the pores of the support media. This allows polymerization of only the small ethylene molecule. Some are more open and can polymerize larger a-olefins such as 1-octene. This explains the broad comonomer and molecular weight distribution associated with traditional linear low density polyethylene (LLDPE) produced by Ziegler catalysts. LLDPE also contains significant amounts of high density polyethylene (HDPE). This fraction is a direct result of excluding larger a-olefins from the sterically confined transition metal site buried in the support media. Analyzing the short chain branching distribution using analytical temperature rising elution fractionation (ATREF) can detect this high density fraction. Figure 5 shows the crystallinity distribution produced by typical Ziegler catalyst systems. Note that the confined catalytic sites that only polymerize ethylene generate some of the highest molecular weight polymer chains due to the faster polymerization of ethylene compared with other large monomers such as hexene or octene.

Fig. 5. Crystallinity distribution produced by typical Ziegler catalyst for a typical LLDPE using ATREF

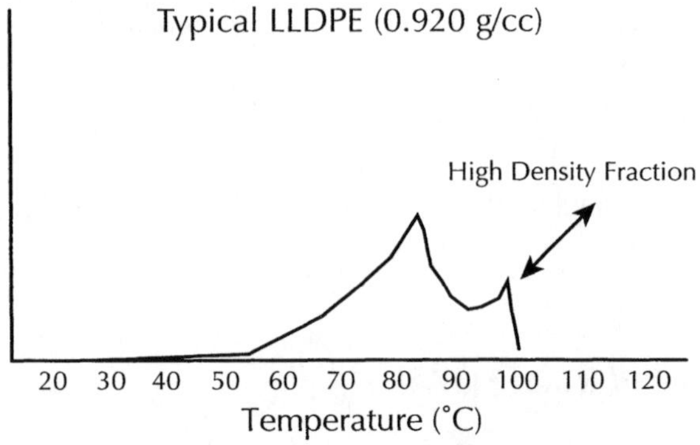

Single-site catalysts

Single-site catalysts differ from conventional Ziegler-Natta catalysts in one significant way. They have only a single active metal site for each catalyst molecule. The activity of polymerization at the transition metal is therefore identical for each catalyst molecule. Polymers produced by single-site catalyst systems exhibit narrow composition distribution. They usually do not contain a high density fraction because each catalyst site produces polymers of similar composition.

Single-site catalyst systems also provide a means for more co-monomer incorporation *(2)*. Low polymer densities are therefore possible for butene, hexene, and octene copolymers. This leads to new terminology that typically describes these polymers using density. The term polyolefin elastomers (POE) refers to single-site products with densities less than 0.880 g/cm^3. POP refers to single-site products with densities of 0.880 g/cm^3 and higher.

For the constrained-geometry catalyst, the open arrangement in the single-site catalyst molecule also opens the metal center for monomer and comonomer insertion. One can therefore design a single-site catalyst that will copolymerize ethylene, butene, hexene, octene, or styrene equally well because of the openness and availability at the transition metal. This technology allows the manufacturer to control the polymerization of olefins selectively with concomitant control of molecular weight (MW), molecular weight distribution (MWD), short-chain branching distribution (SCBD), and comonomer distribution (CD). This then provides increased impact strength and improved toughness, clarity, and organoleptic properties. Linear, single-site catalyzed polymers typically have randomly distributed side-chain branching of Fig. 6 and narrow molecular weight and comonomer distribution.

Fig. 6. Single-site catalyzed polymer with randomly distributed side chains

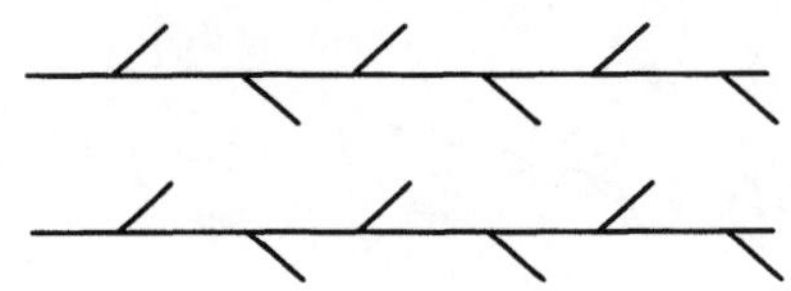

Linear Homogenous Polymer

Under appropriate conditions, CGCT systems can produce polymers not only lacking the high-density polymer fraction of Fig. 7 but also most importantly having a level of long chain branching that imparts unique physical properties to the polymers.

Fig. 7. Crystallinity distribution produced by single-site constrained geometry catalyst (ATREF)

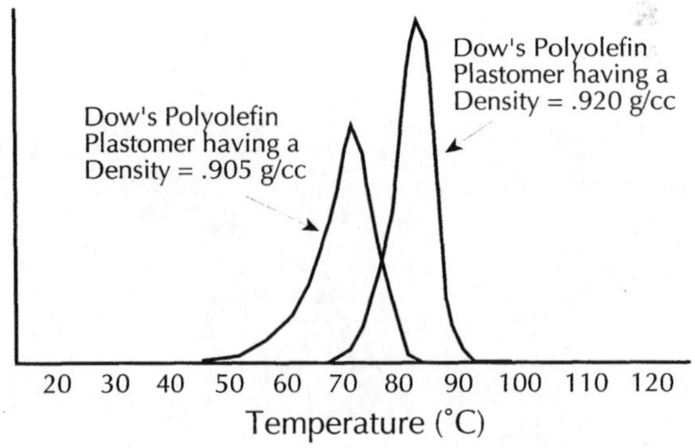

CGCT opens the catalyst geometry to the extent that it allows vinyl- terminated polymer chains to attach or polymerize to a growing chain to form the long chain branches of Fig. 8 *(6)*. Polymers produced with CGCT have narrow molecular weight distributions. Narrow molecular weight distribution typically results in harder processing. The long chain branching in CGCT polymers minimizes this effect. These long chain branches also give higher melt strength that reduces the necking-in of the molten polymer curtain for extrusion coating applications and provides additional bubble stability for blown film applications.

Fig. 8. Polymer produced with Dow's INSITE technology, with randomly distributed side chains and long-chain branching

Dow's AFFINITY resin

A molecular weight distribution—also called polydispersity or Mw/Mn—of approximately 2 is standard for single-site polymers such as a CGCT POP with 0.920 g/cm³ density shown in Fig. 9. Two major benefits result from this narrow molecular weight and comonomer distribution:

Fig. 9. Comparison of the molecular weight distribution of 0.920 density resins

LLDPE Mw = 124600, Mn = 33200, MWD = 3.75
POP Mw = 73800, Mn = 37400, MWD = 1.97

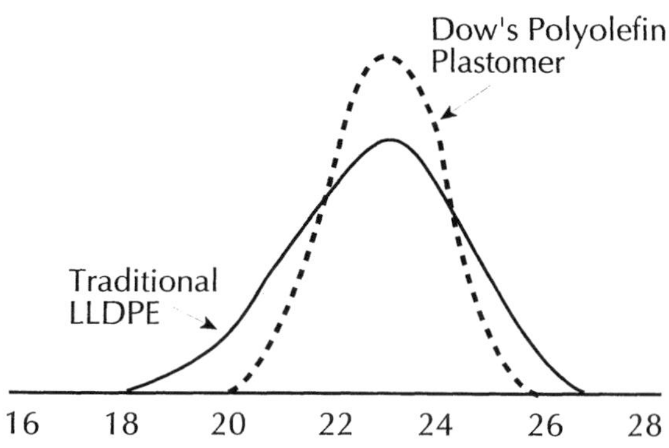

- The absence of the high density and high molecular weight component that yields undesirable high melting points and poor optics in existing conventional multi site linear catalyzed products
- The elimination of the low molecular weight, low density component that imparts high extractables and unwanted blocking tendency in linear polymers.

EXPERIMENTAL

The coating performance of a 7.5 melt index and 0.902 g/cm³ density POP produced by CGCT was compared with LDPE, LLDPE, ethylene acrylic acid copolymer (EAA), ethylene vinyl acetate (EVA), and an ionomer *(7)*. All resins were evaluated on a pilot extrusion coating line having a 3.5-in. diameter primary extruder with a 30:1 L/D single-flight screw, a 2.5-in. and a 2-in. extruder, and a feed block for coextrusion. A 30-in. coat hanger die deckled to 24 in. with a 20-mil die gap and a 6-in. air gap was used for all evaluations. A microprocessor system that included weigh cell feed hoppers for rate calculations and control of coextrusion controlled the extrusion line. A mirror pocket glycol-cooled chill roll was controlled at 57°F. Table I describes the polymers evaluated for this study.

Table 1. Resins evaluated

Designation	Density, g/cm³	Melt Index, g/10 min.
POP	.902	7.5
LDPE	.916	8.0
LLDPE	.923	5.5
EVA	.940	8.0
EAA	.938	10.0
Ionomer	.940	5.0

Monolayer coating evaluations using the primary extruder used 50-lb kraft paper. The polymers were extrusion coated at a 3.5-in. screw speed of 90 rpm (approximately 250 lb/h) at their recommended melt temperature. Neck-in values were determined at 440 ft/min. (1 mil coating thickness).

The substrate speed was then increased until the molten web completely pulled from the die. Any edge tear or draw resonance was noted.

Samples for heat seal, hot tack, and sensory testing were prepared by coextruding a 1.5-mil coating using an EAA (8 melt index and 6.5% acrylic acid concentration by weight) copolymer as the tie layer (20% layer thickness) onto 48-gauge primed polyethylene terephthlate (PET) film and a paper/polyethylene/aluminum foil (.00035 A wettable) substrate. Cast film samples were produced on a pilot cast film line with a 30-in. die. Film samples were tested for physical and optical properties.

Heat seal and hot tack

Lab heat seals were formed on 1-in. wide strips of the substrate and sealed at 40 psi bar pressure with a 0.5-s dwell using heated bars at temperatures of 80°C–150°C on a commercially available hot tack tester. After aging the sealed samples for 24 h at ambient temperature, the force in pounds required to pull the seal apart was determined with the use of a tensiometer at 10 in. per min. crosshead speed and a 90° "T" peel.

Hot tack strength testing used the same hot tack tester. A 1-in. wide sample was clamped at both ends. The top stationary clamp was attached to a load cell, and the bottom clamp was attached to a piston that moved downward at a constant peel speed of 150 mm/s. The folded sample was placed between the seal bars that were then closed for 0.5 s with a 0.2-s delay and 40-psi bar pressure. When the seal bars were released, the piston moved to exert an instant downward force on the seal. The peak force in Newtons/25 mm to break the seal was recorded.

Tear testing

Tear data were generated on 1-mil monolayer coatings on 50-lb kraft paper. A conventional tensile tear tester was used.

Sensory testing

Sensory data were generated using 24 trained panelists to evaluate the film samples in a ranking and paired comparison format. The samples were coded with random three-digit numbers and presented according to a balanced block design to minimize bias. Evaluations were conducted in an air medium and commercial drinking water. The data were analyzed using Friedman Analysis of Variance for ranked data followed by Duncan Multiple Range comparisons *(8)*. T-tests were used for the paired comparisons.

Adhesion data

Slip sheets of different substrates—oriented polypropylene (OPP), aluminum foil, and primed PET—were coated at 1-mil thickness to check adhesion. The adhesion values were obtained by pulling at 90° using a "T" peel configuration on a tensiometer at 10 in./min. crosshead speed.

RESULTS AND DISCUSSION

Properties

Table II shows bulk physical properties. Table III shows cast film properties of the POP resin for thicknesses of 1 and 2 mils.

Table 2. Bulk physical property data

Physical property	POP	LLDPE	LDPE
Octene comoner, %	13.5	7	N.A.
Melt index, dg/min.	7.5	5.5	8
Density, g/cm^3	0.902	0.923	0.916
DSC meltiing point, °C	98	119	103
Vicat softening point, °C	77	99	85
Tensile yield, psi	850	1870	1400
Ultimate tensile, psi	2180	1970	1500
Elongation, %	950	705	580

Table 3. POP cast film properties

Property	1 mil	2 mil
Tensile yield, psi,		
MD	1080	840
CD	840	770
Ultimate tensile,		
psi, MD	7270	5435
CD	4720	4300
Elongation, %	620	750
20° Gloss	46	50
Dart, g	248	452
Puncture, ft·lb/in.^{3}	178	166
Tear, g, MD	88	348
CD	316	580

Fig. 10. Rheology data

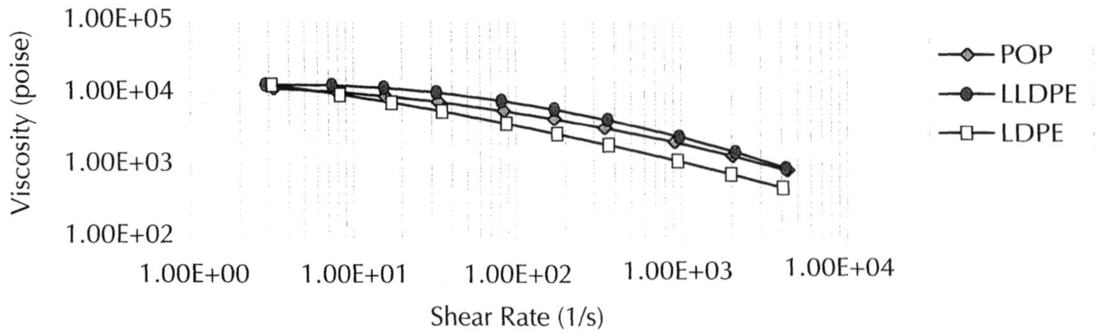

Rheology

Figure 10 compares the shear rheology of POP to the LLDPE and LDPE having the bulk physical properties listed in Tables I and II. The enhanced shear thinning behavior of this POP is due to the presence of long chain branching that improves processability. The LDPE shear-thinned the most because it has more long chain branching and broader molecular weight distribution than the POP.

Extrusion coating performance

Monolayer coating results in Table IV show that the POP had a higher neck-in—4.9 vs. 1.6 in. at 550°F—and required more energy as seen by higher amps to process than the LDPE resin. The POP was easier to process than the LLDPE resin as shown by 8% lower amps when extruded at 600°F. The POP could be drawn down to less than 0.25 mils without any draw resonance problems. The rheology data also supports this.

IV. Coating performance

Resin	Melt Temp., °F	Amps	Neck-in, in.	Min. Thickness mil
POP	550	142	4.9	<.25
POP	600	138	5.3	<.25
LDPE	550	101	1.6	.70
LDPE	600	94	1.6	.50
LLDPE	600	150	4.5	<.25
EAA	550	110	3.0	.42
Ionomer	550	126	1.8	.63
EVA	450	121	4.6	.35

Heat seal and hot tack

Figures 11 and 12 show laboratory heat seal data on the primed PET films. The POP had a lower heat seal initiation temperature than LDPE, LLDPE, EAA, and the ionomer. The plastomer also had the highest ultimate seal strength (>30%) of all the polymers evaluated on the PET film. The POP had a heat seal initiation temperature (2 lb/in.) 18°C lower than LDPE, 26°C lower than the LLDPE, and 5°C lower than the EAA and the ionomer. The seal initiation temperature of the POP was 5°C higher than the EVA (15% vinyl acetate). This suggested that the POP would have equal performance to an EVA containing approximately 12% vinyl acetate.

Fig. 11. Heat seal data: 48 gauge PET/1.5 mil coex EAA (20%)/sealant (80%)

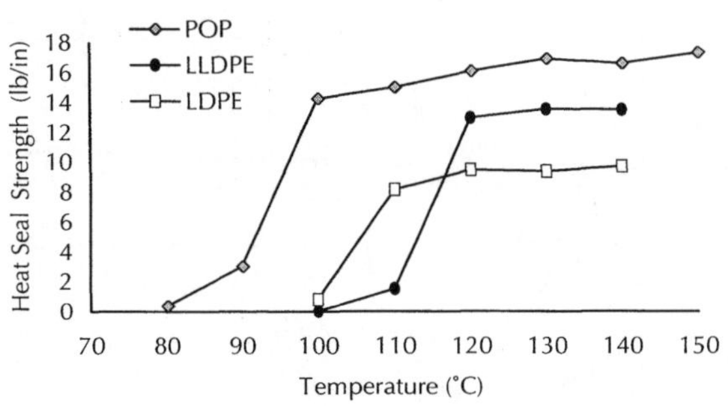

Fig. 12. Heat seal data: 48 gauge PET/1.5 mil coex EAA (20%)/sealant (80%)

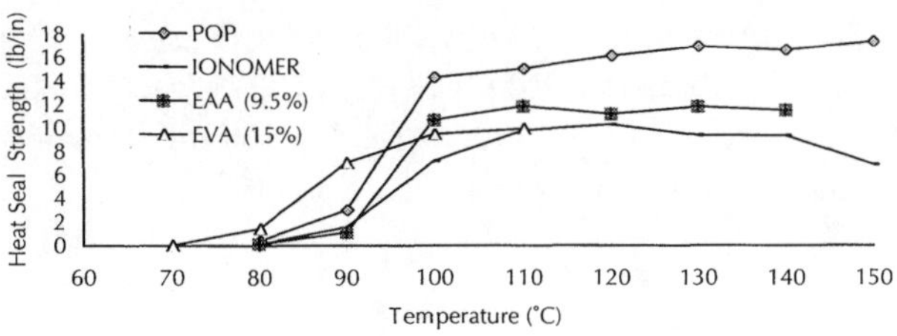

Heat seal performance in Fig. 13 on the aluminum foil substrate showed the POP had lower seal initiation temperature and seal strength than the EAA resin. The seal initiation temperature of the POP was 10°C higher than the EVA on this thicker structure. Adhesion failure at the aluminum foil interface limited the seal strength of all the sealants.

Fig. 13. Heat seal data: (paper/poly/foil (0.00035)/1.5 mil coex EAA (20%)/sealant (80%)

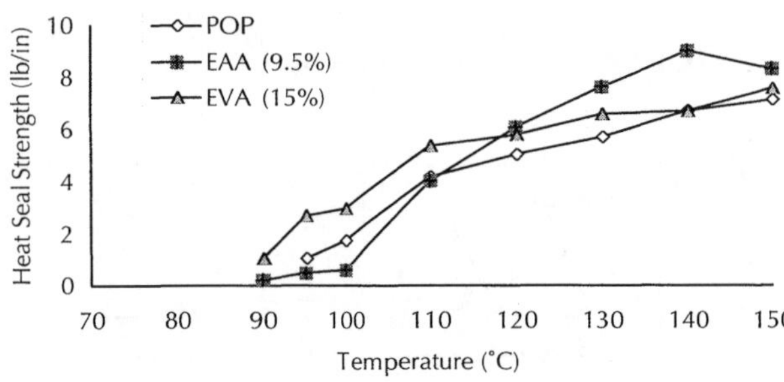

The hot tack performance on the aluminum foil structure in Fig. 14 showed the POP had a higher ultimate hot tack strength (13 vs. 9 N/25 mm) and a broader hot tack window (> 8 N/25 mm) than the EVA copolymer (35°C vs. 15°C). When compared with the EAA, the POP had similar ultimate hot tack strength but a narrower hot tack window. Because hot tack strength strongly depends on substrate adhesion, the same EAA tie layer was used for all sealants to ensure the same adhesion to the substrates.

Fig. 14. Hot tack data: paper/poly/foil (0.00035)/1.5 mil coex EAA (20%)/sealant (80%)

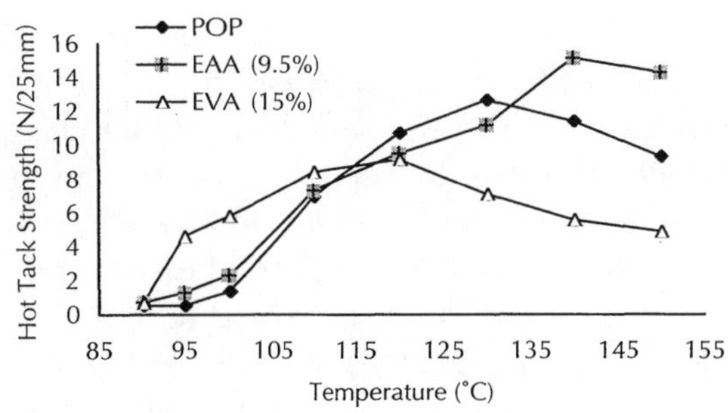

All the resins exhibited much lower hot tack values on the PET film. This is due to the lack of a fast heat transfer medium (aluminum foil) to help in accelerated cooling of the sealant. The ultimate hot tack strength of the POP was much better than the LDPE (5.5 vs. 1.5 N/25 mm). The POP achieved approximately the same ultimate hot tack as the LLDPE resin but at approximately 20°C lower temperature as Fig. 15 shows. The hot tack window of the POP was narrower than the EAA and the ionomer sealants as Fig. 16 shows.

Fig. 15. Hot tack data: 48 gauge PET/1.5 mil coex EAA (20%)/sealant (80%)

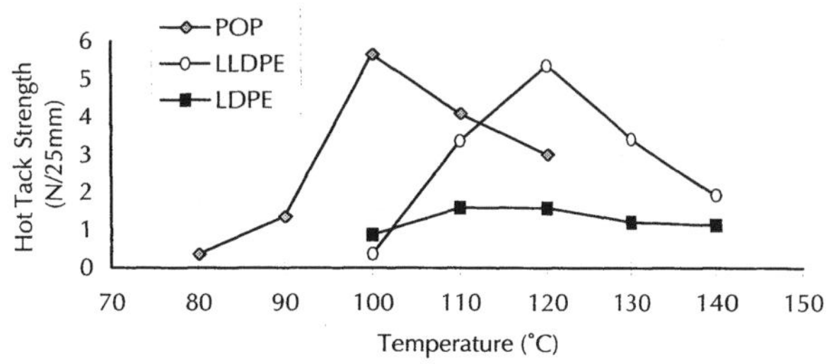

Fig. 16 Hot tack data: 48 gauge PET/1.5 mil coex EAA (20%)/sealant (80%)

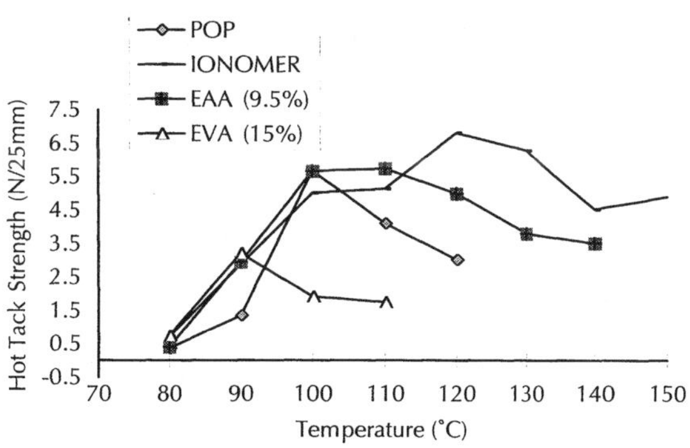

On 50-lb kraft paper in Figs. 17 and 18, the heat seal strength of the POP was twice that of the LDPE (4 vs. 2 lb/in.) and even greater than the LLDPE. The POP initiated a seal at a temperature 20°C lower than the LDPE and LLDPE. All the resins were coated at 600°F. The heat seal strength of the EAA and ionomer (coated at 550°F) was a little better than that of the POP. The EVA (coated at 450°F) had the lowest heat seal strength (2 lb/in.) and did not show an advantage in heat seal initiation. The POP had the lowest heat seal initiation temperature on the paper substrate.

Fig. 17. Heat seal data: 1 mil coating on 50 lb kraft paper at 600°F

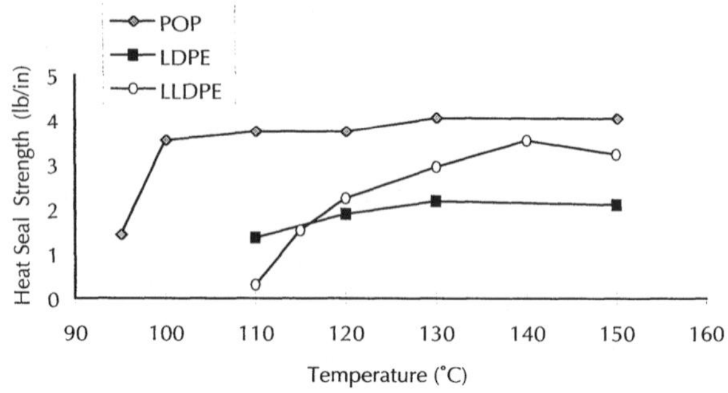

Fig. 18. Heat seal data: 1 mil coating on 50 lb kraft paper

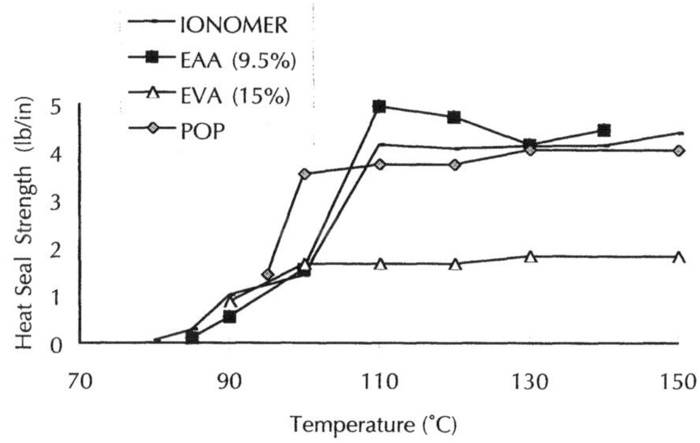

Tear properties

POP extrusion coated onto paper showed excellent tear resistance as Table V shows. The Elmendorf tear values of the POP were more than 2.5 times that of LDPE (373 vs. 133 g) and even higher than the LLDPE (373 vs. 271 g). The tear properties of the POP were also more balanced in the machine and cross directions.

Table 5. Elmendorf tear data using 1 mil on 50 lb kraft paper

Resin	Elmendorf MD	Tear, g CD
Pop	373	366
LDPE	133	141
EVA	152	216
EAA	165	123
LLDPE	271	289
Ionomer	134	173

Table 6. Adhesion data

Resin	OPP adhesion, g	Foil adhesion, g	Primed PET adhesion, g
POP @ 550°F	472	41	36
POP @ 600°F	335	241	636
LDPE @ 550°F	72	41	41
LDPE @ 600°F	123	45	268
LLDPE @ 600°F	77	159	822
EVA @ 450°F	36	54	41
EVA @ 550°F	45	454	341
Ionomer @ 550°F	45	368	318

Adhesion

POP demonstrated excellent adhesion (> 400 g) to the treated high energy industrial grade oriented polypropylene (OPP) of 70 gauge and the primed PET films but only fair adhesion (< 200 g) to aluminum foil as Table VI shows. If higher aluminum foil adhesion is necessary, the POP can be coextruded with an EAA tie layer that will provide excellent aluminum foil adhesion. Previous work also confirmed excellent adhesion to LLDPE films *(9)*.

Sensory results

Taste and odor evaluations shown in Figs. 19 and 20 for the polymers coated on the primed PET showed that the POP contributed the least intense taste and odor compared with the EAA and ionomer.

Fig. 19. Taste intensity ranking

There is a 99% Confidence Level of Difference between 1.35 and 2.66

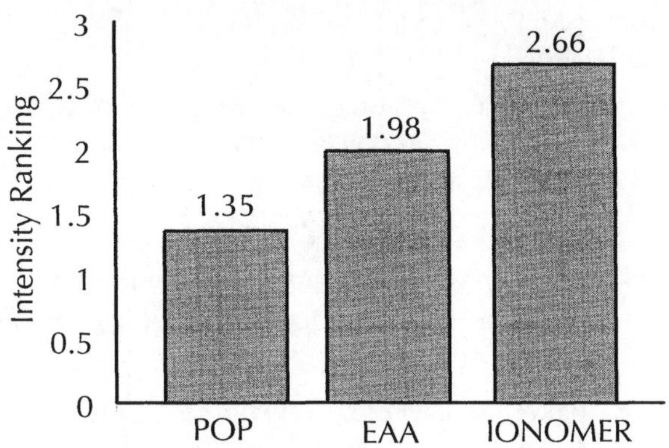

Fig. 20. Odor intensity ranking

1 = Best or least intense
3 = Worst or most intense
There is a 99% Confidence Level of Difference between 1.35 and 2.66

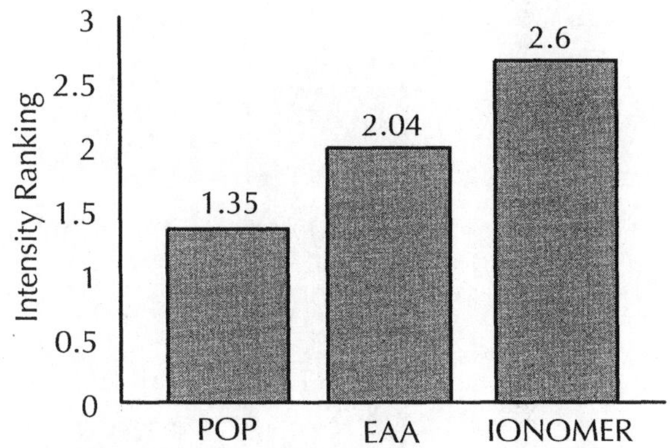

In a separate comparison with the EVA in Figs. 21 and 22, the POP showed the least odor intensity. The EVA contributed the least intense taste to water probably due to lower oxidation of the polymers at lower processing temperatures (450° F vs. 575°F).

Fig. 21. Odor intensity in air (paired comparison)

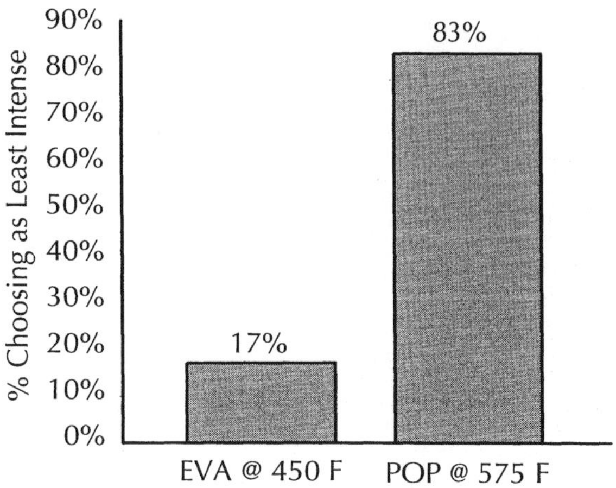

Fig. 22. Taste intensity in water (paired comparison)

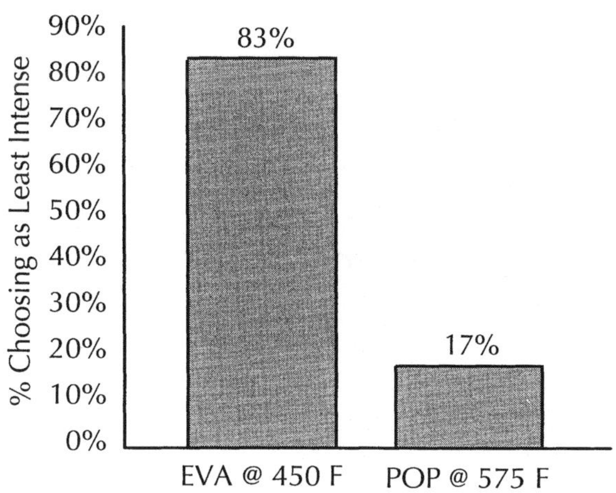

CONCLUSIONS

The performance properties of single-site catalyzed polymers and copolymers make them well suited for coextrusion or monolayer coating and laminating applications. Because tie layers are necessary to achieve the desired adhesion level to some substrates—aluminum foil, unprimed PET, nylon, machine glazed paper, coextrusions will be important to achieve the best combination of cost and performance. POP is also an excellent choice for overall economics compared with more expensive specialty copolymers.

POP can provide flexible coatings that offer good sealing properties and excellent toughness and tear resistance. Their optical properties allow their use in laminating applications that require good clarity. The low modulus of POP should allow their use as coatings for woven and nonwoven scrims for such applications as tarpaulins, tapes, medical applications, and consumer products such as tents and high strength tote bags requiring toughness and flexibility.

The POP products lower seal initiation temperatures, provide excellent seal strengths, and have good hot tack performance. They meet the requirements for form, fill, and seal applications. On aluminum foil-containing structures with outer layers that limit sealing jaw temperatures below 140°C, the performance of POP should compare very favorably to premium sealants especially when cost is a consideration. In nonaluminum foil structures that do not require a broad hot tack window, POP can provide excellent seal strength. An excellent combination of cost and performance with good taste and odor properties makes single-site-based POP a good choice for extrusion coating and lamination applications.

LITERATURE CITED

1. Thayer, A. M., Chem. Eng. News 73:15(1995).
2. Elston, C. T., U.S. pat. 3,645,992 (Feb. 29, 1972).
3. Sinclair, K. B., Polyolefins VIII International Conference, Society of Plastics Engineers, Houston, 1993, p. 1.
4. Devore, D. C., U. S. pats. 5,470,993 (Nov. 28, 1995), 5,556,928 (Sep. 17, 1996), and 5,624,878 (Apr. 29, 1997).
5. Lai, S. Y., Wilson, J. R., Knight, G. W., et al., U. S. pats. 5,272,236 (Dec. 21, 1993) and 5,278,272 (Jan. 11, 1994).
6. Lai, S. Y., Wilson, J. R., Knight, G. W., et al., U.S. pats. 5,272,236 (Dec. 21, 1993) and 5,278,272 (Jan. 11, 1994).
7. Kelley, D. C., Baker, S. L., and Potts, M. W., TAPPI 1995 Polymers, Laminations and Coatings Conference Proceedings, TAPPI PRESS, Atlanta, p. 499.
8. Baker, S. L., Potts, M. W.,and Hughes, M. M., TAPPI 1991 Polymers, Laminations and Coatings Conference Proceedings, TAPPI PRESS, Atlanta, p. 259.

9. Baker, S. L., Goins, J. D. , Kale, L. T., et al., TAPPI 1994 Polymers, Laminations and Coatings Conference Proceedings, TAPPI PRESS, Atlanta, p. 103.

CHAPTER 24

Ethylene Vinyl Acetate Resins for Extrusion Coating

Gary D. Cheney, Equistar Chemicals, LP

INTRODUCTION

Ethylene vinyl acetate (EVA) copolymers are thermoplastic materials that offer outstanding heat seal characteristics, good adhesion to a variety of substrates, and excellent optical properties. These characteristics make EVA copolymers ideal coatings on substrates such as polyester (PET), biaxially oriented polypropylene (BOPP) and nylon for use in cheese wrap, document covers, ID cards, and lidding materials.

EVA copolymers for extrusion coating typically range in melt index from 14 to 35 g/10 min and contain 15-30 weight percent vinyl acetate.

CHEMISTRY

EVA copolymers for extrusion coating are produced by copolymerizing ethylene monomer and vinyl acetate monomer under similar high-pressure/high-temperature conditions used to produce low-density polyethylene (LDPE) extrusion coating resins. Polymerization yields long chains of hydrocarbon molecules with vinyl acetate groups randomly distributed among the ethylene molecules. Figure 1 shows a part of an EVA copolymer chain.

Fig. 1. Part of an EVA copolymer chain

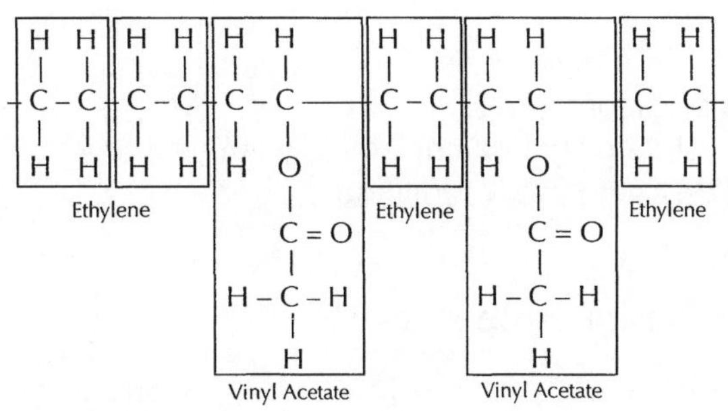

The bulky vinyl acetate units do not allow the chains to pack closely, thus lowering the crystallinity of the polyethylene, which results in the differences between LDPE and EVA copolymers. While the crystallinity of EVA copolymers decreases with vinyl acetate incorporation, the density increases due to the size of the vinyl acetate molecules. Therefore, the normal density and crystallinity relationship of polyethylene is not observed for EVA.

PROPERTIES

Many of the fundamental physical properties and extrusion coating properties of LDPE are affected by incorporation of vinyl acetate. The amount of change in properties is controlled by the amount of vinyl acetate in the copolymer.

Physical properties

The effects of changes in vinyl acetate content in EVA copolymers on various physical properties are shown in Table 1 below:

Table 1. Effect of VA content on physical properties

Increasing VA% Causes an Increase in:	Increasing VA Causes a Decrease in:
Density	Tensile strength
Flexibility	Secant modulus
Clarity	Stiffness
Tackiness	Abrasion resistance
Coefficient of friction	Shore hardness
Impact strength	Resistance to gas and vapor transmission
Puncture resistance	Melting point
Elongation	
Resistance to brittleness at low temperatures	
Resistance to UV radiation	

Adhesion properties

Adhesion is the most important property in extrusion coating. For LDPE, adhesion is obtained by increasing the polarity of the melt curtain through surface oxidation. The surface oxidation is obtained by high melt temperatures (600-625°F) and long air gaps. Treatment of the substrate is also used.

EVA copolymers lack the thermal stability of LDPE. At temperatures above 450°F, EVA copolymers degrade and generate acetic acid. For this reason, the functionality offered by the vinyl acetate groups is the mode for adhesion, not surface oxidation. Adhesion of EVA copolymers to substrates increases with increasing vinyl acetate content. Other adhesion enhancement techniques used when extrusion coating with EVA copolymers are corona treating of the substrate, chemical priming of the substrate, and injecting ozone at the nip to increase surface oxidation of the melt curtain.

APPLICATIONS

EVA copolymers are used as the heat seal coating for document covers and cheese wrap. The properties that make EVA copolymers ideal for these applications are excellent clarity, low-temperature sealing, and good adhesion to many films.

EVA copolymers are the resin of choice in the document cover industry where aesthetics are very important. EVAs offer superior gloss and lower haze versus LDPE, ethylene normal-butyl acrylate (EnBA), and ethylene methyl acrylate (EMA). The optical properties of EVA copolymers improve with the amount of vinyl acetate incorporated into the final product.

As stated in the "Properties" section, incorporation of vinyl acetate decreases the melting point and heat seal initiation temperature of LDPE. The degree to which the melting point is depressed is directly related to the weight percent of vinyl acetate incorporated. Figure 2 shows the effect of vinyl acetate content on melting point.

Fig. 2. Effect of VA content on melting point

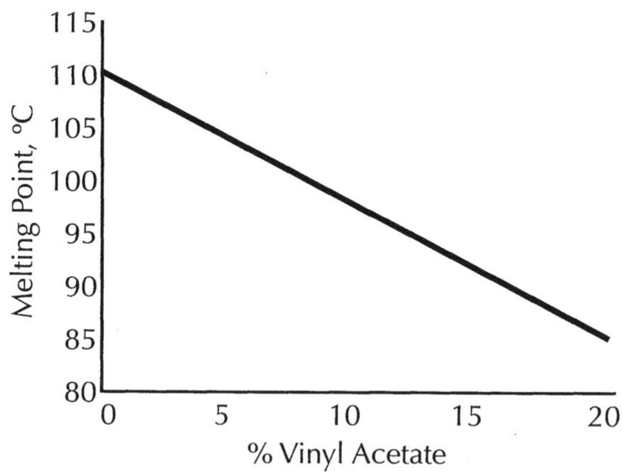

For the same weight percent copolymer, EVA and EMA copolymers have slightly lower melting points than EnBA copolymers.

Adhesion of EVAs to polyvinylidene chloride-Coated polyester and BOPP films is better than LDPE and equal to EnBA and EMA. The adhesion of EVA copolymers to nylon, polyester, and metallized oriented polypropylene films is not as strong as LDPE or EnBA. The adhesion of EVA to these substrates can be increased by chemically priming the substrate or introducing ozone at the nip.

Fig. 3. Recommended temperature profile for EVA copolymers

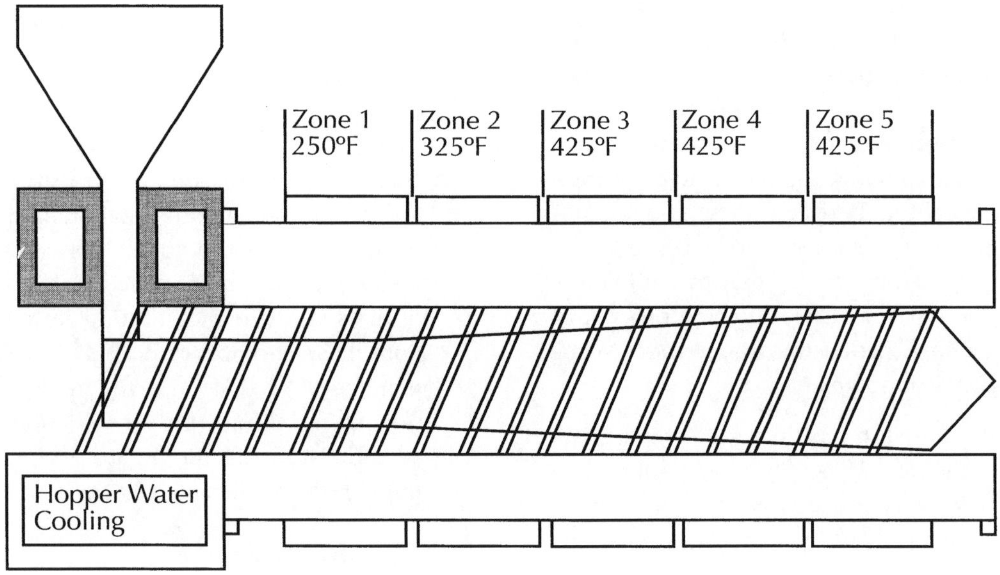

PROCESSING

Processing considerations

Standard extrusion coating equipment used for processing LDPE can be used for EVA copolymers.

As stated in the "Chemistry" section, the pendent vinyl acetate groups make the properties of EVA copolymers different from LDPE. From a processing standpoint, some of the key properties affected are thermal stability, melting point, drawdown, chill roll release, and coefficient of friction.

Because EVA copolymers degrade at temperatures above 450°F, different extrusion temperatures are used than for LDPE. Figure 3 shows a recommended temperature profile to be used when processing EVA copolymers.

The temperatures of the piping following the extruder and the die should be set for 425°F. Because of the low melting points and softness of EVA copolymers, the feed section should be kept at 250°F to prevent bridging of the feed throat. The melt temperature should not exceed 450°F.

Drawdown is a function of melt temperature, air gap and line speed. For EVA copolymers, the melt temperature has the largest effect. The lower melt temperature does not allow EVA copolymers of the same melt index to be run at the low coating weights possible for LDPE. Therefore, EVA copolymers for extrusion coating typically have higher melt indexes than LDPE.

Since surface oxidation is not the primary mode of adhesion for EVA copolymers, a short air gap (less than 7 in.) can be used. The lower air gap will reduce neck-in.

As vinyl acetate content increases, resins become tackier and can result in poor release from the chill roll. Poor chill roll can cause web brakes and adhesion problems. Because of the tackiness of EVA copolymers, a matte surface chill roll is used for most applications. For applications requiring a mirror pocket chill roll, slip is usually added to high vinyl acetate resins to get good release.

Increase in vinyl acetate content causes an increase in the coefficient of friction. This increase in the coefficient of friction can result in blocking. Blocking is the tendency of the surface of a coated substrate to stick to the substrate lying on top of it in the windup roll. A blocked roll causes problems in post extrusion coating converting processes because it is difficult to unwind. Addition of anti-block to the resin can reduce or eliminate the problem.

Scheduling and transition

EVA copolymers used in extrusion coating have lower softening points and higher melt indices than LDPE homopolymers used in extrusion coating. These differences, along with the lower melt temperatures used in extrusion coating EVA copolymers, can result in significant lost production when switching between jobs using LDPE and EVA copolymers. Proper scheduling of products on the extrusion coating line can minimize the lost production time. The following procedure can help maximize production time when extrusion coating EVA copolymers.

Schedule all extrusion coating jobs using EVA copolymers together. If possible, schedule them for Monday if your production is not seven days a week. It is better to start up on EVA copolymers because it is much easier to heat the equipment up to temperatures used for LDPE than it is to cool down to EVA temperatures. Prior to running EVA copolymers, the extruder should be shut down on the lowest density and highest melt index LDPE available.

Use the following steps for the transition into EVA:

1. Set the extruder, piping, and die temperatures as outlined previously.
2. Open the back pressure valve and the deckles completely.
3. Start up on the lowest density and highest melt index LDPE available.
    a) Switch to a dry blend of 70 pounds LDPE and 30 pounds EVA.
    b) Switch to a dry blend of 70 pounds EVA and 30 of LDPE.
4. Transition to 100 percent EVA.

Use the following steps for the transition out of EVA:

1. This should be performed at the temperatures recommended for EVA copolymers.
2. Open the valve and deckles completely.
3. Begin to purge in LDPE.
4. Change the temperature profile to that for LDPE.
5. While the temperatures are rising, change extruder rpms in the following sequence: slow - medium - high - medium - slow.

Resin handling

EVA copolymers are very heat sensitive, with lower softening points than homopolymer LDPE resins. Due to the lower softening points, EVA copolymers are more susceptible to the formation of fines and streamers during resin transfer operations than homopolymer LDPE. To reduce the generation of fines and streamers during transfer, the piping should be run as short a distance as possible and the temperature of the air used to transfer the pellets should be monitored.

REFERENCES

1. "Ultrathene® Ethylene Vinyl Acetate Copolymers for Applications that Demand Flexibility and Toughness" (product literature), Quantum Chemical Corporation, Cincinnati, Ohio.
2. Malin, D.S., "A Fresh Look at Extrusion Coating with EVA Resins," 1984 Polymers Laminations and Coatings Conference Proceedings, TAPPI PRESS, Atlanta, p. 359.
3. Thomas, D., "Extrusion Coating of Paper and Paperboard Equipment and Materials," in EVA for Extrusion Coating, 3rd edn., TAPPI PRESS, Atlanta, 1990.
4. "Enathene® Ethylene N-Butyl Acrylate (EnBA) Copolymers for Extrusion Coating and Laminations" (product literature), Quantum Chemical Corporation, Cincinnati, Ohio.

CHAPTER 25

Ethylene Methyl Acrylate Copolymers for Extrusion Coating

George L. Baker, Chevron Chemical Co.

INTRODUCTION

The ethylene methyl acrylate copolymer (EMA) is one of the most thermally stable of the high-pressure alpha olefin copolymers, making it ideally suited for the high temperatures of the extrusion coating process, as well as molding, sheet, film, coextrusions, and blending applications. The excellent extrusion coating processing characteristics of this material are due in part to the high-pressure autoclave process used for its polymerization. This material is commercially supplied to the extrusion coating industry in monomer contents of 18-24%, in a melt index range of 2-20 g/10 min, and with and without slip and anti-block additives. Other combinations are available for the other applications.

PROPERTIES REVIEW

In the copolymerization of methyl acrylate with ethylene, the methyl acrylate monomer becomes part of the polyethylene main chain and produces additional side branches in addition to those normally present in LDPE homopolymers. These additional side branches serve to reduce the crystalline structure that exists among the homopolymer polyethylene chains. Due to the polarity of the methyl acrylate side groups, a measure of polarity is given to EMA that is not present in a homopolymer LDPE.

Table 1 lists a property comparison of the commercially available 2 melt index (MI), 20% methyl acrylate EMA to a 2 MI, 0.917 density LDPE homopolymer. While both the EMA and the LDPE have similar flow properties by virtue of each having a 2 MI, the EMA melting point by DSC is depressed from 104°C to 83°C as a result of this lower crystallinity. This reduced melting point is the reason that EMA possesses improved heat seal response and will require less heat to melt the pellets in an extruder, hence a more energy-efficient polymer. The reduction of Shore D hardness and flexural stiffness indicates that film and coated items of EMA will have reduced stiffness and be soft to the touch (as compared to LDPE coatings).

EMA has significantly improved stress cracking resistance over LDPE and can be used as is or in blends with LDPE to improve the stress cracking resistance of the extrusion coated product. The surface printability of EMA coatings require about the same level of electronic or flame treatment as LDPE.

The haze of mono-layer EMA film is higher than that of LDPE, but when used as an extrusion coating lamination layer, the contact clarity is equal to that of LDPE. This excellent contact clarity is demonstrated in many of the sparkling snack foods structures where EMA is used in the extrusion lamination of two high optic substrates.

The surface COF of EMA is quite high, allowing it to be used for non-skid coatings. The addition of slip additives can modify the surface to be quite slippery.

The heat seal comparison shows that EMA has significantly improved heat seal characteristics over

Table 1. Property comparison

Resin	EMA Copolymer	LDPE Homopolymer
Melt index, g/10min	2	2
Comonomer content, %	20	0
DSC melting point, °C	83	102
Vicat softening point, °C	59	90
Tensile at break, psi	1620	1850
Elongation, %	720	650
Shore hardness, D scale	33	50
Flexural stiffness, psi	4000	18500
Stress cracking resistance, % failures/2 weeks (Igepal CO-630 @ 50°C)	0	20
Extrusion Coated on to 40# kraft paper		
Printability (after surface treatment)	Excellent	Excellent
Contact clarity	Excellent	Excellent
Coefficient of friction		
(with slip and anti-block additives)	0.2	0.2
(without slip and anti-block additives)	>1	>1
Heat Seal*, lbs/inch of width		
At 71°C (160°F)	0.8	0
At 77°C (170°F)	1.7	0
At 82°C (180°F)	2.4	0
At 88°C (190°F)	2.4	0.3
At 93°C (200°F)	-	1.0
At 99°C (210°F)	-	2.2

*0.5 seconds dwell, 30 psi seal pressure

LDPE. This results in higher packaging line speeds (more bags per hour). EMA is also commercially sealed by radio frequency (R/F) welding methods, due to its increased polarity.

Another attribute of EMA is that the pellets are not hydrophilic. Therefore, EMA pellets can be shipped in the same low cost shipping containers normally used for LDPE and stored in silos without fear of water absorption. The same is true for edge trim and recyclable scrap. This results in reduced handling and storage costs.

EMA is used in many direct food contact applications. It has FDA compliance under Food Additive Regulation 21 CFR 177.1340. It can also be used as the heat seal layer in hydrogen peroxide, aseptic packaging systems and has approval for direct contact packaging of meat and poultry.

The polar nature of EMA results in outstanding adhesion to the very high volume orientated polypropylene (OPP) snack food film (see Table 2). Excellent adhesion is also obtained for substrates that are surface coated with the high barrier PVdC material. EVA does not have the thermal stability to be processed at the heats needed for high adhesions. EAA can be processed almost as hot as EMA and LDPE, where its acid functionality produces excellent adhesion to aluminum foil and to metallized substrates. All materials have their strong points and their resulting places in the market.

Table 2. Adhesion to various substrates

	EMA	LDPE	EVA	EAA
Kraft paper	Excellent	Excellent	Fair	Excellent
OPP (unprimed)	Excellent	Fair	Fair	Fair
OPET (unprimed)	Excellent	Fair	Fair	Fair
Cellophane (PVdC coated)	Excellent	Fair	Fair	Fair
Aluminum foil (unprimed)	Fair	Fair-Good	Fair	Excellent
LDPE film	Inseparable	Inseparable	Inseparable	Inseparable

APPLICATIONS

The very versatile EMA is used in film and molding applications, as well as an impact modifier and as a high loadings master batch carrier base resin. Its low melting point makes it an ideal heat sealing material as well as a low-temperature film fabrication material. The outstanding thermal stability, extrusion coatability, and polar nature make EMA an excellent extrusion coating material that is well suited for the snack foods industry. Over the years, EMA has continued to be the very best material for extrusion coating and laminating to unprimed OPP and OPET substrates, for adhesion purposes. The OPP/EMA/OPP structure is the standard of the snack foods industry. Another construction is OPP/EMA/EAA/ALUMINUM FOIL (where the EMA/EAA are coextruded) to take advantage of the EMA adhesion and the EAA adhesion to ALUMINUM FOIL.

PROCESSING CONSIDERATIONS

EMA copolymers are routinely processed in smooth barrel extruders with coat hanger or key hole dies that are designed for LDPE, and extruded at the same 304-315°C (580-600°F) melt temperatures. This allows the coater to use EMA interchangeably with LDPE without downtime to change heater settings, screws, or dies. The same die gaps, 0.5 mm (approximately 20 mils) and draw distances of 10-17.5 cm (4-7 in.) are also used. EMA is readily compatible with the extrusion coating grades of LDPE and therefore is easily transitioned into the extruder behind LDPE and just as easily purged from the extruder following its use. Water cooling on the extruder throat is recommended to eliminate any bridging potential.

The excellent thermal stability of EMA allows it to be extrusion coated, extrusion laminated, and/or coextrusion coated or laminated at the same high temperatures that are used for LDPE. Thermal degradation in other copolymers can yield corrosive compounds that will damage screws, barrels, adapters, dies, and chill rolls, in addition to generating objectionable odor levels.

Because it is made by the autoclave process, EMA is an excellent extrusion coating resin. It will draw down well and runs in excess of 340 meters/min (1000 ft/min), with neck-in a little more than LDPE. Due to the adhesive qualities of EMA, it is best run on a matte chill when top coating a substrate, to achieve maximum chill roll release. The use of lower chill roll temperatures and a stripper roll also aid in chill roll release for EMA. The lower chill roll temperature will provide improved cooling of the EMA, while the stripper roll allows for longer contact time of the EMA layer on the chill roll. When extrusion laminating, chill roll release is generally not a consideration as the EMA is not against the chill roll.

Coextrusion

EMA is well suited for extrusion coating and lamination coextrusion, working equally well with combining adapters or multi-manifold dies. It has the thermal stability to be processed at the high extrusion temperatures and the polarity needed for adhesion to most of the commonly used thermoplastics. Excellent results have been experienced with most of the materials listed in Table 3.

Table 3. Coextrusion compatibility of EMA to other polymers

LDPE	E
HDPE	E
LLDPE	E
PP	E
PET	E
PVdC	E
Ionomer	E
EMA	E
EVA	E
EAA	E
EVOH	F
Nylon 6	P
PC	F

E = Layers very hard to separate
F = Layers can be separated with some effort
P = Layers can be readily separated

SAFETY

The safety considerations for the handling and processing of EMA are much like that of LDPE. Those skilled in the handling of LDPE can safely handle this material. EMA does not require a hazard warning per the OSHA Hazard Communication Standard 29 CFR 1910.1200. Cool EMA product is expected to cause no eye, skin, inhalation, or ingestion problems; however, material at processing temperatures will cause burns to unprotected body tissue. The higher the temperature is, the more severe the burning will be. Fumes from the hot material can be unpleasant to some individuals and cause nausea and irritation of the upper respiratory tract. Ingested cool material is considered non-toxic.

CHAPTER 26

Acid Copolymers for Extrusion Coating

John D. Vansant, DuPont Co.

INTRODUCTION

Acid copolymer resins are branched ethylene copolymers of either methacrylic acid or acrylic acid. The branched molecular structure makes these resins particularly suitable to the extrusion coating process, as the flow characteristics are similar to those of low-density polyethylene (LDPE). Acid groups are incorporated into the polymer to provide a unique chemical functionality that enhances both the bonding and sealing characteristics. Acid copolymer resins are premium priced compared to polyethylene resins. Their use, however, can create a premium value offering to the converter's end-use customer by enabling longer shelf life and/or greater seal integrity.

PROPERTIES REVIEW

General properties and characteristics

The extrusion coating properties of ethylene-methacrylic acid copolymer (EMAA) and ethylene-acrylic acid copolymer (EAA) are similar except for the difference in molecular weight of the acid groups. The molecular weight of methacrylic acid (86) is slightly higher than acrylic acid (72), due to the additional methyl group. Therefore, there are 1.2 times as many functional acid groups in EAA copolymer as in the same weight percentage EMAA copolymer. For example, a 12% EMAA has the same degree of acid functionality as a 10% EAA copolymer. Since most of the characteristics vary with the actual acid group functionality, this distinction between EMAA and EAA copolymers should be factored into performance comparisons of the sister-like products.

The effect of acidic branching on the molecular structure reduces the crystallinity of the resins proportionately with the number of acid groups present. Hence, the higher the percentage of acid is in the copolymer, the lower the crystallinity is. The crystallinity-dependent properties such as melting point, hardness, and gas barrier are subsequently reduced with increasing amounts of acid comonomer. When acid copolymers are compared to low-density polyethylene homopolymer, the lower melting point (e.g., 95° at 15% EMAA vs. 115°C) can be either an advantage or a disadvantage, depending on the application.

Physical properties of acid copolymers

Adhesion properties

Acid copolymers offer excellent, durable adhesion to metal due to their chemically reactive nature. In general, the adhesion to aluminum varies directly with the percentage of acid groups in the resin. It is the enduring nature of this adhesion that adds significant value over polyethylene homopolymers. High levels of adhesion between LDPE and aluminum can be produced with primers or oxidation, but the bonding can be deteriorated by extended exposure to certain food products such as juice or condiments. With acid copolymers, many of the hard-to-hold products can be packaged without sacrificing long-term durability of the coating adhesion. The level of acid required in the acid copolymer will depend on the product being packaged, and the required shelf life.

The acid functionality that bonds to aluminum will also provide adhesion to certain other materials such as other metals, treated glass, paper, polyurethane, and molten polyamide (nylon). In the case of nylon film, coating with an acid copolymer cannot create a durable bond because the reactive sites of the nylon polymer are turned away from the film surface.

Acid copolymers will also adhere to molten polyethylene, but the level of adhesion decreases as the weight percent acid increases. As a general rule, an acid copolymer containing less than 10% acid should be used for coextrusion with polyethylene. In some cases, acid copolymers are used to adhere to polyethylene films. Corona treatment is generally used to oxidize and increase the wettability and functionality of the polyethylene film surface.

Heat seal and hot-tack properties

Improved heat sealing properties of acid copolymers result from the attractive forces of hydrogen bonding between acid groups, and from the reduction in melting point that is characteristic of short chain branching. Thus heat seal performance is strongly influenced by the acid content of the copolymer. Hydrogen bonding enhances the ability of acid copolymers to seal through contaminants, as the acid groups lock on to those of the opposing sealant web. The improved hot-tack of acid resins is the result of hydrogen bonding between the opposing layers of the heat seal, and within the heat seal layer itself.

Hot-tack in polyethylene resins (and ethylene copolymers not containing acid) depends completely on the viscosity of the polymer, and on the melting and freezing characteristics of those polymers. The unique mechanical structure of linear low-density and metallocene polyethylenes imparts a measure of hot-tack strength, but this hot-tack is limited to the temperature range where melting and freezing take place. Beyond this narrow temperature range, the hot-tack performance is essentially the same as low-density polyethylene or other copolymers such as ethylene vinyl acetate (EVA). As a product of hydrogen bonding, acid copolymers offer hot-tack over a broader temperature range than the linear polyethylenes.

Barrier properties

Acid copolymer resins have moderate moisture barrier properties and relatively high permeation to oxygen and other gases (similar to LDPE). However, when an acid copolymer resin is extruded to aluminum foil, the overall barrier performance of a package can be significantly better than one extruded with LDPE. This is because the superior bonding and toughness properties of the acid copolymer provide an extra measure of protection against pinholing and flex cracking. The sealant performance characteristics must also be considered when looking at package barrier performance.

Mechanical and optical properties

As a general rule, hydrogen bonding between polymer acid groups provides increased toughness compared to LDPE. Mechanical properties such as tensile strength are primarily a function of molecular weight, so differences in MFI must be considered. The mechanical properties of coating resins are not often measured. While film properties data is readily available, differences in crystallinity and orientation make film property values inappropriate for application to extrusion coating. As acid content is increased, the effects listed below can be expected, and performance relative to LDPE can be inferred. Likewise, the effects of increasing MFI are also listed.

Effects of Increasing Acid Content
Higher tensile strength
Lower elongation
Lower modulus
Higher tear resistance
Slightly increased hardness
Higher abrasion resistance
Lower melting/freezing points
Lower sealing temperature
Greater hot-tack
Lower haze
Higher MVTR, Lower OTR
Higher aluminum foil adhesion
Increased solvent solubility

Effects of Increasing Melt Flow Index
Improved drawdown ability
Improved optical properties
Lower melt strength (higher neck-in)
Lower impact resistance

Lower abrasion resistance
Lower flex crack resistance

Processing properties

Appropriate processing conditions will depend on the grade of acid copolymer selected. Grades that are high in acid content and high in melt flow can be processed as low as 220°C (430°F). Grades that are lower in both acid and MFI can sometimes be processed as high as 315°C (600°F). It is important to remember that it is not necessary to oxidize the surface of the acid copolymer resin in order to get good adhesion; higher temperature is not always better. Sufficient time in the air gap is needed, however, both for draw down and adhesion purposes. The air gap should be adjusted *(1)* in order to obtain the require adhesion, while not creating excessive neck-in. For some product grades, a minimum air gap is desired to minimize neck-in, as adhesion will exceed the minimum requirement.

Little more horsepower is required for extrusion than that needed for LDPE, but good cooling must be maintained in the feed casting and in the first barrel zone. First barrel zone temperatures should be maintained between 150 and 220°C. Some screw designs intended exclusively for LDPE might produce too much shearing of the acid copolymer resin. Your machinery or polymer supplier can help you assess your screw design.

Due to the potentially corrosive nature of the organic acid copolymers, equipment should be duplex-chrome or nickel plated at a minimum. Equipment manufacturers have been supplying such plated systems for many years for processing conventional polyethylene resins. Thus, the protective plating recommendation is more of a statement of standard practice than a special need just for acid copolymers. To help avoid corrosion, always do a thorough purge following a run of acid copolymer or ionomer resin. Never shut down the extruder with acid copolymer in the system.

Manufacture

Acid copolymers are manufactured by the same high-pressure process used to manufacture LDPE. Thus the basic molecular structure (long chain branching) and resultant rheological behavior is similar. The presence of hydrogen bonding, however, causes the viscosity change with processing temperature to be greater than that of LDPE.

Commercial extrusion coating products include copolymers with acid levels ranging from 3 to 15 weight percent, and even higher percentage acid products are expected to become available to meet specific requirements in new applications. The molecular weight of these copolymers can be varied easily to cover a broad field of extrusion coating processing needs. With molecular weight represented in terms of commonly used melt flow index value, acid copolymers in the range of 5 to 20 MFI are suitable for most extrusion coating applications. A number of higher MFI grades are available for specialized applications.

APPLICATIONS

Ethylene copolymers of methacrylic or acrylic acid provide value-in-use for extrusion coating applications where improved properties of heat sealability, hot tack strength, oil resistance, or aluminum foil adhesion are desired.

Acid copolymers are frequently specified for their unique aluminum foil bonding characteristics. Often the product being packaged is of an aggressive nature such as with a condiment pouch packaging acidic foods like tomato ketchup. The optimum level of acid in the copolymer will depend on both the level and types of acids, etc., in the food product, and the desired shelf life. The thickness of the acid copolymer layer will also have an impact on the shelf life. In some cases, a layer of acid copolymer is used on both sides of the aluminum foil to prevent the aggressive chemical in the product from permeating through pinholes and attacking the back side. Shelf life pouch studies are often conducted by periodically opening pouches to measure the aluminum foil adhesion and inspecting for signs of aluminum foil corrosion.

The strong, durable adhesion to aluminum foil makes acid copolymers suitable for applications such as liquid (juice) cartons, condiment pouches, toothpaste tube laminates, and communications cable shielding. Premium sealing characteristics provide good hot-tack and sealing through contaminants for packages that have low levels of leakers and for fast packaging line speeds. Numerous sealant applications include sachet packages of condiments, spiced oils, powdered products, alcohol towelettes, and other health and beauty care products.

Keep in mind that, in most packaging machines, it is normal to incur significant fluctuation in seal bar temperature and sometimes line speeds. When specifying heat sealants, make certain to include a margin for error and change of conditions by your customer. While a given sealant specification might perform well under the controlled conditions of a qualification trial, changing conditions of the packaging machinery can result in leaky seals in the future. Thus acid copolymers or ionomers (see chapter on ionomer resins) should be selected where hot-tack over a wider temperature range might be required. Both the sealant material and its thickness must be able withstand such variations.

Frequently, acid copolymers are coextruded with polyethylene resins as a primer to facilitate high adhesion to aluminum, sometimes at very high coating line speeds. Thus, the polyethylene does not need to be oxidized, and can be extruded at much lower temperatures. The resulting product is much less likely to impart an undesirable odor or flavor to the food product being packaged, as the oxidation of the coating is greatly reduced.

PROCESSING CONSIDERATIONS

At elevated temperatures, the acid copolymers are susceptible to anhydride cross-linking of the functional acid groups, with a subsequent increase in viscosity. With sufficient time and temperature, draw characteristics of the coating web can be limited. As cross-linking occurs, a more viscous melt can reduce the ability of the resin to wet the surface of the substrate and adhesion can be diminished. Also, a rough or applesauce-like appearance can occur in part or all of the web. Cross-linking can also cause gel formation that can lead to purging difficulties. Thus it is important to educate the extrusion operators as to the nature of these polymers and stress that the temperatures should not be excessive.

Some grades of acid copolymer are designed to be more resistant to cross-linking and to be more drawable. These grades are of particular value in coextrusion, where the holdup time and nonuniform heating can lead to substantial cross-linking. As a general rule, copolymers of methacrylic acid are more thermally stable than those of acrylic acid. This is not to say that all EMAA grades are more stable than all EAA grades. Resins of both acid types can be successfully processed, as long as there is good temperature control.

Higher acid-containing resins are generally more susceptible to anhydride formation than lower acid products. These anhydrides can crosslink the resin, but can also generate unwanted moisture at high temperature and high holdup times. Therefore, a maximum melt temperature of 260°C (500°F) is recommended for copolymers above 12% acid. For copolymers up to 10% acid, melt temperatures from about 285°C to 305°C (550°F to 580°F) are usually sufficient to achieve excellent adhesion. A variety of grades of copolymers with different molecular weights (implied by melt flow index values) can be selected to meet the speed, drawdown, and neck-in requirements customarily achieved with extrusion coating of polyethylene homopolymers. Resin grades with enhanced drawability are also available. Due to the complex nature of these products and the variety of products available, it is recommended that you consult your resin supplier when developing new specifications.

Extruder barrel temperatures in the initial feed section should be maintained as low as possible to minimize bridging of the low melting point, sticky acid copolymers. First barrel zone temperatures as low as 150°C (300°F) might be feasible with some equipment, while temperatures above 220°C (430°F) are to be avoided. In the second and subsequent extruder zones, where there is positive displacement of melting resin cubes, barrel temperatures can be adjusted as necessary to achieve the melt temperature desired for coating.

Because the acid functionality causes attraction to metal surfaces, purging acid copolymers from the extruder can sometimes be difficult. Normally LDPE or LLDPE is used to purge out the extruder. It is important to vary the rpms *(2, 3)* of the extruder to alternately increase the shear rate at the metal surface and slow down the flow enough for the layers of polymer to adhere to one another. Varying the rpms also changes the flow patterns within the extrusion system.

SAFETY

Good ventilation should be provided when processing acid copolymer resins to contain (evacuate) the polymer smoke. In addition to small amounts of acrylic or methacrylic acid monomer, the polymer fumes will contain normal products of hydrocarbon combustion that are normally found when processing LDPE. These products include small amounts of acrolein and formaldehyde. While a processing temperature limit of 325°C (615°F) is recommended, this is not a sharply defined decomposition temperature. Most of the fumes generated are products of oxidation and, as with any chemical reaction, the rate of oxidative degradation probably doubles with every 10°C increase in resin processing temperature. Thus the melt temperature should never be higher than that required for processing and adhesion. Often it is possible to process the acid copolymer resins below 290°C, and rarely is it necessary to exceed 305°C. As you should with any type of smoke, avoid breathing the polymer fumes, and optimize the performance of ventilation hoods.

For any polymer processing, MSD sheets should be consulted and made available per OSHA requirements. Compliance with the OSHA limits for airborne levels of chemicals produced by polymer decomposition is the responsibility of the polymer processor (converter).

REFERENCES

1. Antinov, V., and Soutar, A.M., "Foil Adhesion with Copolymers: Time in the Air Gap," TAPPI 1991 Polymers, Laminations and Coatings Conference Proceedings, TAPPI PRESS, Atlanta.
2. Soutar, A.M., "'Disco' Purging: A Technique," TAPPI 1996 Polymers, Laminations and Coatings Conference Proceedings, TAPPI PRESS, Atlanta.
3. Bulletin H-02368-1, 1991, "The Disco Purge Procedure for Extrusion Coaters," DuPont.
4. Hoh, G.K.L., Sadik, S.A., Gates, J., "Aluminum Foil Adhesion of Ethylene/Methacrylic Acid Copolymer Coatings," TAPPI 1989 Polymers, Laminations and Coatings Conference Proceedings, TAPPI PRESS, Atlanta.
5. Hester, D.L., Sadik, S.A., Trouilhet, Y., "Using Acid Copolymers as Coextrusion Tie Layers," TAPPI 1989 Polymers, Laminations and Coatings Conference Proceedings, TAPPI PRESS, Atlanta.
6. Soutar, A.M., "Hot Tack Testing: or Is It?" TAPPI 1996 Polymers, Laminations and Coatings Conference Proceedings, TAPPI PRESS, Atlanta.
7. Hammond, F.M., "Processibility and Performance Properties of High Performance Ethylene Acrylic Acid Copolymers," TAPPI 1987 Extrusion Coating Short Course Notes, TAPPI PRESS, Atlanta.
8. Kelley, D.C. and Baker, S.L., "EAA Extrusion Coating Sealants for Non-foil Packaging," TAPPI 1996 Polymers, Laminations and Coatings Conference Proceedings, TAPPI PRESS, Atlanta.

CHAPTER 27

Ionomer Resins for Extrusion Coating

John D. Vansant, DuPont Co.

INTRODUCTION

Ionomers are distinctive polyolefin copolymers formed by metallic ion cross-linking of ethylene acid copolymers, of either acrylic acid or methacrylic acid. The combination of ionic and hydrogen bonding imparts unusual solid-state and molten characteristics to the ionomers. The cross-linking is thermally reversible; therefore, the beneficial effects of high molecular weight on melt rheology, toughness, and other end-use properties can be achieved without sacrifice in melt processing. With acid copolymers, the product properties are a function of acid content, polymer chain length, and branching. With ionomers, secondary variables such as the type and amount of metallic ion permit a much wider range of resin properties than normally achieved with olefin polymers. To fully understand the issues discussed in this section, the user should first read the previous chapter on acid copolymers.

Most applications for ionomers in extrusion coated structures take advantage of the exceptional heat sealing characteristics. Some applications also benefit from the grease, abrasion, and chemical resistance characteristics. The use of ionomer resins spreads across the full range of packaging applications, wherever challenging seal requirements exist.

PROPERTIES REVIEW

General properties and characteristics

Ionomers are very similar to low-density polyethylene in many characteristics important in processing and in end uses. With melting points varying by grade from 82°C to 100°C (180°F to 212°F), ionomer resins can be easily melted from free-flowing pellets to viscous melts that can be extruded as thin webs at high elongation rates. When compared with LDPE, the metallic ion cross-linking in ionomers has an important effect on molten resin properties. At low molten temperatures, the ionic structure causes high resistance to flow, or high melt viscosity. Then at elevated temperatures, the cross-linking is broken to yield melt viscosity much lower than for conventional polyethylenes of seemingly equivalent molecular weight.

Physical properties

Adhesion properties

The same acid functionally found in acid copolymers is also available for adhesion of ionomers to aluminum, nylon, and other materials. The amount of acid available for bonding is that acid in the polymer that has not been reacted with the metal salt, or "neutralized."

When discussing adhesion, it is important to make a distinction between true adhesion and peel strength. Ionomer resins are very stiff by nature. Even though the adhesive forces can be quite high, peel strength is often less than for a similar acid copolymer resin. When peeling a resin layer from another material, higher stiffness will change the nature of the peel so that the stress is concentrated to a sharper area of the interface. The result is a cleaving action that causes lower measured peel values. Peel values alone should not be overemphasized in setting specifications. In many cases, the resultant heat seal strength should be tested when setting peel strength specifications.

It is normal for the adhesion of ionomers to aluminum foil to increase or "age up" over the first few days after coating. Occasionally, the stiffening of the ionomer over time will lead to an aging down of peel strength. This happens when the initial value of adhesion is relatively low. When establishing specifications for any given structure, adhesion should be measured green (right off the line) and then checked periodically over at least four weeks. As a rule of thumb, green peel values of 5 Newtons/25 mm (500 g/in.) will normally age up, while green values less than 3.5 Newtons/25 mm can actually decrease in peel value. Very thick coatings (greater than 35 microns or 22 lb/ream) or film laminations should be watched carefully for aging effects.

For applications involving coextrusion with polyethylene, peel values will be lower as the acid comonomer and metal salt contents are increased. By properly selecting the grade and process conditions, delamination-type peelable seals can be obtained. In some cases, polyethylene-based adhesive resins will not be suitable for coextrusion with the ionomers, as heat seal and hot-tack strength will be limited by the peel and cohesive strength of the tie resin.

Ionomers are often specified for extrusion onto aluminum foil. If the product being packaged is aggressive in nature, sufficient acid must be available in the polymer to fight off the chemical attack. For ionomers, the product resistance and shelf life will generally depend on the amount of unneutralized acid in the polymer. For more explanation of product resistance and testing requirements, see the corresponding section on acid copolymers. Never assume that a given polymer will provide the needed protection without consulting your polymer supplier or performing shelf life testing.

Heat seal and hot-tack properties

Where hydrogen bonding between acid groups enhances the hot-tack properties of acid copolymers, the additional attractive energy provided by ionic cross-linking extends the hot-tack of ionomers over a much wider temperature range. Hydrogen bonding is responsible for the enhanced ability of acid copolymers to seal through contaminants, as the acid groups can lock on to those of the opposing sealant web. As with hot-tack, the addition of ionic bonding further enhances the ability of the ionomer to seal through contaminants. Thus heat seal performance is strongly influenced by both the acid content and the ion content of the copolymer. Figure 1 shows a comparison of 9% MAA acid copolymer (10 MFI) with a low acid/low neutralized zinc ionomer (5 MFI) and a high acid/higher neutralized ionomer (14 MFI).

Fig. 1. A comparison of 9% MAA acid copolymer with a low acid/low neutralized zinc ionomer 1 and a high acid/higher neutralized ionomer 2

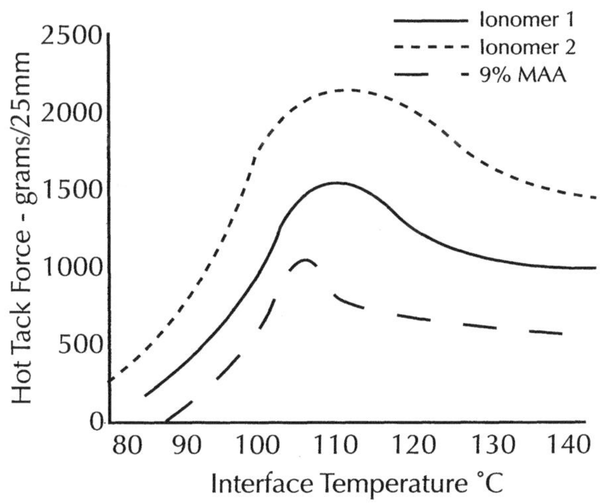

When compared to linear low-density-type polyethylene (LLDPE) and metallocene polyethylene resins (MPE), the hot-tack of both acid copolymers and ionomers is effective over a wider heat sealing temperature range. This is because the polyethylene resins depend completely on the freezing process to obtain hot-tack, since they lack the strong chemical attractive forces that benefit the acid copolymers and ionomers. The extrusion coating grades of MPEs and LLDPEs have less impressive hot-tack curves than the blown film grades because the viscosity is lower. Figure 2 shows a comparison of hot-tack strength for a common extrusion coating ionomer grade of low acid and low neutralization, compared to a common extrusion coating grade metallocene polyethylene. Even for this low-end performing ionomer, a substantial difference in hot-tack is shown. Note the wide range of temperature under which the ionomer can operate and still provide a given level of hot-tack. Thus, acid copolymers or ionomers should be selected for applications where hot-tack over a broader range of sealing temperature is required.

Fig. 2. DTC hot tack force ionomer vs. MPE

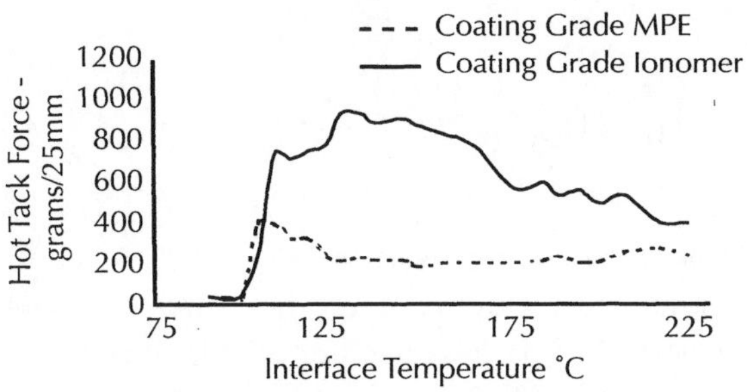

.5 second heat seal dwell time
.1 second actual delay time

Keep in mind that, in most packaging machines, it is normal to incur significant fluctuation in seal bar temperature and sometimes line speeds. When specifying heat sealants, make certain to include a margin for error and change of conditions on the packaging machinery. While a given sealant specification might perform well under the controlled conditions of a qualification trial, changing conditions of the packaging machinery can result in leaky seals in the future. Both the sealant material and its thickness must be able to withstand such variations. Premium sealant materials can often pay for themselves in reduced complaints and claims made at a later time.

Mechanical and optical properties

As was the case with acid copolymer resins, blown film property values should not be applied to extrusion coating applications. In general, ionomers are tougher, stronger, and stiffer than LDPE and acid copolymers. The toughness, stiffness, and strength properties of ionomers typically age up over a four-week period of time following coating. This is due to a reorientation or clustering of the ions.

The following relationships will be helpful in understanding the properties of ionomers:

Higher % Acid
Stiffer
More oil resistant
Better optical properties
Higher tensile strength
Higher impact strength
Lower seal initiation temperature

Higher % Unneutralized Acid
Higher aluminum adhesion and nylon coextrusion adhesion
Higher flex durability
Higher tear strength
Lower notch sensitivity

Higher % Neutralization
Stiffer
More oil resistant
Better optical properties
Higher tensile strength
Higher impact strength
Tougher
More abrasion resistant
Lower notched tear strength
Higher/broader hot-tack
Less blocking

Processing properties

Appropriate processing conditions will depend on the grade of ionomer selected. Grades that are high in acid content and high in melt flow can be processed as low as 220°C (430°F). Grades that are lower in acid and MFI can sometimes be processed as high as 315°C (600°F). It is important to remember that it is not necessary to oxidize the surface of ionomer resins in order to get good adhesion to aluminum; higher temperature is not always better. Sufficient time in the air gap *(1)* is needed, however, both for drawdown and adhesion purposes. The air gap should be adjusted in order to obtain the required adhesion while not creating excessive neck-in. For some product grades, a minimum air gap is desired to minimize neck-in because adhesion will exceed the minimum requirement.

Typically, more horsepower is required for extrusion of ionomers than for LDPE but substantially less than that required for LLDPE or metallocene polyethylenes. Most extrusion coating lines will have sufficient horsepower to run ionomers at reasonable rates. Good cooling must be maintained in the feed casting, and in the first barrel zone. First barrel zone temperatures should be maintained between 150°C and 200°C. The second zone, however, normally must be around 50°C hotter than the first zone. This helps the melting process, reduces the horsepower requirement, and actually reduces the shear heat buildup on the screw.

Screw designs intended exclusively for LDPE can produce too much shear heating of the ionomer and will also be limited in achievable output. Typically, for a given screw design, there is an rpm above which the extruder cooling cannot keep up and bridging will occur. Your machinery or polymer supplier can help you assess your screw design.

Due to the potentially corrosive nature of ionomers, equipment should be duplex-chrome or nickel plated at a minimum. Equipment manufacturers have been supplying such plated systems for many years for processing conventional polyethylene resins. Thus, the protective plating recommendation is more of a statement of standard practice than a special need just for acid copolymers. To help avoid corrosion, always do a thorough purge following a run of acid copolymer or ionomer resin. Never shut down the extruder with acid copolymer in the system.

Manufacture and chemistry

Ionomers are produced by reacting acid copolymers with metal salts. This reaction introduces an ionic cross-linking, thus the name ionomer. Figure 3 illustrates this molecular structure. The circle on the left is the ionic cross-link, while the circle on the right is an unneutralized acid group.

Fig. 3 Ionomer structure

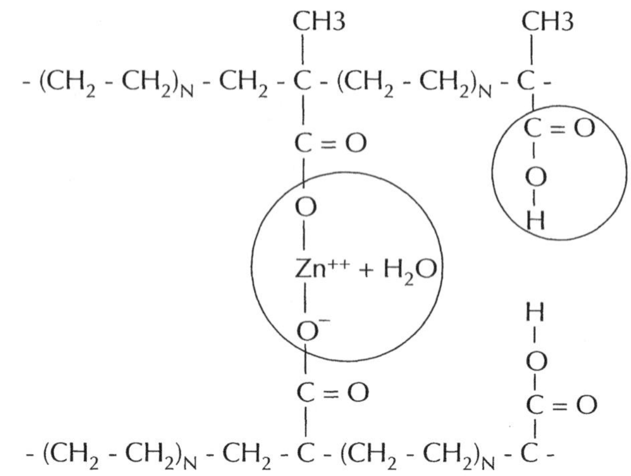

Acid copolymers used to produce ionomers are manufactured by the same high-pressure process that is used to manufacture LDPE. Ionomers are normally produced by further reaction of the acid copolymer resin with metal salts. The presence of hydrogen bonding in the acid copolymer causes the viscosity change with processing temperature to be greater than that for LDPE. The additional ionic bonding present in ionomer resins further intensifies this effect. Ionomers are thus characterized by very high melt viscosity at lower temperatures. Keep in mind that the standard 190°C melt flow index of an ionomer cannot be used for rheology matching.

Fig. 4 Melt viscosity vs. temperature

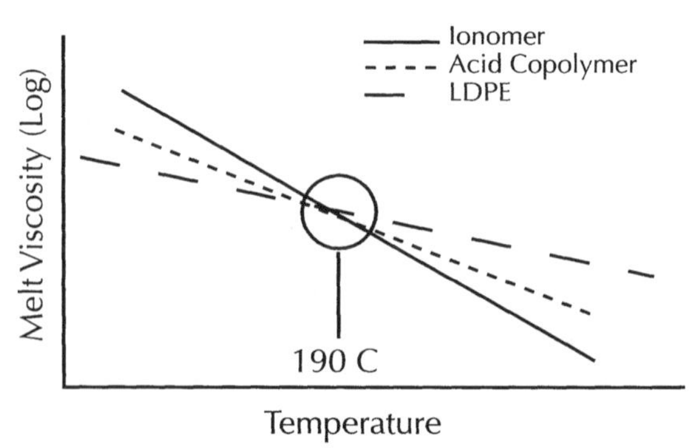

Two basic families of ionomers have been used for many years in commercial extrusion coating applications. The first of these, introduced in 1963, were the sodium ion cross-linked ionomers based on ethylene-methacrylic acid copolymers. They continue to be used for extrusion coatings where abrasion resistance, high resistance to greases and oils, and excellent heat sealing characteristics with extraordinary hot tack strength are needed. A second family of ionomers is made by cross-linking EMAA or EAA copolymers with zinc ions. The zinc ionomers provide a combination of high, durable adhesion to aluminum and excellent heat sealing characteristics, with high hot-tack strength, over a broad temperature range.

Although the commercial grades of ionomers cover a wide spectrum of molecular weights, resins for extrusion coating are characterized by melt flow index values from about 2-20 MFI. Ionomers are made from acid copolymers that range typically from 5-15% acid before neutralization. With varied degrees of metallic ion cross-linking, tailored solid and melt properties are achieved for the different grades of ionomers.

APPLICATIONS

Ionomer resins are used in many extrusion coating applications where sealing requirements are critical. The unique ionic melt properties provide a high degree of heat sealability over a broad temperature range, with excellent seal strength, extraordinary hot tack strength, and sealability through contamination. In some applications, the ionomer's superior abrasion and grease resistance is needed. The combination of high hot-tack strength and low seal initiation temperature make them uniquely qualified for packaging at very high speeds. Hot-tack is also important for packaging of heavy products in vertical form fill seal applications to prevent mooning, or partial opening of the bottom seal under the weight of the product. It also plays a significant role in sealing through product contamination.

For fine powdered products, an acid copolymer might be a better sealant selection because it holds a much lower static charge than ionomers. Although some powdered product applications benefit from the greater seal through contamination capability of the ionomer, this can be a trade-off because more contamination can be attracted to the ionomer surface. Acid copolymers will generally have sufficient ability to seal through the contaminant without the static attraction, and thus provide a better overall performance.

Packaging machinery variables will often dictate the use of ionomer sealant resins. Applications that span a variety of packaging machines and vintages benefit from the resin's ability to seal over a wide range of conditions. Often setup time and packing setup waste is significantly lower than with sealant resins that have a narrower sealing temperature range. Packaging machines are notorious for poorly designed and poorly maintained seal bar temperature controls. The use of ionomer sealants can result in substantially fewer leakers, and thus lower overall cost to the product or food producer.

The majority of extrusion coated ionomer applications are on paper and foil laminates. These numerous applications range from dry mixes to medical devices. Coatings on film are used for snack packaging, sometimes with fat free cookies and cakes that require exceptional moisture barrier and seal integrity. Ionomer sealants are also found in non-flexible packages such as spiral wound cans and other food cartons. Ionomers are commonly used in coextrusion extrusion coatings, where the resin's special properties contribute value to complex, multilayer packages.

PROCESSING CONSIDERATIONS

At elevated temperatures, zinc ionomers are susceptible to anhydride cross-linking of the functional acid groups, resulting in an increase in viscosity. As with acid copolymers, processing can be affected. Ionomers, however, are more resistant to this type of cross-linking than their acid copolymer cousins. It is still best, however, to keep the processing temperature as low as possible.

Ionomer resins have chemically bound water associated with the ionic cross-linking. In manufacture, the resins are dried to a level of moisture that is suitable for extrusion coating, and then are shipped in moisture-proof packages. Ionomers in opened packages exposed to humid environments for more than eight hours can absorb sufficient moisture to cause bubbles in the molten web or die buildup. Partially used boxes of resin should be tightly re-sealed, preferably by heat sealing of the moisture-barrier liner bag.

Processing temperatures suitable for extrusion coating ionomers range from 230°C to 315°C (450°F to 600°F), depending on the grade. Melt temperature above 315°C (600°F) should be avoided. Unlike polyethylene coating, better substrate adhesion is not achieved as melt temperatures are elevated over 315°C (600°F). It is important to prevent overheating of feed hoppers and feed-section barrel zones to minimize sticking of the low-melt ionomers on the initial section of the extruder screw. This causes bridging or starved-feed conditions.

Extruder barrel temperatures in the initial feed section should be maintained as low as possible to minimize bridging of the low-melting-point, sticky ionomer resins. First barrel zone temperatures as low as 150°C (300°F) might be feasible with some equipment; temperatures above 220°C (430°F) are to be avoided. In the second and subsequent extruder zones, where there is positive displacement of melting resin cubes, barrel temperatures can be adjusted as necessary to achieve the melt temperature desired for coating.

To avoid bridging problems, good cooling must be maintained in the feed casting, and in the first barrel zone. First barrel zone temperatures should be maintained between 150 and 200°C. The second zone, however, normally must be around 50°C hotter than the first zone. This helps the melting process, reduces the horsepower requirement, and actually reduces the shear heat buildup on the screw. Due to the sticky nature of the molten resins, screw rotation should be maintained when ionomers are in the system. The third and subsequent zones should be adjusted to obtain the desired melt temperature. Normally the valve should set completely open, as back pressure is not require to obtain the needed output temperature. Excessive shearing should be avoided.

Because the acid functionality causes attraction to metal surfaces, purging ionomers from the extruder can sometimes be difficult. This difficulty can by overcome if the following practices are maintained.

- Introduce polyethylene into the system with temperatures maintained as used with ionomer coating.
- After polyethylene has completely filled the melt system, raise temperatures to reduce viscosity of residual ionomer, which changes more with temperature than polyethylene. Temperatures commonly used in polyethylene coating can be used for this resin transition period.
- When high temperatures are reached, extrude at fast screw speeds for short periods of time. It is important to vary the rpms *(2, 3)* of the extruder to alternately increase the shear rate at the metal surface and slow down to allow adhesion to the purging resin. The subsequent high shear rates will further reduce viscosity of the molten resin and helping to purge ionomer from the system. Without utilization of high temperatures and periods of fast purge rates, complete elimination of viscous ionomer from the extrusion system can require excessive resin scrap and downtime. Using high rpms for periods longer than 2-3 minutes at a time can result in a quicker purge-out, but will result in much higher purge waste.
- Continue purging until the web is visually free of troublesome gel, when viewed at both high and low rpm.

The transition into ionomer from LDPE is normally simple, but the following steps can be helpful:

- Before introducing the ionomer to the extruder, change the temperature settings to those suitable for processing the ionomers and allow the actual temperatures to reach their setpoints.
- Introduce the ionomer and randomly vary the rpm *(3)* until the ionomer is well established. The viscous ionomer melt should easily displace and purge polyethylene from the system.
- To minimize stagnation of ionomer in the ends of externally deckled extrusion dies, the deckles should be adjusted to approximate positions while polyethylene is still in the system.

Extrusion coating equipment suitable for processing low-density polyethylene is used very successfully to process ionomers. Duplex-chrome or nickel-plated surfaces on the screw and in all melt flow sections have been found suitable for resistance to the corrosion that is encountered with mild steel equipment. To help avoid corrosion, always do a thorough purge following a run of acid copolymer or ionomer resin. Never shut down the extruder with acid copolymer in the system.

Because of the inherent chemical functionality of ionomer resins, the need for air-gap oxidation to attain adhesion is not required when coating ionomers onto paper or aluminum foil. In general, close positioning of extrusion dies to the pressure and chill-roll nip will minimize web neck-in, lessen heat loss of the web and result in good substrate adhesion even at very high line speeds. When adhesion is insufficient, or edge tearing (snapping) occurs, it is beneficial to increase the air gap as the processing speed is increased *(1)*.

SAFETY

Good ventilation should be provided when processing ionomer resins to contain (evacuate) the polymer smoke. Ionomer resin fumes contain small amounts of acrylic or methacrylic acid monomer and can contain acetic acid. The polymer smoke or fumes will also contain the products of hydrocarbon combustion normally found when processing LDPE. These products include small amounts of acrolein and formaldehyde. While a processing temperature limit of 325°C (615°F) is recommended, this is not a sharply defined decomposition temperature. Most of the fumes generated are products of oxidation and, as with any chemical reaction, the rate of oxidative degradation probably doubles with every 10°C increase in resin processing temperature. Thus the melt temperature should never be higher than that required for processing and adhesion. Often it is possible to process the acid copolymer resins below 290°C, and rarely is it necessary to exceed 305°C. As you should with any type of smoke, avoid breathing the polymer fumes and optimize the performance of ventilation hoods.

For any polymer processing, MSD sheets should be consulted and made available per OSHA requirements. Compliance with the OSHA limits for airborne levels of chemicals produced by polymer decomposition is the responsibility of the polymer processor (converter).

REFERENCES

1. Antinov, V., and Soutar, A.M., "Foil Adhesion with Copolymers: Time in the Air Gap," TAPPI 1991 Polymers, Laminations and Coatings Conference Proceedings, TAPPI PRESS, Atlanta.
2. Soutar, M., "'Disco' Purging: A Technique," TAPPI 1996 Polymers, Laminations and Coatings Conference Proceedings, TAPPI PRESS, Atlanta.
3. Bulletin H-02368-1, 1991, "The Disco Purge Procedure for Extrusion Coaters," DuPont.
4. Soutar, A.M., "Hot Tack Testing: or Is It?" TAPPI 1996 Polymers, Laminations and Coatings Conference Proceedings, TAPPI PRESS, Atlanta.

CHAPTER 28

Extrudable Adhesive Resins for Coextrusion Coating

Dean A. Zimmerman and Dan Ward, Equistar Chemicals, LP

INTRODUCTION

Extrudable adhesive resins, commonly referred to as tie-layer resins or tie-resins, are used to bond two dissimilar materials that would otherwise have very poor adhesion. For example, polyolefins have very poor adhesion to polar substrates such as nylon and ethylene vinyl alcohol copolymers (EVOH). However, by adding functionality to the polyolefin, it is possible to achieve adhesion to EVOH, nylon, paper, metal, and polyethylene terephthalate (PET) while still maintaining adhesion to polyolefins. As a result, multilayer structures can be produced that take advantage of the unique features of each type of material resulting in a final product that has properties that would be impossible to achieve without tie-resins.

Extrudable adhesive resins can either be reactive or nonreactive. Examples of potentially reactive polymers include those resins grafted with acids, anhydrides, epoxides, or silanes. Examples of nonreactive adhesive resins include copolymers of ethylene and vinyl acetate (EVA), methyl acrylate (EMA), or n-butyl acrylate (EnBA). The focus of this chapter is on reactive resins, namely, anhydride modified polyolefin (AMP) tie-layers.

MECHANISM OF ADHESION

Adhesion between the tie-layer and substrate can occur either by chemical bonding or non-chemical bonding. Chemical bonding involves the formation of covalent or hydrogen bonds while non-chemical bonding includes chain entanglement, mechanical interactions (i.e., rough surface) or Van der Waals' forces.

In order for the AMP to form a chemical bond to the substrate, there must be a labile hydrogen that is capable of reacting such as in hydroxyl (OH), amide (NHCO), or amine (NH_2) groups. The anhydride ring opens and forms an ester linkage with the functional group of the substrate with no by-products. The reactivity of the functional group also plays a role in adhesion. A reactive group that is in the backbone of the resin (i.e., a copolymer) compared to being grafted is not as reactive since mobility is restricted.

Excellent adhesion can be obtained without chemical bonding due to chain entanglements or polar interactions. For example, a functionalized HDPE adheres well to HDPE due to the ability for chain entanglement to occur. The likelihood of two resins to chain entangle can be obtained from the solubility parameter. Two resins that have similar solubility parameters are expected to have better adhesion (such as EVA and LDPE) than two materials (such as HDPE and nylon) which have very different solubility parameters. Also, the polarity of the resins affects the ability of the resins to adhere. A non-polar resin such as polyethylene will not adhere well to a polar resin, although having two polar resins does not guarantee adhesion.

TIE-LAYER SELECTION

Before selecting a tie-layer resin, the following questions should be answered:

- What are the materials to be bonded?
- What is the minimum adhesion requirement?
- What is the process including secondary operations?
- What are the processing conditions (temperature, line speed, etc.)?
- Will the tie-layer be in direct food contact?
- What is the end-use environment (solvents, moisture, temperature)?
- What are the end-use requirements (stiffness, barrier, etc.)?
- What are the economics?

Before selecting a tie-layer, it is important that the tie-layer supplier and processor review these questions since tie-layers are often optimized for certain resins or processes. Minimum adhesion requirements are also critical to produce the most economical product. While maximum adhesion is often desired, it can be expensive and it is important not to over-engineer the product.

To maximize tie-layer performance, the proper base resin, MI, graft type, and graft concentration must be selected. Tie-layer suppliers have developed extensive information on specific applications and can assist in selecting the right tie-layer. The grafted resin will have similar properties to the base resin, which should be selected to maximize performance. The tie-layer should be an integral part of the multilayer system and can provide additional benefits besides adhesion. For example, HDPE has good stiffness, moisture barrier properties, and thermal stability while EVA has good heat sealing characteristics and clarity. The adhesion is also affected by the polarity and crystallinity of the base resin, especially in cases where no chemical reaction takes places as with polyvinylidene chloride (PVdC). The MI should be selected so that the layers in contact have similar viscosities to prevent flow instabilities that can lead to waviness or other imperfections.

While there are many possible grafts, anhydrides are the most common due to their high reactivity and ease of use. AMP chemically bond to a variety of substrates to provide excellent adhesion including paper, metals, nylon, and EVOH while maintaining excellent bondability to most polyolefins. The adhesion between a substrate and AMP is a function of the concentration of the anhydride (up to a given value) and is also strongly dependent on the physical properties of the AMP. High graft levels usually result in a higher priced resin; therefore, it is important to select a resin with the proper amount of graft. "High-performance" tie-layers have been developed that take advantage of key physical properties of the base resin that can improve adhesion.

PROCESSING GUIDELINES

Resin handling and safety

Grafted anhydrides will react with moisture and convert to the less reactive diacid form. It therefore is recommended that AMP be stored in sealed bags or liners and in places to avoid exposure to high humidity. Resin drying generally is not required if the liner or bag is sealed. Moisture in the resin will be evident by bubbles or foam in the extrudate and a significant decrease in adhesion. Your tie-resin supplier might have specific recommendations on drying resins.

General processing considerations

An AMP typically processes like the non-functionalized base resin. For instance, an anhydride functionalized LLDPE will process at similar pressures, output rates, and neck-in levels as an unmodified LLDPE with the similar viscoelastic properties. However, manufacturers should be knowledgeable of several key tie-resin processing parameters that are required to maximize adhesion and reduce the incidence of problems such as gels.

First, AMP require a certain amount of heat, pressure, and contact time to ensure adequate bonding to the barrier resins and/or the coating substrate. The amount of heat in the tie-resin and thus bond strength can typically be improved by increasing tie-resin melt temperatures, increasing chill roll temperatures, preheating the substrate, reducing line speeds, or by modifying other parameters used to increase the temperature, pressure, and contact time between the layers.

Second, AMPs are prone to stick to metal surfaces and thermally degrade when subject to high tem-

Table 1. Relationship between shear force and resin viscosity at various process locations

	Die Wall		Center		Die Wall
Shear Force	High	Intermediate	Low	Intermediate	High
Resin Viscosity	Low	Intermediate	High	Intermediate	Low

peratures for long time periods that can lead to cross-linking or oxidation gels. To avoid this problem, extruders should not be shut down or started up with AMP, and an extruder should never be idled with AMP for a significant period of time without lowering temperatures. Many tie-resins can withstand fairly high processing temperatures, provided the screw is not stopped. Special temperature considerations, however, should be given to tie-resins based on EVA, which are not as thermally stable as PE, EnBA, or EMA.

Blending tie-resins with less expensive polymers can be an effective method for reducing cost. However, the secondary polymer dilutes the functional component in the tie-resin, which in turn reduces the number of potential effective bonds. If the number of bonds formed drops below critical levels, package failure, and/or delamination will occur. In addition, the likelihood of delamination or gel formation significantly increases if there is incomplete mixing between the tie-resin and secondary polymer.

Minimizing gel formation

AMP can act as a purge by chemically bonding to polymers and other materials left in the extruder and die from previous runs. The resulting purge gels are often erroneously attributed to a defective AMP since they clear up when the AMP extruder is shut off. A complete purge of the system is necessary to resolve these purge gels.

Guidelines for minimizing gel formation when using AMP include:

- Follow all processing guidelines and supplier recommendations.
- If gels are present, completely purge the extrusion system beginning and ending with the material most likely to form gels.
- Shutting down or idling an extruder containing AMP can lead to gels upon restart.

Extreme caution should be used when blending AMP with pigments, additives, and/or other polymers. Additives such as titanium dioxide (TiO_2) can chemically interact with the anhydride and lead to gel agglomerates.

Melt viscosity considerations

As with any coextrusion process, it is necessary to choose resins with compatible viscoelastic properties. Failure to do so will result in quality defects and processing difficulties. For best results, the lowest viscosity resin (highest MI) should be located in the region of highest shear rate that is at the die walls. (Table 1) The inside layers of the coextrusion should have melt viscosities that are equal to or slightly lower than the outside layers. If the tie or inner layers have lower melt viscosities than the outer layers, wavy surfaces, grainy appearance, poor layer uniformity, and even voids in the web can develop as the lower viscosity resins migrate to the area of highest shear rate.

The melt or melt flow indices of the various polymers can be used as a general predictor of relative melt flow properties. However, the melt index is indicative of flow at one shear rate and the resin processor should refer to a complete viscosity/shear rate curve (often available from the resin manufacturer) when screening component resins in a coextrusion.

Factors affecting adhesion strength when using tie-resins

Most commercial tie-resins are designed to run on conventional extrusion lines and do not require special equipment modifications. Adhesion performance of a tie resin will be affected by equipment design factors such as die land length, feedblock design, and extruder length which influence the contact time, temperature, pressure, and degree of oxidation of the polymer and substrate. Additives can also affect the adhesion. Slip agents, such as erucamide, and inorganic antiblock can decrease adhesive bond strength.

Adhesion of an AMP to a given substrate will generally improve if the substrate is primed, treated, or preheated immediately before the coating stage. Flame treatment can be beneficial for coating of paper with AMP. Priming can be essential for tie-layer adhesion to substrates such as oriented PET films. General practices for improving polyethylene bond strength to substrates such as increasing the air gap are also applicable for improving tie resin bond strength to a given substrate. Adhesion of the tie-layer to the substrate or adjacent layers can generally be improved by increasing the tie-layer thickness. Typically, increasing the temperature will lead to an improvement in adhesion as long as the resin degradation temperature is not reached. Adhesion is also a function of line speed and will decrease as the line speed increases. The AMP has less time to react, but also the interfacial stress is increased-which decreases the adhesion.

APPLICATIONS

The use of reactive polyolefins in extrusion coating and extrusion laminations is growing rapidly. Much of the growth has occurred in paperboard coating applications such as gable-top and other juice cartons where the reactive polyolefin is required to coat an EVOH or nylon barrier layer onto the paperboard. The use of extrudable tie-resins is also growing in snack food, meat, and cheese laminations involving coated or printed polymer films. In these applications, the tie-resin coextrusion offers economic, processing, and end-use performance advantages over adhesive laminations.

Anhydride and acid modified resins are well known in the industrial coatings industry for their ability to bond to oxidized metal surfaces. In packaging applications involving aluminum foil or metallized films, acid-copolymer or acid-modified resins are generally preferred over AMP due to their ability to bond at lower temperatures and faster line speeds. However, these acid/metal hydrogen bonds can be susceptible to long-term attack by organic acids present in certain packaged products resulting in package failure or delamination. In these cases, AMP might be a suitable alternative, provided the process and/or substrate can be modified accordingly.

SUMMARY

Reactive extrudable adhesive resins are used in a variety of coextrusion processes to bond dissimilar materials which allows for final products with properties that otherwise could not be obtained. The choice of the tie-layer resin is often the last consideration when developing a new packaging structure and is often based on the lowest cost resin which meets the minimum adhesion levels. However, the tie-layer can add value to the structure by increasing the moisture barrier, eliminating the sealant layer, improving physical properties, or adding to the overall package integrity. In addition, the lowest cost tie-resin might not always be the most economical, as higher performance tie-resins can result in down-gauged layers or higher processing speeds. Therefore, it is important to review processing and performance requirements with the tie-resin supplier early in the package development process to ensure that the best product is chosen.

CHAPTER 29

Extrusion Coating Polypropylene

Ray Edwards

ABSTRACT

Polypropylene extrusion coating formulations offer property advantages over polyethylene in grease-proofness, abrasion resistance, and heat resistance. This chapter will discuss applications where these property advantages are important such as in coating release sheet, woven scram, and spiral-wound cans. Processing conditions for adhesion to paper substrates require a short air gap and melt temperature of about 590°F. New product development was required in tougher coatings and coatings with better adhesion to label stock.

INTRODUCTION

Polypropylene formulations for extrusion coating became commercially available in 1964. Prior to this time, polypropylene showed draw resonance and excessive edge weave at speeds of only 100 or 200 ft/min. The low coating speeds made polypropylene unacceptable for most commercial coating applications.

In about 1964, polypropylene formulations were developed that could be extruded at commercially acceptable speeds and coated at low weights. It is now possible for polypropylene to be extrusion coated at speeds greater than 1000 ft/min at coating weights as low as 8-10 lb/ream.

DISCUSSION

Polypropylene is more crystalline and has a higher melt point than polyethylene. This difference in properties contributes to polypropylene having advantages over polyethylene in grease-proofness, abrasion resistance, and heat resistance. Some examples of polypropylene use are:

1. Release sheet
2. Coating woven scrim and nonwoven products such as "Typar" spunbonded polypropylene
3. Beverage bottle labels
4. Aseptic packaging
5. Spiral-wound cans
6. Retort pouches.
7. Dielectric laminates.

Let us examine the properties of polypropylene that make it useful in some of the applications mentioned above.

HEAT RESISTANCE

The heat resistance of polypropylene is indicated by a response called the vicat softening point.

Coating	Vicat Softening, °F
LDPE	190
HDPE	235
pp	270

The higher the vicat softening point is, the higher is the heat resistance of the coating. The higher heat resistance of polypropylene has led to its use in the manufacture of linoleum floor tiles, decorative laminated table and counter tops, and reheatable food trays. Polypropylene shows very little adhesive attraction to other polymers, and the property also contributes to its use in release-type applications.

ABRASION RESISTANCE

The abrasion resistance of coatings can be determined by the Taber Abrasion Test as shown.

Coating	Weight Loss
LDPE	0.018 g/1000 cycles
HDPE	0.012 g/1000 cycles
PP	0.008 g/1000 cycles

In this test, a Taber Abraser Model 174 or equivalent is used to abrade the test specimen. The specimen weight is determined before and after abrasion for 1000 cycles. The results are reported as the weight loss per 1000 cycles.

The abrasion resistance of polypropylene makes it desirable for coating beverage bottle label stock. In the filling and distribution of beverages, the labels rub against each other and against the handling equipment. The abrasion resistance of polypropylene helps the labels retain their high gloss and remain aesthetically appealing until the product is purchased.

The beverage bottle labels have been traditionally laminated with biaxially-oriented polypropylene film. The labels can be extrusion coated more economically.

GREASE RESISTANCE

Grease resistance can be an important property in many packaging applications. The following data were generated using military tests Mil-B-121B.

Coating	Time to Fail (hours)
LDPE	4-5
HDPE	8-10
PP	14-16

In this grease-proofness test, a 4-in. by 4-in. test specimen is folded diagonally to form right angles in the center of the specimen. At the apex of the right angles, 5 g of sand is placed and then 1.5 ml of the test reagent is added to the sand. The reagent is a mixture of turpentine, calcium chloride, beacon grease 325, and red dye. The prepared specimen is held in place on white filter paper with a 50-g weight. The samples are checked for staining on the white filter paper every two hours until failure is observed.

The grease resistance of polypropylene makes it suitable for use in spiral-wound cans, packaging for food products, and animal foods.

OTHER PROPERTIES OF INTEREST

Polypropylene also shows good stress crack resistance, curl resistance, and stiffness-all of which can be important properties in packaging applications.

HEAT SEALING

The heat resistance of polypropylene can be advantageous in some applications but, in those applications requiring a heat seal, polypropylene requires higher heat sealing temperatures and/or longer heat sealing times than polyethylene. The polypropylene seal will also tend to be brittle unless the seal can be cooled rapidly. Therefore, the impulse sealers with their ability to cool the seal generally yield better seals than do heated bar sealers. New polypropylenes have been developed to overcome these heat seal deficiencies.

PROCESSING CONDITIONS

Screw design

Equipment used for processing low-density polyethylene is satisfactory for extension coating grade polypropylenes. However, it is desirable to have an extruder with an L/D of at least 24:1, with 28 to 32:1 being more desirable. The screws compression rations should be from 3.5 to 4.0:1 for better mixing. Mixing sections are not suggested.

If it is necessary to maximize the extrusion rate, then an investment in a new screw might be desirable. Screws designed specifically for polypropylene consider the low coefficient of friction between polypropylene and the barrel wall. Due to the low CoF of PP against the barrel, a long feed section is required to obtain the desired output. There are several equipment manufacturers that have the technical expertise to build extruder screws for polypropylene.

Die design

The same die design normally used for polyethylene is generally suitable for polypropylene. While shorter land lengths of 0.3-0.5 in. and narrow die openings of about 15-20 mils have been found to be optimum, longer lands and wider openings can be used. The shorter lands and narrower openings are important in the processability of polypropylene at faster line speeds and lighter coating weights.

CONDITIONS

When processing polypropylene in extrusion coating, its relatively low viscosity can result in some problems. If low output rates are used, the melt pressure inside the die will be very low and unstable. In turn, this will result in coating weight variations across the web and will yield rather poor drawdown characteristics. Therefore, it is suggested the output rate be at least 10 lb/in. of die width. Even when high output rates are used, the drawdown of polypropylene is not equivalent to low-density polyethylene. When attempts are made to produce thin coatings at high line speeds, polypropylene will generally streak out and tear off at the edges. Relatively fast line speeds (900-1000 ft/min) can normally be obtained if the output rate is sufficient to yield a coating thickness of about 1 mil or greater.

The following conditions compare polypropylene to low-density polyethylene.

PROCESSING CONDITIONS

Variables	Polypropylene	Polyethylene
Screw speed, rpm	100	100
Drive current, amps	120	150
Output, lb/h	325	375
Melt temperature, °F	580	620
Melt pressure, psi	1200	1800
Maximum line speed, ft/min	700	1500
Minimum coating weight, lb/ream	7.5*	4.0
Neck-in, in./edge	1.6	0.9
Air gap, in.	2 3/8	7

*Pin holes are a potential problem at this coating weight with polyethylene.

ADHESION

The adhesion of polypropylene to most substrates will not normally be as good as that obtained with an equivalent of polyethylene. When polyethylene is processed at 620°F through an air gap of 6-8 in., it will oxidize and the adhesion is enhanced. However, polypropylene cannot be processed at the melt temperature needed for oxidation because of the extremely low viscosity that would result. Therefore, the next best approach is to use the shortest air gap possible so that the molten polypropylene will have the minimum opportunity to cool to better wet the substrate. While liquid primers will improve the adhesion to paper, flame treatment or preheating of the substrate with a hot roll is much easier and should do just as well. Polypropylene formulations are also available that have built-in primers that will enhance the adhesion to paper substrates.

NEW PRODUCTS

Most polypropylene formulas for extrusion coating applications have a flow rate in the range of 10-90 dg/min. This compares to a melt index range of about 3-20 dg/min for polyethylenes.

Polypropylene, as mentioned earlier, does not stick to paper as well as polyethylene. Development efforts have been directed toward solving this adhesion problem. Products are now available that show improved adhesion to printed label stock and to paper. A polypropylene with good adhesion to aluminum foil has recently been developed.

SUMMARY

Polypropylene for extrusion coating applications has been available for about 25 years. Formulations available today can be extrusion coated at speeds in excess of 1000 ft/min. Polypropylene can generally be extrusion coated on equipment designed for polyethylene, but screw and die design can be improved for better performance. Applications for extrusion-grade polypropylene include release sheet, spiral-wound cans, retort pouches, woven bags, and tarpaulins. New product development has resulted in lower flow rate formulations and toward products with improved adhesion.

CHAPTER 30

Extrusion Coating Polyesters

Ray Edwards

ABSTRACT

Polyesters (PET and glycol-modified PETs) are enjoying increasing acceptance as extrusion coating materials. This acceptance is due to their advantage in cost over aluminum foil in applications requiring flavor retention or moderate oxygen barrier properties and their advantages in physical properties (higher melting temperatures) over polyolefins in dual ovenable board applications. The chemical and physical nature of polyesters require processing considerations unlike those required to process a low-density polyethylene (LDPE). This paper will address some characteristics of polyesters and outline the considerations required to successfully process these materials.

Polyesters are characterized by an intrinsic viscosity (IV) that is an indirect measurement of the average molecular weight. To ensure adequate performance, the IV must be protected by adequately predrying the polyester at elevated temperature with desiccated air. Modification to the screw design used to process polyolefins might also be necessary because of significant theological differences.

The IV and the level of crystallinity of the coated polyester determine the performance of the finished product. Higher IV coatings demonstrate increased modulus, elongation, toughness, reduced brittleness, and incidence of pinholes. Increased levels of crystallinity improve the barrier properties of the polyester coating; however, it significantly reduces modulus, toughness, and elongation. Polyesters then are generally characterized by their need for predrying, rheological differences that can require screw design considerations, and physical properties in coated products that differ materially from typical olefin products.

POLYESTERS FOR EXTRUSION COATING

When polyethylene terephthalate (PET) was introduced as an extrusion coating material approximately 10 years ago, the skeptics doubted its potential because of the problems and processing requirements characteristic of PET. Today PET forecasts for use in the extrusion coating of dual ovenable board for 1990 total approximately 15 million pounds. In addition to the strong growth enjoyed by PET in the dual ovenable board market, PET and glycol-modified PET offer considerable potential for growth in other packaging applications.

What advantages do these polyesters offer over other extrusion coating materials available? Heat resistance and grease resistance, two characteristics critical for the demanding ovenable board application, are probably the best known physical properties of PET. Other advantages of PET include a moderate oxygen barrier as well as flavor and vitamin retention superior to that of polyolefins. While glycol-modified PETs do not, in most cases, have the heat resistance of PET, the material formulated for extrusion coating offers similar oxygen barrier and flavor and vitamin retention characteristics. In addition, this glycol-modified PET offers significant improvements in heat sealability.

However, as mentioned earlier, PET and glycol modified PETs do require additional processing considerations above those needed for polyolefins, for example. These include the need for drying with desiccated air before processing, the use of a screw design somewhat modified from that typically used for polyolefin extrusion, and the use of air pins with a minimized air gap to control neck-in. We will further discuss these and other requirements.

But first, let's review the basic chemistry of polyesters since that's the key to their difference. PET and glycol-modified PET can be manufactured by first producing a monomer in a reaction known as esterification.

$$\text{Glycol + Acid} \underset{\text{Hydrolysis}}{\overset{\text{Esterification}}{\rightleftharpoons}} \text{Ester} + H_2O \uparrow \text{removed}$$

A combination of glycols is used in the manufacture of glycol-modified PET. The production of polyester is completed by polymerizing the monomers in one of or a series of polycondensation reactions to build a material of the desired average molecular weight. Water is a by-product of this reversible reaction. It follows logically that esters and polyesters are susceptible to hydrolytic degradation in the presence of moisture at moderate temperature.

Polyesters are characterized by their intrinsic viscosity, which is an indirect measurement of the polymers' average molecular weight. Briefly, inherent viscosity is determined by dissolving 0.5 g of polyester in a solution of phenol in tetrachloroethane at a 60:40 mixture by weight. The resulting mixture is introduced into a modified Wagner viscometer having a 0.5-mL bulb and a 0.55-mm capillary at 25°F. The rate of flow of material through the capillary is determined, and the rate is compared with that of a solvent blank. Intrinsic viscosity is an extrapolation from inherent viscosity. A test method thoroughly describing the entire procedure is available upon request. This method involves the use of toxic solvents, and safety precautions are required.

It's important to minimize IV drop during an extrusion coating operation since toughness and pinholing will directly correlate. Excessive IV drop during processing will make the coated board susceptible to cracking and failure in subsequent scoring and forming operations.

We will review polyester processing from the extruder hopper to the coated board.

Drying the hydrolytically unstable PET is of primary importance for the extrusion coater. The moisture content must be reduced to and maintained at 0.005% by weight during extrusion. This is a standard requirement to which the massive PET bottle industry adheres; it can be attained with any of several desiccant dryers which provide a dew point of -20°F to -40°F at temperatures of 300°F or more. The extruder hopper must be sized to allow a four- or five-hour holdup in a continuous process, and the blower must deliver at least 1 ft^3/min per pound per hour of output. Drying conditions will vary for polyesters other than PET but, in any case, it is important to diligently check and recheck regeneration cycles, air heaters, air filters, and especially dew point temperature.

Screw design is also a major concern in polyester extrusion. Polyethylene was the first material to be extrusion coated, and it is not unusual for a polyethylene screw to be the only one that a processor has available. The massiveness of a typical coating extruder and the difficulty and expense associated with frequent screw changes make it desirable to use a "hybrid" all-purpose screw. We might find future operations dedicated to polyesters but, until then, a compromise might be necessary.

PET differs from polyethylene in that it melts at a higher temperature and over a more narrow temperature range. A square-pitch, single-flight screw suitable for polyethylene can have a deep short feed and extremely long transition (possibly one-half the total length of the screw) followed by a metering zone. An evolution of design has seen a shallower and longer feed and a shorter transition with metering essentially unchanged. This screw is capable of handling several materials in addition to polyethylene.

Barrier screws are in common use, but they can be somewhat more difficult to generalize. In addition to length and depth of feed and metering, we must now consider implications of barrier flight clearance and possibly mixer design for low-density polyethylene versus PET. Again, it is preferable to tend toward shallower and longer feeds for PET. Barrier flight clearance, which can be somewhat tighter for low-density polyethylene, could generate too much frictional heat with PET. We will not attempt to arrive at an optimum design today, but it is fair to say that the increasing importance of PET in the current coating markets will influence future screw designs.

If a polyester is extruded on a polyethylene screw, the two problems to expect are pumping air into the melt and controlling the melt temperature. In each case, the condition is created by the use of a screw not suited for polyesters. Internal screw cooling, higher zone temperatures in the rear of the extruder, and lower screw speeds ordinarily eliminate air in the melt. Excessive melt temperatures, which increase neck-in, edge weave, and breaks in IV can also be controlled with high zone temperatures in the rear of the extruder, reduced screw speeds, and external barrel cooling.

Let's look at comparative extrusion conditions for low-density polyethylene versus PET on laboratory-size extrusion equipment. In this comparison, these conditions will vary for polyesters other than PET.

The equipment used includes a 2-in., 28:1 L/D extruder with a screw having 10 flights of feed at 0.353 in., four flights of transition, and 14 flights metering at 0.106 in. The die has a maximum deckled width of 42 in. with a die opening of 20 mils.

	LDPE	PET
Barrel zone 1, F	400	530
Barrel zone 2, F	480	580
Barrel zone 3, F	560	600
Barrel zone 4, F	620	620
Barrel zone 5, F	620	620
Adapter zones, F	620	620
Melt temperature, °F	640	635
Screw speed, rpm	117	115
Output, lb/h	140	177
Motor load, % of max.	38	56
Melt pressure	1500	450
Air gap, in.	8.5	5.0
Neck-in, in/6dge	1.2	>3.0
Line speed, ft/min	600	400
Barrel zone 3, F	560	600
Barrel zone 4, F	620	620
Barrel zone 5, F	620	620
Adapter zones, F	620	620
Melt temperature, °F	640	635
Screw speed, rpm	117	115
Output, lb/h	140	177
Motor load, % of max.	38	56
Melt pressure	1500	450
Air gap, in.	8.5	5.0
Neck-in, in/edge	1.2	>3.0.
Line speed, ft/min	600	400

Polyesters require higher rear-zone temperatures than polyethylene does. The 485°F melting temperature can also be expected to increase drive motor loads as much as 50% at similar screw speeds. Dies designed for low-density polyethylene work well with PET with openings at 20 mils and die land lengths of 1 in.

The neck-in problem when extrusion coating PET is greater than normally experienced with low-density polyethylene and is controlled by the die gap, the melt temperature, the final IV of the coating, and the air gap. Reduced die openings in the range of 15-25 mils are best suited for PET. A larger die opening results in increased neck-in, which requires reduced line speeds to maintain web control.

Although increases in melt temperature improve adhesion of a polyester to the substrate, higher melt temperatures result in additional neck-in and reduced viscosities, which will yield an extrudate that is more difficult to control. The IV of the polyester coating exhibits a similar trend since lower IVs offer improved adhesion to the substrate but, again, an increase in neck-in is experienced and lower line speeds are required.

A reduction in air gap will minimize neck-in by decreasing the length of the unsupported web and by affording the air pins a good opportunity to do their job. Air pins are tubes that direct jets of air onto the web surface on the bead approximately 1 in. from the edge. This jet of air acts to pin and partially quench the polymer coating to the substrate. As neck-in is controlled, edge weave is eliminated.

Adhesion of PET does not compare favorably with that of polyethylene. Factors which affect adhesion of a polyester to a given substrate include priming, melt temperature, air gap, and the final IV of the product. Flame alone, or in combination with corona treatments, has been found to be effective in improving PET adhesion to paper. Other types of polyesters will also exhibit improved adhesion to films or aluminum foil when similar treatments are used. Higher melt temperatures offer improved wettability with all polyesters; however, it has a negative effect on neck-in. Short air gaps will improve neck-in and maximize adhesion. PET has a crystalline melting point of 485°F and must be coated at melt temperatures of 580°F to 630°F to achieve required processability and adhesion. Progressively hotter melt temperatures will tend to decrease IV with accompanying reductions in toughness and a higher probability of pinholing in the finished coated product.

PET will usually not stick to a chill roll if the roll's surface is maintained below 80°F. A fast quench will ensure minimum crystallinity and thus a more ductile coating. Crystallinity in the finished coating can slightly improve the barrier properties, but it also creates a less ductile coating and can cause cracking in score lines during fabrication. The use of a chilled water system will normally maintain roll temperatures suitable for polyester.

A PET coating exhibits physical properties desirable for many applications. These include an oxygen barrier of 100 cm^3 mil/m^2 24h ATM and a moisture barrier of 4 g/mil/100 $in.^2$/24 h. It has a maximum service temperature of about 450°F. PET makes an excellent coating for ovenable board, where it is seeing major use. Glycol-modified PETs are being considered for packaging applications requiring improved oxygen barrier, heat sealibility, and flavor/vitamin retention. Polyesters are available that can be overcoated onto other polyesters as well as some polyolefins and offer heat sealability to paper and paperboard, PVC, polycarbonate, PET, and other glycol-modified PETS. Polyesters are providing the marketplace with many new packaging opportunities.

In conclusion, it must be emphasized that the successful use of polyesters in extrusion coating will depend on recognizing that they are significantly different from low-density polyethylene. Polyester must be pre-dried with desiccated air, and proper attention must be given to screw design and extrusion conditions. However, this effort should be justified because polyesters have demonstrated a solid growth potential in extrusion- coated packaging applications.

Please note:

Neither Eastman Kodak Co. nor its marketing affiliates shall be responsible for the use of this information or of any product, method, or apparatus mentioned. Users of this information must determine its suitability and completeness for individual use, for the protection of the environment, and for the health and safety of employees and purchasers of products. No warranty is made of the merchantability or fitness of any product, and nothing herein waives any of the seller's conditions of sale.

CHAPTER 31

Ethylene Vinyl Alcohol (EVOH) Resins for Coextrusion Coating

Ronald H. Foster, EVAL Company of America

INTRODUCTION

The advances made in the development of barrier resins and equipment technology over the last ten years have fueled the trend toward coextruded multi-layer structures to replace the traditional poly-paper-aluminum foil-poly laminate. Today, the converter has a choice of barrier polymer resins that can provide protection from the effects of oxygen, solvents, flavor loss, odor, etc. These resins, when combined with various substrates and other functional polymers, provide a multi-layer structure that has definite cost/performance advantages over traditional forms of packaging.

The use of barrier polymers in coextrusion coated multi-layer structures offer the following advantages:

- Elimination of the loss of barrier properties due to "pinholing" or flex cracking of the aluminum foil barrier layer
- Ability to prevent "flavor scalping" by putting the barrier layer in direct contact with the product
- Cost reduction by the ability to use only the amount needed for protection and the elimination of costly second and third production steps
- Ability to provide a clear structure if desired.

Ethylene vinyl alcohol copolymer (EVOH) has fast become the barrier layer of choice for packaging food products, cosmetics, chemicals, etc. Due to its superior gas barrier properties and its resistance to solvents, greases, and flavor loss as well as its excellent processing properties, EVOH has replaced many of the traditional barrier materials

The general properties of EVOH and specific information pertaining to their use in coextrusion coating applications are presented in this chapter.

GENERAL

Even though the first synthesis of EVOH occurred over 50 years ago *(1)*, these resins were not fully commercialized until 1972 and were not used in substantial quantities in the United States until 1984. Since that time, the use of EVOH resins has gained popularity and, in 1995, over 30,000 metric tons (66,000,000 lb) were used in a variety of packaging applications such as bottles, flexible films, tubes, formed containers, and coated substrates.

EVOH copolymers are highly crystalline in nature and their properties are highly dependent upon the relative concentration of the comonomers. Figure 1 shows the relationship between comonomer concentration and moisture/gas barrier properties.

Fig. 1. Comonomer concentration vs. barrier properties

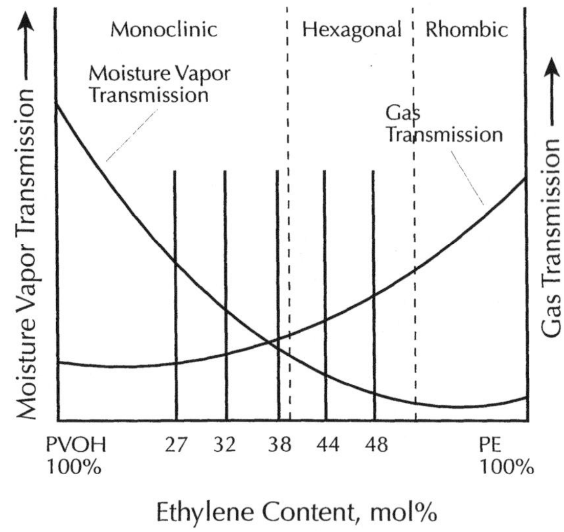

By combining the exceptionally high barrier properties of polyvinyl alcohol (PVOH) with the processability of polyethylene, EVOH resins offer both superior barrier properties and ease of processing.

EVOH resins, unlike polyolefins, are characterized by the mol% ethylene in the copolymer. Typical commercial grades of EVOH range from 27 mol% to 48 mol% ethylene.

Figure 2 shows the general relationship between ethylene content and resin properties.

Fig. 2. Commercial grades of EVOH

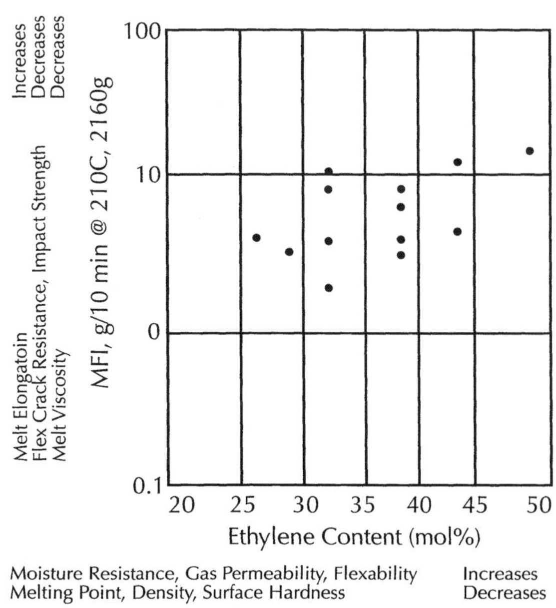

Because of the presence of a hydroxyl group in the polymer chain, the barrier properties of EVOH resins are moisture sensitive. As the moisture content of the EVOH layer increases, the barrier properties decrease (Table 1).

Table 1. Oxygen permeability (cc 25 μ/m2 24 hr atm at 20°C)

Ethylene Content (mol%)	65% RH	85% RH	100% RH
27	0.15	0.94	18.91
38	0.54	1.73	10.23
44	1.17	2.60	7.90

This effect is much more pronounced with gases, due to the small molecule size, than it is with larger molecule substances such as solvents, greases, flavors, etc.

Through the proper use of other resins such as polyethylene and polypropylene in conjunction with EVOH, the relative humidity (moisture content) of the EVOH layer can be controlled. For most packaging applications, these techniques allow the relative humidity of the EVOH layer to be controlled between 65% and 85%.

MANUFACTURING PROCESS

EVOH copolymers, as the name implies, are copolymers of ethylene and vinyl alcohol.

Ethylene-Vinyl Alcohol Copolymer

$$\text{---}(CH_2 \text{ -- } CH_2)_m \text{ ---- } (CH_2 \text{ - } \underset{\displaystyle OH}{\underset{|}{CH}})_n \text{ ---}$$

EVOH is produced by fully hydrolyzing an ethylene vinyl acetate random copolymer.

There are four basic steps in the production of EVOH. These are:

- Polymerization
- Saponification
- Coagulation
- Drying.

Figure 3 is a flow diagram of the EVOH manufacturing process.

Fig. 3. Flow diagram of EVOH manufacturing process

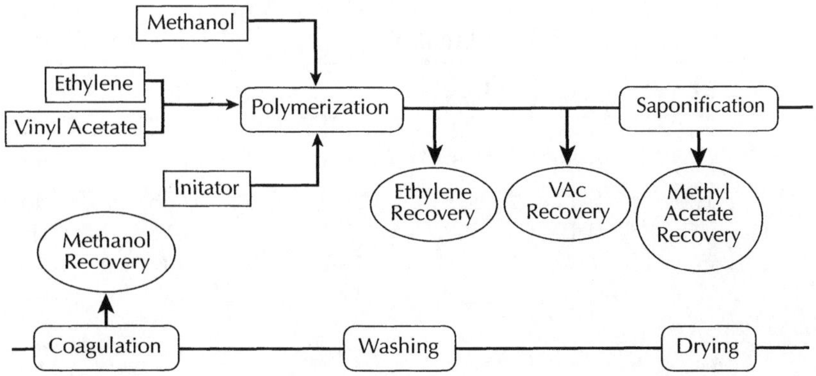

Two raw materials, ethylene and vinyl acetate, are used to produce EVOH. Methanol is used as a solvent.

In the polymerization section, ethylene and vinyl acetate (in a methanol solution) are copolymerized to ethylene vinyl acetate using an initiator. Unreacted ethylene, vinyl acetate, and methanol are separated from the ethylene vinyl acetate solution and sent to the purification section for recovery and recycling.

In the saponification section, the ethylene vinyl acetate is hydrolyzed with caustic soda to convert it to ethylene vinyl alcohol copolymer. Methyl acetate is formed as a by-product of the reaction and is separated and sent to the purification section together with methanol. The methyl acetate is converted to acetic acid for sale as a by-product.

The ethylene vinyl alcohol is then coagulated and formed into solid pellets. Impurities are washed out, and the pellets are dried to a low moisture level before packaging.

PROPERTIES

Polymer

EVOH resins are commercially available in a broad range of properties which allows the converter the choice of selecting the best resin for his particular process. Table 2 shows the broad range of properties available.

Table 2. EVOH properties

Property	EVOH-27	EVOH-32	EVOH-38	EVOH-44	EVOH-48
Melt index, g/10 min at 210 °C	3.9	1.9-10.5	3.8-6.0	3.5-13.0	14.7
Density, g/cm^3	1.20	1.19	1.17	1.14	1.12
Melting point, °C (°F)	191 (376)	183 (361)	175 (347)	165 (329)	158 (320)
Crystallization temperature, °C	167	161	151	142	136
Glass transition temperature, °C	72	69	62	55	49

Table 3. Gas transmission rate

	cc 25μ/m² 24 h atm				cc mil/100 in.² 24 h atm			
	O_2	N_2	CO_2	He	O_2	N_2	CO_2	He
EVOH-32	0.31	0.03	1.0	291.2	0.02	0.002	0.06	18.8
EVOH-38	0.39	0.08	2.3	341.3	0.03	0.005	0.15	22.0
EVOH-44	1.2	0.12	3.5	391.5	0.08	0.008	0.23	25.3
PVDC	2.3	0.19	17.4	423.2	0.15	0.01	1.1	27.3
Oriented nylon	55.2	21.7	205.2	3596	3.6	1.4	13.2	232.0
Oriented PET	71.3	7.1	303.8	2790	4.6	0.46	19.6	180.0

23°C (73°F), 65% RH

For coextrusion coating applications, the most common type of EVOH used is EVOH-32 (melt index 10.5) and EVOH-44 (melt index 13.0).

Gas barrier

EVOH resins are best known for their outstanding barrier to the permeation of gases such as oxygen, carbon dioxide, nitrogen, and helium. The use of EVOH copolymers in a packaging structure enhances flavor and quality retention by preventing oxygen from penetrating the package. For packages utilizing Modified Atmosphere Packaging (MAP) technology, EVOH resins effectively retain the carbon dioxide, oxygen, or nitrogen used to enhance the product.

Table 3 compares the gas transmission properties of EVOH copolymers with other commonly used packaging materials.

Hydrocarbon and solvent resistance

EVOH resins have a very high resistance to hydrocarbons and organic solvents. The weight percent increase in EVOH resins after immersion for one year at 20oC is 0% for solvents such as cyclohexane, xylene, petroleum ether, benzene, and acetone, 2.3% for ethanol and 12.2% for methanol.

Table 4 shows the performance of protective gloves containing an EVOH layer and Table 5 reflects the resistance of EVOH to various fuels.

Table 4. Protective glove performance

Breakthrough Time in Minutes (1,440 minutes = 24 hours)

Chemical/Solvent	Time	Chemical/Solvent	Time
Acetone	1,440	Carbon Disulfide	1,440
MEK	1,440	Sodium Hydroxide	1,440
Acetonitrile	1,440	Ethyl acetate	1,440
Methylene chloride	1,440	Sulfuric acid 93%	1,440
Benzene	1,440	N hexane	1,440
Nitrobenzene	1,440	Toluene	1,440

Gloves: 2.7 mils thick (LLDPE/Tie/EVOH/Tie/LLDPE)-ASTM F739

Table 5. Resistance to fuels

Fuel/Component	Permeation (g 25µ/m2 24 h atm at 40°C)			
	EVOH-32	EVOH-27	Nylon	HDPE
REF Fuel C	0.014	0.012	0.232	3,224
CM15	9.45	7.13	77.5	2,604
MTBE15	0.011	0.009	0.186	2,573
Toulene	0.016-	--	--	-4,665
Iso octane	0.046	--	--	372

REF Fuel C: 100% hydrocarbon
CM15: REF Fuel C + 15% methanol
MTBE15: REF Fuel C + 15% MTBE

These properties make EVOH resins an excellent choice for the packaging of oily foods, edible oils, agriculture pesticides, and organic solvents.

Flavor, fragrance, and odor protection

The food industry is becoming increasingly aware of the importance of flavor barriers in containing natural flavors and preventing extraneous flavors from contaminating the product. Some food products are notorious for their ability to absorb extraneous flavors or aromas. Another important flavor consideration is commonly called "flavor scalping." This phenomenon is where some plastics, such as polyethylene, selectively absorb certain flavor constituents from the product.

EVOH resins provide excellent protection to the absorption of d-Limonine, the primary flavor constituent in citrus juices, as shown in Fig. 4. Where food products are concerned, aroma barrier is somewhat analogous to flavor barrier. Food products must be protected from outside aromas in the distribution chain, grocery store, and home. Other food and non-food products such as garlic, agricultural chemicals, pesticides, insecticides, or perfumes can be highly aromatic. In this case, the desire is to retain the aroma in the package and not let it escape into the surrounding environment.

Fig. 4. d-Limonene sorption

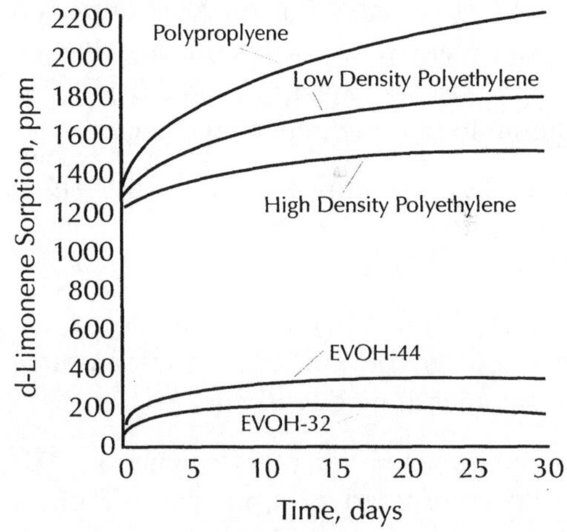

As shown in Table 6, EVOH can provide protection to aroma migration.

Table 6. Aroma permeability measured by gas chromatography

Film Type	(g/m² 24 h 100 ppm at 23°C, 75% RH)				
	Allyl Sulfide	Acetic Acid	Ethyl Pheny Acetate	Beta Pinene	Ethyl Acetate
HDPE/Nylon/EVA (30µ)	<0.00008	0.92	0.00099	<0.001	0.03
HDPE/EVOH/EVA (30µ)	<0.00075	0.035	>0.002	>0.001	0.004
PP/PET/PP (25µ)	------------	-------	>0.01	>0.09	>0.005
PVDC Ct'd OPP (25µ)	<0.0068	1.98	<0.0099	<0.09	0.34

Table 7. Typical properties of EVOH resins

Property	Unit	EVOH Resin Grades				
Ethylene content	mol%	27	32	38	44	48
Melt index at 210°C	g/10 min	3.9	3.8	3.8	13	15
Tensile strength, break	kg/cm^2	730	730	520	470	380
Tensile strength, yield	kg/cm^2	960	790	660	600	440
Elongation, break	%	200	230	280	380	330
Elongation, yield	%	6	8	6	7	3
Young's modulus	kg/cm^2 x 10^6	32	27	24	21	20
Rockwell hardness	M	104	100	93	88	-
Tabor abrasion	mg	-	1.2	1.2	2.0	2.2
Electric resistance						
P_s (surface)	Ω x 10^{15}	-	4.7	-	1.2	-
Dimensional stability, 140°C, 1 h						
MD	%	-	-2.7	-6.9	-1.6	-
TD	%	-	-0.9	1.4	-1.2	-

Mechanical properties

Although not intended to use as structural resins, EVOH resins have high mechanical strength, elasticity, surface hardness and excellent weatherability. EVOH is highly antistatic in nature, preventing the buildup of dust and/or static charges. These resins have a very high gloss and low haze, resulting in outstanding clarity characteristics.

Table 7 lists typical properties for EVOH.

Processability

EVOH resins are thermoplastic polymers easily processed on conventional fabrication equipment without special modifications.

Depending upon the requirements, EVOH resins can be coextruded with all types of polyolefins, nylon, polyvinyl chloride, and polyester. Downstream processing such as thermoforming, vacuum forming, and printing is easily accomplished with EVOH resin-containing structures.

Regulatory approvals

Depending upon the resin grade, EVOH resins have U.S. FDA approval for direct food contact as specified in the Code of Federal Regulations, Title 21, Section 177.1360. EVOH resins can also be used in high-temperature laminates (retortable applications), providing a functional barrier is used to separate the EVOH layer from the food product.

EVOH resins are considered non-toxic according to the U.S.P. Class VI test protocol.

APPLICATIONS

Throughout the world today, the primary use for EVOH resins is food packaging. Multi-layer structures containing EVOH provide a cost-effective alternative to traditional forms of packaging such as glass, metal, or aluminum foil. In addition, they also offer other advantages such as convenience of use and source reduction of packaging materials.

The use of EVOH for coextrusion coating found its start in the manufacture of packages for juice products. It has evolved into the coating of various substrates for both food and non-food products. Traditionally a coated/laminated poly-paper-aluminum foil-poly structure has been used for packaging such products as:

- Juices
- Milk
- Snack foods
- Dry mixes.

In addition to the coating/laminating process being a rather expensive process, converters faced another problem. Aluminum foil has been increasing in cost and decreasing in availability. Aluminum foil also has the inherent disadvantages of "pinholing" and flex cracking. For these reasons, converters actively sought an alternative to the poly-paper-aluminum foil-poly structure.

Dry mixes

A coextrusion coated structure such as shown in Fig. 5 has replaced the paper-aluminum foil laminate for many dry products such as:

- Marinade
- Spices
- Drink mixes.

This structure not only offers the economics of a one-step method of conversion, but also permits product inspection for metal contamination and excellent machinability.

Fig. 5. Dry mix structure

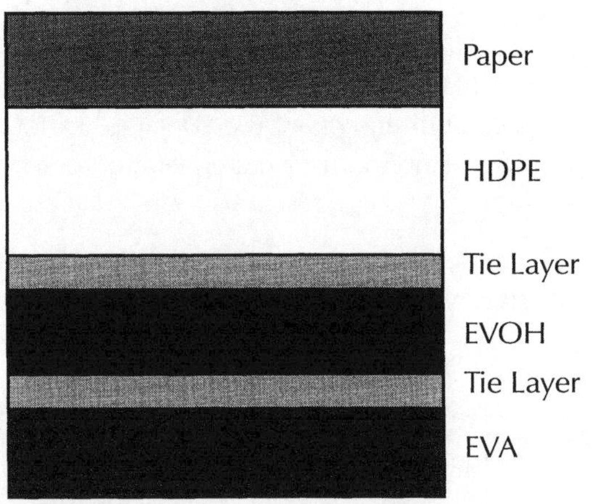

Liquid packaging

Probably the most spectacular growth in coextruded liquid packaging materials has occurred in the use of cartons. Most of this growth has been with aseptically filled containers such as the "Brik Pack." However, many products such as fruit juices, vegetable juices, and milk are packaged in "gable top" containers ranging in size from 12 oz to 96 oz. With the advances made in aseptically packaged particulate products, items such as soups and sauces are now being seen in both the single-serve and larger containers.

Gable top containers

Gable top barrier containers for packaging fruit juices, fruit drinks, and other juices are available utilizing non-aluminum foil structures. Typical structures such as shown in Figs. 6 and 7 utilize only 0.2-0.4 mils of EVOH. These containers offer not only an economic advantage over aluminum foil barrier containers, but also increased shelf life. Also by putting EVOH in direct contact with the product (Fig. 7), "flavor scalping" can be eliminated.

Fig. 6. Non-foil barrier structure for juices

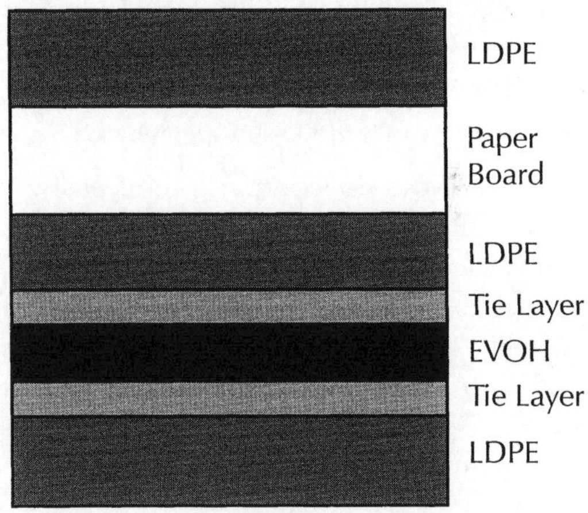

Fig. 7. Anti-flavor scalping barrier structure for juices

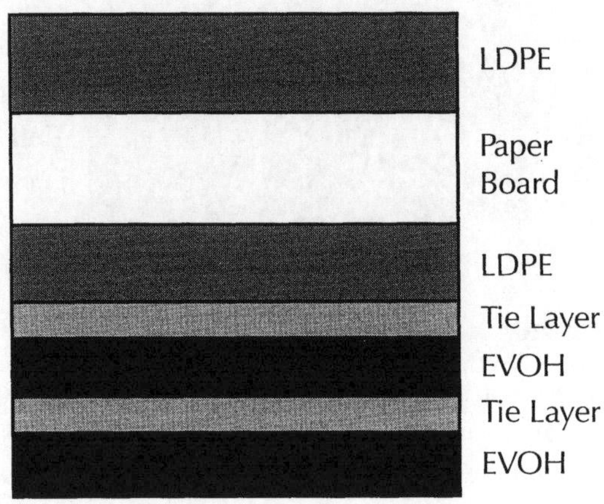

Aseptic single-serve container

One advantage that aluminum foil has in the aseptic single-serve container is its use as a sealing medium. Most aseptic packaging lines use induction sealing which, in turn, requires the use of a aluminum foil layer. Equipment manufacturers are presently producing sealing systems that would not require the aluminum foil layer for sealing.

Pouches

Bulk products such as those used for the hotel, restaurant and institutional (HRI) market have traditionally been packaged in No. 10 metal cans. These cans are not only heavy and expensive, but are also difficult to handle and dispose of. Plastic pouches in sizes from 32 oz to 110 oz are rapidly replacing this cumbersome package.

Many of these pouches are produced by straight blown film coextrusion or by laminating three films such as biaxially oriented nylon (BON), EVOH, and linear low density polyethylene (LLDPE) together to form the structure. An alternative to these methods would be to coextrusion coat the EVOH and LLDPE to a BON film substrate as shown in Fig. 8.

Fig. 8. HRI pouch structure

Other food products

Many other food products such as snacks, baked goods, candy, etc. are packaged in a range of laminated structures. Materials such as metallized films, biaxially oriented polypropylene, aluminum foil, and others are used to provide the necessary properties. With the advances made in coextrusion technology, it can be said that "most structures produced by lamination can be produced by coextrusion coating" and most likely with less expense. Typical structures might include:

- OPP - Tie layer - EVOH - Tie layer - LDPE - EVA (baked goods)
- OPET - Tie layer - EVOH - Tie layer - EVA (coffee)
- PVDC Ct'd BON - Tie layer - EVOH - Tie layer - LLDPE (lid stock).

Non-food products

In Europe, Japan, and Latin America, the use of small sachets or pouches for the packaging of a variety of products is commonplace. These products include:

- Medical supplies
- Liquid soap
- Shampoo
- Dry and liquid chemicals.

These packages use a variety of substrates such as paper, PET, OPP, or nylon. Depending upon the barrier requirements of the package, various barrier, and/or non-barrier materials are coated on the substrate.

Other non-food packaging utilizing coextrusion coating with EVOH includes:

- Photographic base paper
- Release paper
- Protective sheets.

As can be noted from the above discussion, the use of EVOH in a coextrusion coated structure requires the use of an adhesive layer (tie layer) to gain adequate bonding strength to the other polymers or substrates. Commercially available adhesive resins such as Admer® (Mitsui Petrochemicals), Tymor® (Morton Chemical), Bynel® (Dupont), or Plexar® (Quantum Chemicals) are suitable for use with EVOH resins.

PROCESSING CONSIDERATIONS

EVOH resins are thermoplastic polymers. As with other thermoplastics, they are subject to thermal degradation. EVOH resins are also inherently hydrophilic. When processing EVOH resins, care

Table 8. Minimum and maximum extrusion temperatures, °C (°F)

EVOH Grade	Melting Point	Minimum	Maximum
EVOH-27	191 (376)	200 (392)	240 (465)
EVOH-32	183 (361)	190 (374)	240 (465)
EVOH-38	175 (347)	180 (356)	240 (465)
EVOH-44	165 (329)	170 (338)	250 (480)
EVOH-48	158 (320)	165 (329)	250 (480)

should be taken to ensure the resin is kept dry and that the proper extrusion conditions and equipment design is used.

Moisture absorption

EVOH resins are hygroscopic and will absorb moisture when exposed to the atmosphere. The amount of moisture that will be absorbed and the speed of absorption will depend upon the temperature and relative humidity of the environment. Depending upon the type of fabrication process being used, an increase in the moisture content of EVOH resins can cause processing difficulties. Foaming, voids, and gels can occur at increased moisture levels. EVOH resins are shipped in moisture proof containers. Under normal humidity conditions, a container can be left opened while being used for periods up to 24 hours. If high humidity conditions exist and the container is left open for prolonged periods of time, moisture can be absorbed.

If moisture is absorbed, the resin can be dried in a desiccant-type, circulating air dryer at temperatures of 90-110°C (195-230°F) for two to three hours. Caution should be taken against over-drying EVOH resins to prevent thermal degradation.

Thermal degradation

EVOH resins are not overly sensitive to thermal degradation; however, like other polymers, they are subject to thermal breakdown under extreme conditions. As thermal degradation is both temperature- and time-dependent, care should be taken to minimize residence time and maintain the proper temperature in the extrusion system.

Extrusion temperatures

Table 8 indicates the minimum and maximum extrusion temperatures for various grades of EVOH resins.

Coextrusion with polymers such as polyesters, nylons, polycarbonate, or polypropylene can cause the EVOH resin melt to come in contact with melt streams that are hotter than the recommended maximum temperature. This contact is usually for a short period of time and adverse effects are not encountered.

Dies

EVOH resins are suitable for use in either multi-manifold or feedblock-type die designs. Special die designs are not necessary, although care should be taken to streamline the flow channels as much as possible. Dies that minimize the amount of wetted surface area, minimize residence time, and have no dead spots are the most suitable for EVOH resins.

Melt path

EVOH resins have an affinity to metal surfaces. If the EVOH extrusion system consists of concave or convex parts, acute angles, etc., EVOH resins can easily become restricted in these dead spot areas. Also, in a passage of fixed diameter, retention can occur on the wall if the diameter is too large for the flow. Retained resin exposed to heat for a prolonged period of time will degrade.

Equipment

Extruder

The size of the extruder should be determined by the amount of EVOH desired in the package structure. Do not oversize the extruder, as this causes the speed to be reduced and increases residence time.

Table 9. Changeover sequence

Resin Before Change	Resin After Change	Sequence
LLDPE, LDPE, PP	EVOH resin	Direct
EVOH resin	LLDPE, LDPE, PP	Direct
Nylon	EVOH resin	Nylon to LDPE to EVOH resin
EVOH resin	Nylon	EVOH resin to LDPE to nylon

The use of a "disco" purge procedure is recommended.

Screw

A transition zone of 20-25% of the total length in a 24-32 L/D screw is recommended to minimize gel formation and provide a homogeneous melt. The feed zone should be 30-40% of the total length, and the metering zone should be 35-50%.

The desired compression ratio for EVOH resins is 3.0-4.0. A larger compression ratio can increase shearing of the resin and raise melt temperatures.

Dies

Either a feedblock system or a multi-manifold die can be used for coextrusion coating with EVOH. If the processing temperatures and viscosities of the various resins used are not significantly different, the feedblock system is the most suitable. A multi-manifold die does a very good job of stabilizing the flow when the viscosities of the resins are quite different.

The air gap is typically 80-150 mm. Due to the use of extrudable adhesives in EVOH coextrusion, surface oxidation is not necessary to promote adhesion. Therefore, the air gap should be as small as possible to prevent excessive neck-in.

Extruder shutdown

The following procedure is recommended when the extrusion operation is stopped:

Shutdown Time	Procedure
Up to 30 minutes	Maintain temperature settings, screw rotation can be stopped.
Up to 3 hours	Maintain temperature settings, operate screw slowly.
More than 3 hours	Change to another polymer, preferably LDPE, and follow shutdown procedures in accordance with commercial practices.

Polymer changeover

Table 9 indicates the recommended purge sequences to be taken for changeovers involving EVOH resin.

SAFETY

EVOH resins are not hazardous as defined by the U.S. Occupational Safety and Health Administration (OSHA) under its Hazard Communication Standard (HCS), 29 C.F.R. 1910.1200.

EVOH is listed in the Toxic Substance Control Act (TSCA) inventory of chemical substances as ethylene-vinyl-acetate-vinyl alcohol copolymer. The CAS RN Number is 26221-27-2.

EVOH is not considered hazardous pursuant to Title III of SARA and is not considered subject to the annual reporting requirements specified by Section 312 and 313 of Title III of SARA and 40 CFR, Part 372.

GENERAL REFERENCES

1. Roland, J.R. Jr., U.S. Pat. 2386347 (1945).
2. Salame, M., Coatings and Plastics Preprints of the 167th Meeting of the American Chemical Society, 34(1):515 (1974), ACS, Div. of Organic Coatings and Plastics Chemistry.

3. Ikari, K., "Oxygen Barrier Properties and Applications of Kuraray EVAL Resins," Proceedings of the 1982 Annual International Conference on Coextrusion Markets and Technology.

4. Sonsino, S., Packaging Design (1990).

5. Satas, D., Coating Technology Handbook 43(399):405 (1991).

6. Robertson, G., Food Packaging (1993).

CHAPTER 32

Color and Additives for Extrusion Coating

Joseph Turick

ABSTRACT

Color and additive concentrates used in extrusion coating and lamination are exposed to the most demanding environment of all polyolefin extrusion processes. This chapter discusses the various types of organic and inorganic colorants available for extrusion coating. Examination of the properties of these materials attempt to explain how factors such as heat stability, dispersion, and FDA concerns are used in formulating for a particular application. The information also reviews flame retardants, ultraviolet stabilizers, and other additive systems.

This chapter provides an understanding of why certain problems are experienced in the extrusion of colorants at 316°C and how they can be resolved or minimized.

INTRODUCTION

While many plastic products are produced clear or uncolored, a considerable amount is also colored or modified by additives. As in other types of plastic processes, extrusion coating and lamination uses colorants and additives to provide different colors, opacity, surface modification, and specific types of performance for a multitude of end uses. No doubt, many individuals in the industry who purchase and use these materials lack a basic knowledge of what the products are, how they work, and what specific concerns there should be when buying or using them.

This chapter reviews the types of colorants and additives used in extrusion coating/lamination, the basics of how they perform, and concerns associated with their use. Pigments and additives are chemicals in powder form and, in most instances, are unusable in extrusion coating processes. While these raw materials are blended or compounded together in a variety of systems, it is pelletized concentrates that are most widely used for extrusion coating. This chapter also discusses these concentrates.

COLOR

Color is described as a psycho-physiological phenomenon. It is the interaction of radiant energy (light source) on an object and an observer. Light falling on an object is modified and a portion of the visible spectrum is reflected to the eye, sending a signal to the brain. This information, provided the observer has normal color vision, is interpreted subjectively in most cases and objectively only when the observer has color experience. It is not necessary to be a color expert when using colorants, but it can prove quite beneficial to possess at least some basic knowledge of the subject when discussing color concentrate requirements or problems with a supplier.

Think of color as being three-dimensional. First, there is hue, most commonly called color-red, blue, green, etc. Second, chroma indicates the intensity of the color. Is it a weak color versus a strong color? The stronger color will yield more hiding of the substrate than a weaker color, if usage level and coating weight are equal. Third, lightness indicates the degree of light reflected and depends generally on the amount of white and black pigments contained in a coating.

Light falling on an object is absorbed, scattered, transmitted, and reflected. This is shown in Fig. 1 and explains why objects have color and opacity. Light entering an object scatters in all directions and, in addition to absorption, provides opacity. No scattering to little scattering is observed as transparency or translucency. As scattering is increased, the object becomes more opaque to the point where no light is transmitted. The portion of light that is not absorbed and reflected from the surface to the eye yields the color we observe.

Fig. 1. Light travel in a pigmented film

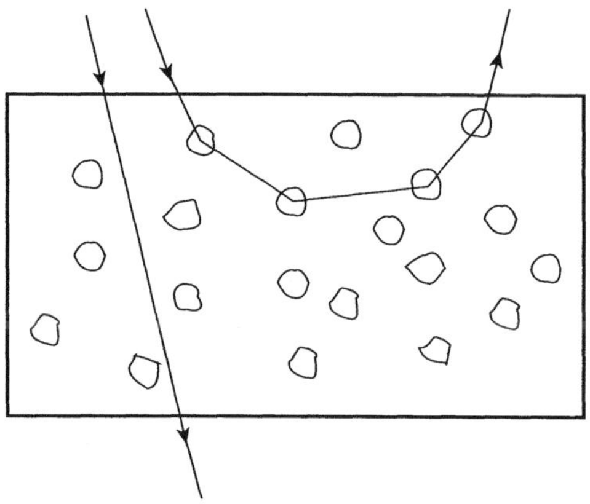

We are able to obtain the color we require for a material by using one or more colorants. Colorants are chemical compounds that absorb and reflect certain wavelengths of light. A colorant absorbing light in the blue-green portion of the spectrum and reflecting in the red would be seen as a red color (Fig. 2). Black, absorbing all light and reflecting little, has a very low reflective value (Fig. 3). White, reflecting most light, will have overall very high values (Fig. 4).

Fig. 2. Red reflectance curve

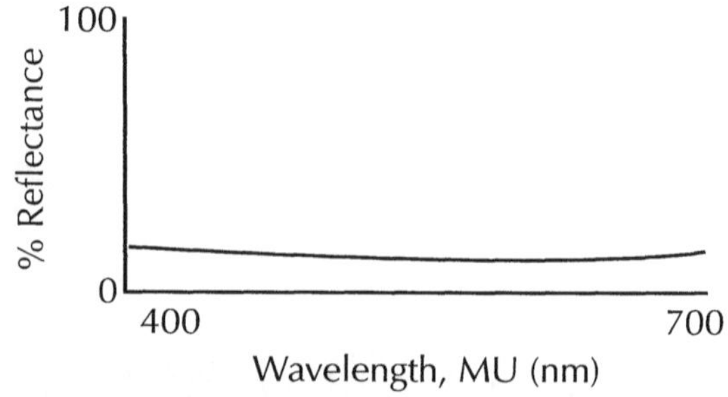

Fig. 3. Black reflectance curve

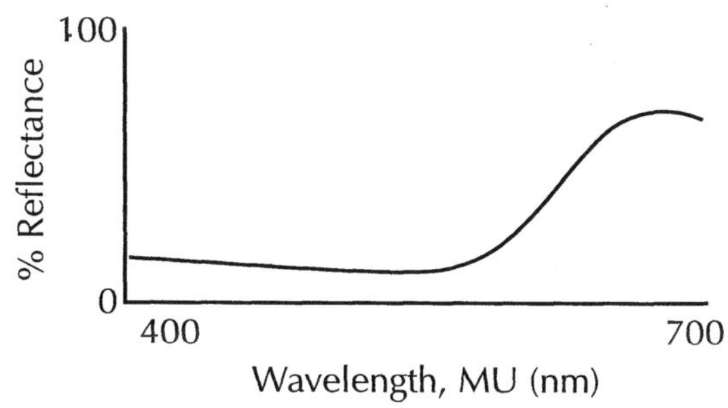

Fig. 4. White reflectance curve

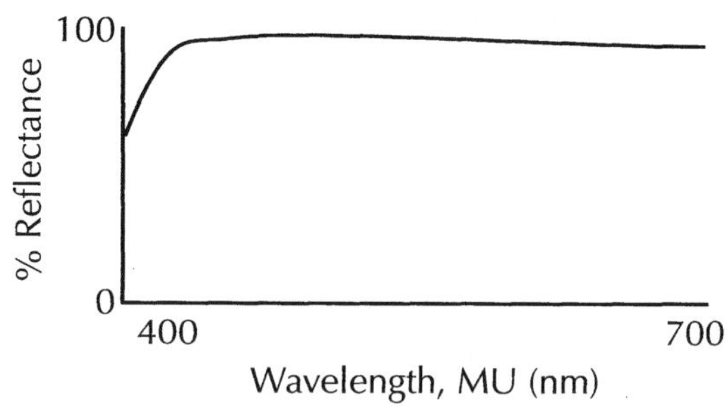

Reflectance values can be measured with a spectrophotometer. These measurements and other types of data make it possible to develop numerical values to calculate color differences. The use of instrumentation can provide a mathematical description of a color and allows measurement of a color, be it a match or a production color, against a target (desired color) or a standard.

Colorants are available as dyes or pigments. Dyes are organic compounds that are soluble in a resin. Since there are no discrete particles to refract and scatter light, dyes do not provide opacity. Used in clear resins such as general-purpose polystyrene and polycarbonate, intense transparent colors are obtained. If opacity is needed, an opacifier such as titanium dioxide is added. Because of their solubility, with few exceptions, dyes cannot be used in polyolefins since they will migrate or bleed to the resin surface where they will rub off. Consequently, for extrusion coating of polyolefins, we are only interested in pigments.

Pigments are either organic or inorganic. A variety of pigment properties such as color, cost, heat stability, strength, light-fastness, FDA concerns, chemical reactivity, etc., must be considered by a colorist when formulating for a specific end user. There are scores of pigment families and types from which a colorist can choose, but the selection process is simplified considerably in the extrusion coating field. Processing temperature of 316°C and excellent dispersion requirements eliminate the majority of pigments available in a color matcher's palette. What remains is shown in Table 1.

Table 1. Colorants for extrusion coating

Inorganic	Organic
Titanium dioxide	Carbon black
Iron oxides	Pthalocyanine blue/green
Cadmium	Quinacridone red
Ultramarine blue	Vat pigments
Aluminum flake	Diazo pigments
Pearlescents	

I have only provided the basics on color theory and colorants, hopefully providing some insight as to the cause-and-effect relationships in the manufacture of color coatings. Anyone wishing to acquire a more in-depth knowledge on the subject should find a number of excellent books available.

CONCENTRATES

Now that we know something about pigments and how they perform, how do we obtain them? Colors for extrusion coating are purchased in two forms. The type least used is a compound. A compound is a product formulated with pigments and a base resin with a particular melt index, density, and processing properties suitable for a specific application. With knowledge of the coating weight to be used on a given substrate, it is designed to be used as is, without reduction with natural resin.

The second and most widely used form is a concentrate that typically contains 50% pigment. It is in a resin carrier and in pellet form. The base resin can be low-density polyethylene (LDPE), high-density polyethylene (HDPE), polypropylene (PP), or a copolymer. Selection of the appropriate resin is based on the resin type and melt index with which it will be mixed. A concentrate is designed to be let down or mixed with a resin to yield a required color and opacity. Letdown ratios for extrusion coating generally range from 3:1 (33%) to 10:1 (10%).

Properly formulated and manufactured, color concentrates should incorporate well into the letdown resin and have good to excellent pigment dispersion. Incorporation is the mixing of the concentrate and the resin and ideally provides uniform color and opacity. This is achieved by selecting a base resin that is compatible and has a melt index and density suited for the letdown resin or resins. For example, an LDPE-based concentrate, with melt index (MI) of 9, can be used in resins with an MI less than 9 and in other resins such as HDPE and some copolymers. Using such a concentrate in a 12-MI resin will probably yield a nonuniform coating with streaks. It is possible for the extruder operator to obtain some improvement in such a situation by making temperature adjustments in the barrel profile or increasing back pressure to effect more mixing. Back pressure can be increased by valving down if a valve exists or changing to heavier screen packs. Either of these actions, however, will increase head pressure and reduce output. There will also be instances of poor incorporation due to poor mixing. This is attributable to older equipment with worn screws and barrels or a screw that is not of the best design for the task at hand.

Formulations for an acid copolymer such as ethyl acrylic acid (EAA) must consider the percent comonomer. In many extrusion situations, an LDPE concentrate incorporates well into a low-acid (3.5%) resin. Generally, with a higher-acid resin (6-9%), a copolymer-based concentrate shows improved compatibility, giving better incorporation. It is less prone to promote gel formation and has less negative impact on the intrinsic adhesion values of the copolymer.

DISPERSION

Discrete pigment particles exposed to heat, moisture, and packing have a definite tendency to lump together and form agglomerates. Agglomerates, in an extreme case, can be larger than the thickness of a coating. This results in a sandpaper finish. Even smaller agglomerates are visible to the eye as specks or gels and exhibit a poor surface appearance. Since the letdown concentrate has not maximized the pigment surface area of discrete particles, optimum opacity and/or color will not be achieved. Finally, screen pack life will be shortened due to rapid buildup of pigment on the screen. Good dispersion can only be achieved by the concentrate manufacturer selecting the best pigments available and compounding with high shear to break up the agglomerates. There is nothing you can do if you have received a batch of concentrate with poor pigment dispersion except to return it to the supplier.

ADDITIVES

In a sense, even pigments are considered additives that provide opacity and color. Generally, when we refer to additives, we think of a material that modifies a resin's physical properties, aids processing, or adds a type of performance not characteristic to the resin. There are a variety of additives available for use in plastics, and Table 2 lists those commonly used in extrusion coating.

Table 2. Additives

Slip additives
Anti-block
Ultraviolet stabilizers
Flame retardants
Chill roll release agents

SLIP AND ANTI-BLOCK

Slip additives typically are fatty amides added to resins to reduce the coefficient of friction (CoF). CoF is the sliding friction between a film surface and another surface or itself. A fatty amide is incompatible with a resin. After extrusion, it exudes to the film or coating surface. Proper CoF values are especially important to converting and packaging operations where the polyolefin coating must slide over forming mandrels.

Wound rolls and stacked coated board have a tendency to stick to or block adjacent surfaces. Anti-block additives are added to the resin to provide a minutely roughened surface that lessens the blocking tendencies. Anti-block additives are generally some form of silica.

Slip and anti-block concentrates are commonly used in the blown- and cast-film industry but, to a relatively minor degree, also in extrusion coating. Generally, extrusion coating resins are available with the required levels of additives. You should be aware that, if additional levels of slip and anti-block are required, they can be added via a concentrate.

ULTRAVIOLET STABILIZERS

Solar radiation in the ultraviolet (UV) region of 290-400 manometers (nm) is absorbed by plastics and is responsible for the destructive process of photodegradation. This degradation is observed by discoloration, embrittlement, and loss of physical properties.

It is actually chemical groups such as carbonyl and peroxides called chromophores that absorb the UV energy. Chromophores are formed by thermal oxidation during extrusion and by impurities which include catalyst residuals, processing contaminants, and additives in the resin.

The absorption of UV energy leads to broken chemical bonds generating free radicals. The radicals react with other polymer molecules forming hydroperoxides and more free radicals. The entire cycle continues and results in polymer degradation unless halted.

The degradation process can be reduced considerably, extending outdoor life and performance for plastic materials and coatings. Applications include shipping sacks, lumber wrap, tarpaulins, and materials used in construction and agriculture. Protection is obtained by using UV stabilizers of one or more types and perhaps an anti-oxidant.

Table 3. Types of UV stabilizers

ABSORBERS-Compete with chromophores by preferentially absorbing UV radiation and converting it into harmless low level heat or fluorescence.

Type	- Benzophenone
Advantages	- More effective in thick cross sections than thin sections - Low cost - FDA approved (polyolefins) - Good compatibility
Characteristics	- Light yellow color

QUENCHERS-Take over the UV energy absorbed by chromophores and dissipate heat.

Type	- Nickel phenolate complex
Advantage	- Effective in natural and pigmented systems - Has antioxidant activity - Better than absorbers for use in thin cross sections
Characteristics	- Green in color - Also acts as radical scavenger and peroxide decomposer

RADICAL SCAVENGER-Reacts with free radicals, thus interfering with the degradation process.

Type	- Hindered amine light stabilizers (HALS)
Advantages	- More effective in natural and pigmented systems than other types of UV stabilizers at lower usage level - Low volatility - No color.

*NOTE: Hindered amines are more reactive to the presence of other additives such as sulfur-type antioxidants, halogenated flame retardants, and certain pigments. Exposure to agricultural chemicals, which include pesticides and fumigants, can adversely affect HALS

ANTISTATS

Unfilled plastics are nonconductive materials and are susceptible to the accumulation of electrostatic charges. These charges are generally caused by friction during the extrusion process or through contact with other materials during handling. Static charges can cause problems in high-speed packaging operations. Cling or the attraction of a surface to itself or other surfaces can cause wrinkles or hang-ups. Employees exposed to electrostatic discharge can experience a shock, ranging from barely noticeable to quite severe. Sparks can cause explosions in areas containing volatile gases and damage sensitive electronic components.

Antistat agents are chemicals that are used with plastics. They fall into two types: external and internal. An external antistat is a solution in water or a solvent that is topically applied by dipping or spraying. Aside from specific handling procedures, the disadvantage with such an application is that the antistat can be wiped or washed off leaving the plastic unprotected. An internal antistat migrates to the plastic surface and if wiped replenishes itself on the surface.

The layer of antistat on the polymer surface is hygroscopic and attracts moisture from the atmosphere to form a conductive path that minimizes the accumulation of a static charge. It is this mechanism that makes antistatic performance so dependent on the relative humidity of the environment.

The amount of antistat used in a given application depends on a variety of factors which include resin type, thickness, FDA concerns, and, most important, the required degree of performance. Performance parameters are static decay time at a particular percent, relative humidity, and surface resistivity. These values are obtained by instrumental testing.

FLAME RETARDANTS

A variety of extrusion products such as tarpaulins and building and packaging materials are potential candidates for flame-retardant additives. Flame-retardant additives do not prevent a plastic from burning. Presence of flame-retardant additives inhibit ignition and flame spread, and provide self-extinguishment if a flame and kindling temperature is removed or reduced.

There are many chemicals available for the formulations, but only a few can be used for extrusion coating. Extrusion temperatures of 316°C and requirements for excellent dispersibility eliminate many additives that are used quite successfully in injection molded products.

Flame-retardant concentrates for polyolefin film are typically synergistic combinations of antimony trioxide and a halogen. Chlorinated paraffin is used extensively in LDPE film but, since it is temperature-restricted to about 205°C, it is not used for extrusion coating. Here the halogen is an aromatic bromine that can withstand 316°C.

The amount of a flame-retardant concentrate needed is dependent upon the degree of performance required and cross section part thickness. Extrusion coatings of relatively thin cross sections require heavy loadings of flame-retardant additives. A commonly used letdown ratio would be 4:1 to 6:1.

CHILL-ROLL RELEASE AGENTS

A chill-roll release agent is usually required when coating with a mirror-surface chill roll. This additive is typically available in the coating resin selected and reduces sticking of the coating surface to the smooth metal surface. Pronounced sticking causes separation of the bond between the polymer coating and the substrate.

Chill-roll release concentrates are available when performance of the resin is inadequate. These additives can also be included in the color concentrates used.

MORE ON OPACITY

The amount of opacity required for a given application is typically considered critical. Generally, either the substrate is not adequately covered or unnecessary cost is incurred by using too much concentrate. Comparative evaluations of competitive concentrates, such as a 50% loaded white concentrate, have evoked comments such as one being more opaque than the other or vice versa.

This chapter already discussed the importance of pigment dispersion and concentrate incorporation for optimum opacity. If we assume two concentrates are of like quality, then opacity will be equal. If one concentrate is superior in dispersion quality, most likely it will provide more opacity at the same usage level. More often the answer to an observed difference in opacity is a mechanical rather than a physical one.

Concentrated pellets are available in different configurations: strand cut, underwater, hot die face, etc. They can be cylindrical, round, flat, or cubed and of varying size. A volumetric feeder, metering in a strand-cut pelletized concentrate at a particular setting, will require a new setting if a change is made to a concentrate with a different pellet configuration or size. While the specific density of the two concentrates are probably equal, the bulk density will most likely differ and require feeder adjustments to obtain equal amounts of pigment or additives in the final product.

CONCLUSION

There are many factors to consider when formulating a color or additive concentrate. A concentrate that performs well for one job or resin might be inappropriate for another. When contemplating a new application or resin change, talk to the concentrate supplier. It might only take a few minutes to discuss it, but it could save hours of scrap generation and downtime.

CHAPTER 33

Purging Extruders with Functionalized Polyolefins

Maged G. Botros, Equistar Chemicals, LP

ABSTRACT

This chapter discusses the use of functionalized polyolefins such as Plexar PX909 (produced by Equistar Chemicals, LP) as purge products for extruders. This functionalized LDPE-based compound will be referred to as PE-F. Special emphasis will be directed toward purging extruders of polar residues such as residual polyamides, ethylene vinyl alcohol copolymers (EVOH), oxidized polyethylenes, or color concentrates adhered to the internal parts of the extruder. Several characterization techniques were used to compare the purging efficiency of functionalized polyolefins to other commercial purge products and to low-density polyethylene (LDPE). These techniques include Fourier Transform Infrared spectroscopy (FTIR), INTEC gel analysis technique, and optical microscopy. The FTIR technique is used to determine quantitatively the amount of residue (nylon or EVOH) purged versus time and the total nylon concentration removed at the end of purging. The INTEC technique is used to determine the amount and the size of gel particles in a LDPE film extruded after cleaning the extruder. The optical microscopy measurements determine the compatibility of the purge compound with nylon or EVOH remaining in the extruder. It has been concluded that purge products made of functionalized LDPE are at least twice as efficient in purging nylon or EVOH from extruders compared to LDPE and other commercial purging compounds. The high purging efficiency of functionalized polyolefins-based purge products is attributed to their ability to chemically react with residual polar compounds remaining in the extruder.

INTRODUCTION

The importance of regularly purging an extruder to remove residual resins is well known. Residual material subject to heat and shear degrades over time and contaminates fresh polymer passing through the extruder. As a result, gels and other imperfections could appear in the finished product. Periodic cleaning by mechanical means is most frequently employed by wire brushing or scraping. This is a very tedious and time-consuming process and can result in unnecessary abrasion or damage to processing equipment. On the other hand, purging residual resins from the internal parts of the processing equipment with suitable purge compound is more convenient. By properly selecting the purge product, this operation is often very effective and rapid. However, purging the extruders of barrier resins such as nylon or EVOH or other polar compounds often requires hours of downtime when using conventional purging materials. In contrast, new purge products based on functionalized PE have been found to reduce purging time by two-thirds or more and clean the extruder of these contaminant residues extremely well.

These new products were designed to maximize end product quality and to minimize maintenance downtime. Since purging time is shorter, less purging compound is required; productive machine time is increased due to less downtime, and less product has to be scrapped because of gels and other imperfections. Savings in money as well as time are thus achieved.

SOME COMMERCIAL CLEANING/PURGING TECHNIQUES

U.S. patent number 4,056,403 discloses a solvent composition comprises a halogenated hydrocarbon and a monohydric alcohol capable of cleaning polyurethane foam-generating equipment.

U.S. patent number 4,226,817 describes the use of an acrylonitrile polymer composition containing ethylene carbonate and optional water as an effective purge material for startup and shutdown of melt spinning operations employing hydrated acrylonitrile polymer melts.

Japanese patent number 91-313489/43 discloses a cleaning resin composition for hollow moldings comprising polyolefin resin, PMMA, natural salt of sulfonic acid and magnesium carbonate, magnesium hydroxide, aluminum hydroxide, etc. It is claimed that the discoloration of the weld line of the discharged parison by a preceding colored resin can be effectively removed.

Japanese patent number 91-344311/47 describes a detergent for resin molding machines, comprising copolymer of methyl methacrylate and acrylate ester. The detergent is claimed to effectively remove fouled resin on the inner wall of heat cylinder for a resin molding machine.

Japanese patent number 92134A151 describes a plastic composition for cleaning extruders. The cleaning compound contains inorganic powder such as silica sand, calcium carbonate, magnesium carbonate, or slaked lime of particle sizes about 100 μm.

Japanese patent number 87-254555/36 covers a composition for a compound suitable for purging thermoplastic resins of extrusion or injection molding machines. The purge compound is composed of a mixture of aromatic polyester such as terephthalic acid or isophthalic acid and a compound containing water of crystallization such as calcium chloride, barium chloride, magnesium sulfate, etc. The aromatic polyester used has a higher melt viscosity than the resin to be purged.

U.S. patent number 4,670,329 describes a method and a composite article for cleaning molds. A moldable compound of the type which has the inherent capability of cleaning contaminating stains and residues in molds is deposited on a carrier sheet to form an uncured mat for placement in a mold to be cleaned. When placed in a mold and subjected to a molding operation, the moldable compound coating of the uncured mat will melt and flow into conformity with the interior configuration of the mold and cure in that conforming configuration. The cured mat is then capable of being ejected in as a reinforced unitary structure from the mold.

U.S. patent number 4,808,355 describes a method for changing the resin in an injection molding machine using a nozzle touch plate.

EXPERIMENTAL

Materials

Some characteristics of the resins and purge compounds used in these tests are given in Table 1. The purge compounds A, B, and C are commonly used for cleaning extruders when running polyolefins or polyolefin-based colorants. They are polyolefin-based purge products filled with inorganic scrubbing agents and other fillers. Purge compound D is a LDPE resin.

Table 1. Characteristics of resins and purge compounds

Designation	MI (g/10 min.)	Density (g/cc)
LDPE	1.00	0.918
PE-F	1.20	0.920
Purge A	5.00	Polyolefin-based
Purge B	2.00	Polyolefin-based
Purge C	2.00	Polyolefin- based
Purge D	0.90	0.919

EVOH MFI = 1.5 at 210°C (32% ethylene)

Fourier transform infrared spectroscopy (FTIR)

The FTIR spectrometer (Nicolet 60SX), purged with dry nitrogen to eliminate water vapor and equipped with a DTGS detector, was used to record the infrared spectra. Film samples collected during purging were scanned for nylon-6 bands in the IR spectrometer. Typically 100 scans were signal averaged at 4 cm^{-1} nominal resolution with double precision. Band absorption, area, and position measurements were obtained using standard Nicolet software.

INTEC Model 4000 gel analyzer

A sample of 450 ft of LDPE film of 2.5-mil thickness was collected using the Killion extruder each time after purging. A 300-ft length was scanned for total and differential gel counts using this technique. Gel sizes were categorized as follows:

V: Very small (1-3 mil)

S: Small (3-5 mil)

M: Medium (5-7 mil)

L: Large (7-10 mil)

MICROSCOPY

An Olympus SZH stereoscope was used to obtain low magnification pictures of the resin collected at the onset of purging nylon or EVOH from the Killion extruder using several purge formulations. It was also used to examine the final LDPE films extruded after purging the extruder of EVOH.

Ash test

In this test, a 2-g sample of the extrudate was collected during purging color concentrates. This sample was then burned at 800°C in a muffle furnace. The amount of ash remaining after 5 minutes was determined and calculated as percent ash in the sample.

TEST METHODS

To determine the effectiveness of PE-F in purging polar residues from extruders, the following procedure was used:

1. A 1-in.-diameter Killion extruder and an 8-in.-wide flat film die was purged for 45 minutes with low density polyethylene (LDPE).
2. LDPE film was then produced, using this extruder, and collected as a control.
3. Polar compounds were run on the Killion extruder for 30 minutes.
4. PE-F were then run through the Killion extruder for 30 minutes.
5. Samples of the extrudate were collected for Fourier transform infrared spectroscopy (FTIR) and ash tests.
6. LDPE films were again produced using the same die and extruder. The number of gels and contaminants were determined with the INTEC gel analyzer and optical microscopy.
7. This procedure was repeated to compare the efficiency of functionalized LDPE-based purge products to that of other purging materials.

RESULTS AND DISCUSSION

The efficiency of a purge compound is usually based on the mechanism(s) by which it operates. Several mechanisms could be considered when designing a purge compound, e.g., flushing, scouring, penetration, and chemical reactions. Other less efficient purge compounds depend only on one or two of the above mentioned mechanisms and usually were not as efficient in cleaning extruders. The flushing mechanism is the simplest and the least efficient mechanism. It involves running a low-cost resin, e.g., LDPE through the extruder to replace the previously used resin. Scouring is based upon the addition of abrasive additives to base resins. This factor increases the purging efficiency as compared to the flushing effect. The penetration effect is defined as the ability of the purge compound to penetrate between metal and residues to encourage release and to reach the dead spots in the processing equipment. The chemical effect, on the other hand, appears to be one of the most important mechanisms in purging. Its main function is to react with the polar polymeric residues remaining in the extruder. This interaction increases the compatibility of the non-polar olefinic purge compounds and the polar contaminants. This compatibility factor increases the interfacial thickness and therefore enhances polymer intermixing. Various authors *(1-11)* have studied such systems to overcome the incompatibility of non-polar/polar resins. Functionalized polyolefin-based purge compounds can have a high purge effeciency by reacting with polar resins such as nylon, EVOH, oxidized polyethylenes, and color concentrates during purging.

Purging polyamides and EVOH

Several analytical tools including FTIR, INTEC gel analysis, and optical microscopy were used to determine the efficiency of LDPE-F compared to other commercial purge compounds. The purge efficiency is judged by the following:

a. The rate at which the purge compound can clean the extruder
b. The amount of nylon or EVOH residue remaining in the extruder after constant purge time
c. The extent of reaction observed between a purge compound and nylon/EVOH
d. The gel content in an LDPE film extruded after cleaning the extruder with a purge compound.

In the first test to determine the effectiveness of several purging compounds, the test method described earlier was used on an extruder which has been extruding nylon-6. The results of the gel analysis on the LDPE film produced after the purge attempts are shown in Fig. 1. Film produced after purging with LDPE-F exhibits the lowest percentage total gel compared to the films produced after purging with the other products. It is shown that functionalized LDPE-based purge products lead all the other materials in efficiency of removing gels; after 30 minutes of purging, it has only 43% of the gels found with the best of the other purging compounds, Purge D. Figure 2, on the other hand, illustrates the effect of functionality on how rapidly nylon residues can be removed from the Killion extruder. Even when the length of time for purging with LDPE was doubled (from 30 min to 60 min), the LDPE could not match the results of purging for 30 minutes with PE-F. The total gel level in a LDPE film produced after 60 minutes of purging with LDPE was 60% greater than that of a similar film produced after 30 minutes of purging with PE-F. In other words, better purging was obtained in half the time with functionalized LDPE-based purge products.

Fig. 1. % total gel in LDPE film produced after purging with various compounds

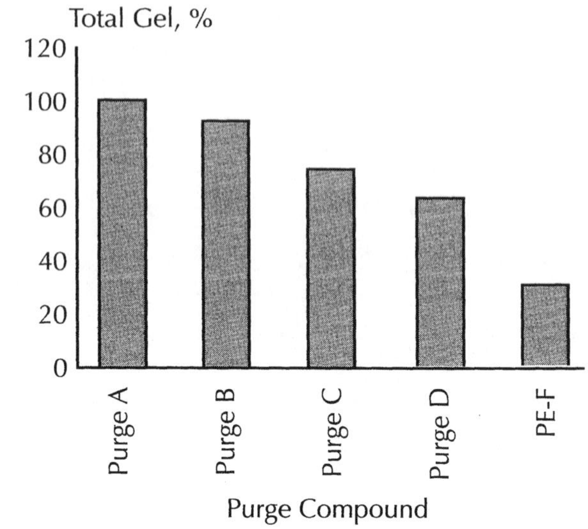

Fig. 2. % total gel in LDPE film produced after 30 and 60 minutes purging of nylon by LDPE and a 30-minute purge by PE-F

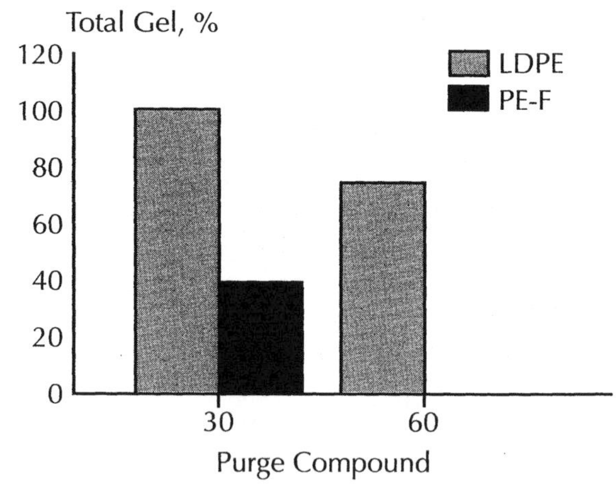

The amount of nylon purged can be determined by FTIR, which measures the precise concentration of various components in a sample. For example, Fig. 3 gives the IR spectrum of nylon-6 with the four bands which were used throughout this study. The band assignments are as follows *(12, 13)*:

NH Stretching Modes:

Secondary amides (bonded NH) (trans-): 3300 cm^{-1}

Secondary amides (bonded NH) (cis- and trans-): 3085 cm^{-1}

CO Absorption:

Secondary amides [Amide I]: 1640 cm^{-1}

NH_2 Deformation:

Secondary, non-cyclic amides only [Amide II]: 1550 cm^{-1}

Fig. 3. FTIR spectrum of nylon-6

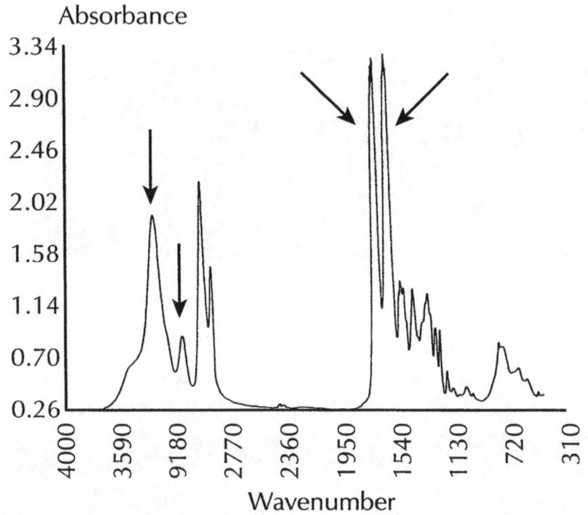

Figure 4 shows the IR spectrum of LDPE and nylon-6. This sample was taken from the extrudate collected after purging nylon from the Killion extruder with LDPE as a reference purge compound. It is clear from this figure that the previously assigned bands are well resolved and can be used for the analysis. As previously stated, samples of the purging compound contaminated with nylon were collected out of the extruder for FTIR analysis. Nylon concentration in the extrudate was calculated using the identified spectral bands. With such data, it was possible to calculate that the PE-F type purge products have removed twice as much nylon from the extruder as LDPE in 30 minutes. The extruder screw was then removed and the nylon remaining was scraped off and measured, showing that twice as much nylon remained in the extruder after purging with LDPE as compared to PE-F.

Fig. 4. FTIR spectrum of LDPE + nylon-6

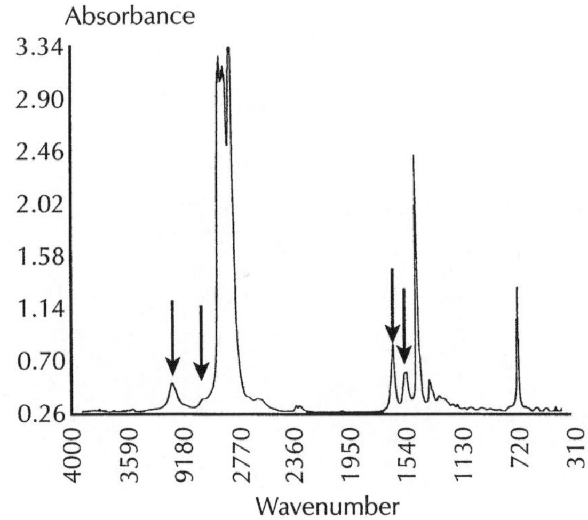

The purging efficiency of PE-F products can be seen visually using optical microscopy. The micrographs in Figs. 5 and 6 are of the extrudate collected while nylon was purged with LDPE or PE-F. As shown in Fig. 5, LDPE is not compatible with nylon. In the absence of any interaction, LDPE is unable to remove residual nylon from the extruder. Figure 6 shows almost complete homogenization of nylon and PE-F.

Fig. 5. Optical micrograph of poorly mixed nylon and LDPE

Fig. 6. Optical micrograph of the through mixing of nylon and PE-F

The effectiveness of functionalized LDPE-based purge products in terms of the quality of the film produced after purging nylon was determined by measuring the total gel levels using the INTEC equipment. Film produced after nylon was purged out of the extruder with PE-F has 60% less gels than the film produced after purging the extruder with LDPE.

INTEC gel analysis technique was also used to determine the efficiency of PE-F in purging EVOH from the extruder. Data shows that the LDPE film produced after purging an extruder running EVOH with PE-F type product had nearly 70% fewer gels than film produced after the extruder was purged with LDPE.

Optical micrographs taken of LDPE film produced after the extruder was run with EVOH and then purged show the results even more dramatically. In Fig. 7, the bright white line is EVOH on top of the final LDPE film product made after the extruder was presumed to be purged of EVOH by LDPE. Figure 8 shows the final LDPE film product after an extruder was run with EVOH, then purged with LDPE-F and then run with LDPE to produce the film. No EVOH is apparent in the final product.

Fig. 7. Optical micrograph of LDPE film produced after purging the extruder of EVOH with LDPE. The white line is EVOH extruded onto the film.

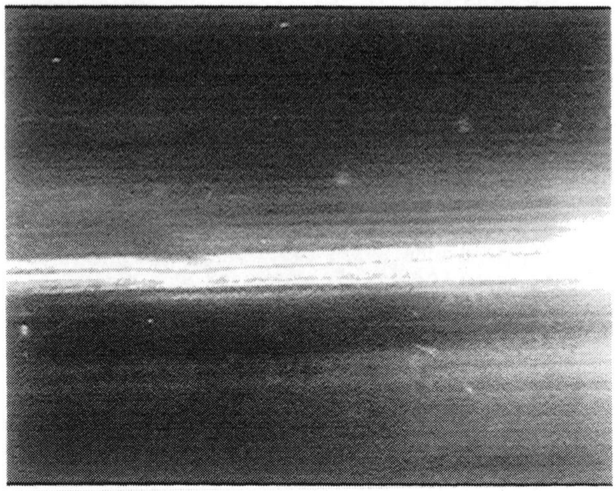

Fig. 8. Optical micrograph of LDPE film produced after purging the extruder of EVOH using PE-F. No EVOH is visibly apparent.

Cleaning the extruders of degraded polymers

Functionalized LDPE-based purge products (PE-F) have also proven to be efficient in cleaning extruders of degraded polymers. A blown film line was contaminated with oxidized high-density polyethylene. Several commercial purge products were used to clean the extruder with no apparent improvement. The line was then purged with PE-F for 30 minutes with the result that over 70% of the gel particles were removed. Figure 9 shows the % total gel in a LDPE film produced before and after cleaning the extruder with PE-F and compared to historical control values.

Fig. 9. % total gel in LDPE film produced before and after cleaning the extruder with PE-F

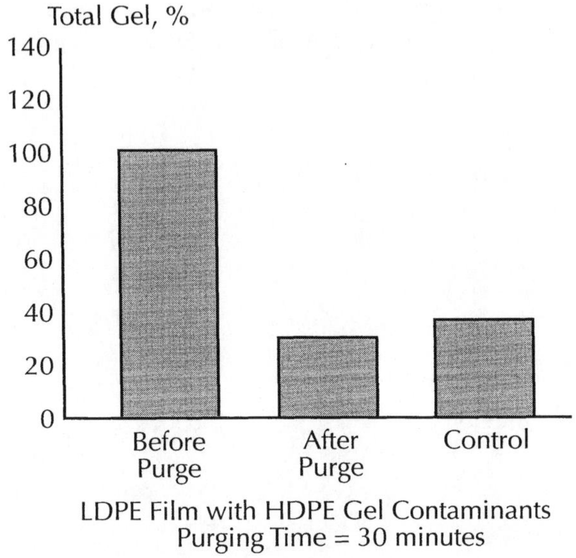

Purging color concentrates

Extruders contaminated with color concentrates were hard to clean using conventional methods. This is because colors are usually polar compounds that tend to stick to the internal parts of the extruders. To determine the efficiency of purging extruders of color concentrates using PE-F type compounds, the following experiment was designed. A white concentrate (50% by weight TiO_2 in LDPE) was run through the Killion extruder. PE-F as well as other commercial purge compounds including LDPE were used to clean the extruder. The purge efficiency was determined using the ash test. In this test, the ash obtained from the extrudate collected during the first 15 minutes of purging was determined. This was then compared to the amount of ash obtained from the color concentrate. Figure 10 shows that PE-F product had removed over 80% of the residual TiO_2 color concentrate compared to 20% for Purge Compound B and less than 10% for the LDPE purge. This again is attributed to its ability to chemically interact with polar residues.

Fig. 10. Purging white concentrate with PE-F and other purge compounds

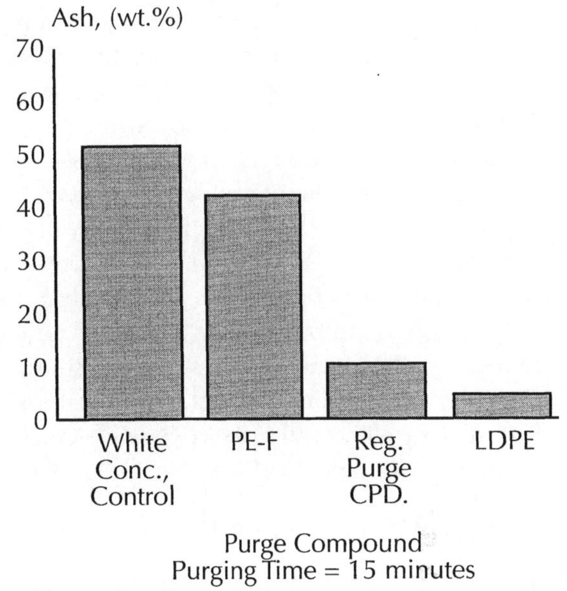

PROCESSING FUNCTIONALIZED LDPE-BASED PURGE PRODUCTS:

The following procedures are recommended for processing PE-F purge products:

1. Machine operating and temperature setting should remain unchanged when PE-F is added to the hopper.
2. For best results, purge over a cycled range of rpms.
3. It is recommended to purge the line until no gels or imperfections appear. The time required is dependent upon the severity of contamination.
4. For blown film dies, allow PE-F to "pancake" on top of the die orifice if a bubble cannot be blown because of film imperfections. For blow molding and flat film dies, allow the material to drool from the extruder.
5. Functionalized LDPE-based purge products should not remain in the system when the extruder is shut down. Use LDPE to remove PE-F from the extruder before shutting it down.
6. When a line is restarted, the resin scheduled for processing can be used to flush out any residual LDPE.

CONCLUSIONS

Several analytical techniques including FTIR, INTEC gel analysis technique, and optical microscopy were used to evaluate the effectiveness of purging compounds. Tests on extruder output showed that a functionalized PE-based purge product (Plexar PX909) was at least twice as efficient as four commercial purging compounds in purging polar residues such as EVOH and polyamide barrier resins, as well as oxidized polyolefins and color concentrates. The high purging efficiency was attributed to its ability to chemically react with polar residues remaining in the extruder. Laboratory and field experience show that this product will usually provide adequate purge of nylon and EVOH within 15-30 minutes.

ACKNOWLEDGMENT

The author would like to thank Paul Jackson at Equistar Chemicals, LP for helpful discussions.

REFERENCES

1. Modic, M.J. and Pottick, L.A., SPE 1992 ANTEC Conference Proceedings, Society of Plastics Engineering, p.1427.
2. Saito, *et al.,* U.S. Pat. 4,429,076.
3. Shiraki, *et al.,* U.S. Pat. 4,628,072.
4. Shiraki, *et al.,* U.S. Pat. 4,657, 971.
5. Gelles, R., Modic, M., Kirkpatrick, J., "Modification of Engineering Thermoplastics with Functionalized Styrenic Block Copolymers," SPE 1988 ANTEC Conference Proceedings, Society of Plastics Engineering, p. 513.
6. Gilmore, D. and Modic, M., "Engineering Polymer Blends with Styrenic Block Copolymers," SPE 1990 ANTEC Conference Proceedings, Society of Plastics Engineering, p. 1228.
7. Benabdallah, S.H., 1992 ANTEC Conference Proceedings, Society of Plastics Engineering, p. 1431.
8. Van Gheluwe, Favis, P., Chalifoux, B., J. Mat. Sci. 23(11):3910 (1988).
9. Kamal, M.R., Hozhabr, S., Arghyris, L., Lambla, M., SPE 1992 ANTEC Conference Proceedings, Society of Plastics Engineering, p. 2677.
10. Setua, D.K., Lim, S., White, J.L., SPE 1992 ANTEC Conference Proceedings, Society of Plastics Engineering, p. 2686.
11. Bellamy, L.J., The Infra-red Spectra of Complex Molecules, John Wiley & Sons, New York, 1958.

CHAPTER 34

Processing and Converting Tips for Cast Film And Extrusion Coating

Thomas Bezigian, PLC Technologies

ABSTRACT

Extrusion is the fundamental polymer processing technique used today to convert raw plastic resin into a usable finished shape. The basic laws of extrusion physics are well known and have been well documented. However, despite this knowledge of basic physics and chemistry, a certain skill that can only be learned over time is required to operate a manufacturing line efficiently and productively. This chapter reviews some of the finer points of extrusion processing that have been learned through years of experience and relates those tips to the cast film and extrusion coating processes.

INTRODUCTION

Extrusion processors must be concerned with many variables including substrates, plastic resins, material handling, chemical primers, corona and flame treaters, extrusion variables, web handling, and, of course, finished product attributes. The nature of the work involved in plastics engineering spans many disciplines including chemistry, chemical engineering, mechanical engineering, industrial engineering, physics, and more. The skills therefore required to produce a good quality product are quite diversified and take years of trial and error to master.

This paper will provide insight into the following areas:

1. The effect of temperature profile on adhesion
2. Output variations due to temperature profile
3. Output variations due to additive addition
4. A comparison of priming methods
5. Laminating vs. coextrusion
6. Die streaks
7. Slitter dust.

THE EFFECT OF TEMPERATURE PROFILE ON ADHESION

Quite often in extrusion coating, the most important product property is adhesion of the coating to the substrate. Adhesion is the state in which two materials join together and do not separate, and is the sum of chemical and physical adhesion. Chemical adhesion is a result of the type of polymer, oxidation of the polymer melt in the air gap, and treatment of the substrate surface, either by chemical priming, corona, or flame treatment. Physical adhesion is a result of the viscosity of the polymer melt, the porosity of the substrate, and nip pressure used in the laminator.

Increasing temperature has two beneficial effects. First, oxidation of the melt curtain in the air gap increases with increasing melt temperature. Second, melt viscosity is an exponential function of temperature, thus higher temperature increases penetration of the melt into the substrate, thereby improving adhesion.

A comment on the effect of temperature profile on adhesion must be made. How the melt temperature is achieved is as important as the actual melt temperature itself. For products that are melt temperature sensitive, such as flexible packaging products that require good adhesion but low odor, melt temperature can be problematic. Though methods exist to run low melt temperatures without sacrificing adhesion, not all converters have the required equipment (ozone generators) and thus must find alternative solutions.

At temperatures high enough to yield good adhesion, odor problems in the finished package result in off-quality product. At lower temperatures, odor is reduced to an acceptable level, but adhesion is reduced.

There is a little known fact that at a given melt temperature, say 320°C (608°F), higher temperatures in Zone 2 of the extruder give better adhesion to the substrate, regardless of the type of substrate. The question here really is: Why does this happen?

As mentioned above, melt temperature does two things to the polymer being extruded. First, it determines the viscosity of the polymer, which in turn determines how much the molten plastic will "bite" into a porous substrate such as paper or nonwoven materials. The more fluid the plastic is, i.e., the lower the viscosity or the higher the melt temperature is, the more the pressure roll forces the plastic into the porous matrix of the substrate. Secondly, the higher the melt temperature is, the more oxidized the surface of the polymer is when it reaches the substrate at the nip point. This is important because adhesion is both a physical and chemical phenomena. The oxidized sites on the polymer surface are like magnets that seek out oxidized sites on the surface of the substrate. The more magnets, so to speak, there are, the better the adhesion is to any substrate, porous or nonporous. In this way, high melt temperatures act as a pseudo-chemical priming, improving adhesion. Chemical or physical bonding alone will result in inferior adhesion to any given substrate.

Many, if not all, converters assume that if the last zone of the extruder, the adapter, piping, and die are set at the desired melt temperature, good adhesion will result. By and large this is true, but in special circumstances, such as this one, it is not. The extruder takes solid plastic and, through friction between the plastic and the barrel, melts the polymer, and then pumps it out the die. Much of the energy absorption and melting occurs in the transition section of the screw. This is the portion of the screw where the root diameter increases to the final metering depth.

This section of the screw, usually located at Zone 2 on the temperature control panel, is where most of the melting and chemical changes occur to the plastic during extrusion. Pressure at this stage of extrusion can easily exceed 35 MPa (5000 psi) and temperature in the screw is more than 210°C (400°F). When the pellets of plastic enter the extruder, there is air (oxygen, nitrogen, etc.) on the surface of the pellet. The presence of oxygen under these extreme temperature and pressure conditions causes chemical changes in the polymer, which we affectionately call "oxidation." If these conditions are too severe, the changes we see are gels and decomposed polymer. If they are appropriate, we observe good melt quality, adhesion, and the like. If less than optimal, we can observe poor adhesion.

In setting up the process, one must observe where the zone temperatures stabilize. This is where the process "wants" to be, i.e., how the screw was designed to operate. The problem can be solved by maintaining the melt temperature where desired, but varying the Zone 2 temperature to achieve the optimum balance of low odor and good adhesion.

OUTPUT FLUCTUATIONS DUE TO INCORRECT TEMPERATURE PROFILE

A very common problem in the extrusion industry is improper setting of the rear zone temperature. This particular process setting is extremely critical in determining output stability. It can be said that the primary driving force in the extrusion process is the coefficient of friction (CoF) between the plastic and metal within the extruder *(1)*. When the CoF becomes too small, output will also drop off accordingly. Though much information has been written on this subject, it might be helpful to examine once again the basic physics behind this phenomenon.

Let's begin with two extreme examples. First, let's assume that we are extruding a plastic that has a CoF of zero (the minimum value) to the barrel and one (the maximum value) to the screw. In this case, as the screw rotates, the plastic will slip on the barrel and stick to the screw. When this happens, the extruder output is zero because, as we said, the plastic is sticking to the screw, not generating any pressure, and thus not moving forward.

Next, let's assume that the situation is reversed. In this case, the CoF between the plastic and barrel is one, and between the plastic and screw is zero. Now the plastic slips on the screw and sticks to the barrel. When the screw flight comes around, it wipes the plastic that is stuck on the barrel and pushes it down the screw channel immediately. In this situation, the output is at a maximum, though residence time and melt quality are at a minimum.

These two extremes show us a lot about how the extruder works. In actual practice, the CoF values are generally between zero and one, and we do not experience either condition. If we explore the reasons for varying CoF, we would develop a list as shown in Table 1.

Table 1. Causes of CoF variation in an extruder

1. Surface finish of the barrel and screw
2. Polymer type
3. Screw speed
4. Rear zone temperature
5. Additives

Because of historical reasons, limited drive motor torque, etc., high rear zone temperatures are frequently used. High temperatures mean low viscosity, which in practical (not theoretical) terms has the same effect as low CoF. Experimental data show maximum "CoF" to occur at about 10-20°C above the DSC melting point of the polymer *(2)*. Though not cast in stone and unchangeable, I recommend that Zone 1 be set at 20-30°C higher than the DSC melting point for optimal feeding, maximum output, reduced surging, and good gauge control.

OUTPUT FLUCTUATIONS DUE TO IMPROPER ADDITIVE ADDITION

The use of certain additives, particularly oils or low melting waxes, has a similar effect on the extrusion process as does excessive rear zone temperature, and is detrimental to the stability of the extrusion process.

High rear zone temperatures reduce the viscosity of the polymer melt in the extruder because of the logarithmic function of viscosity versus temperature of the non-Newtonian polymer melt. Similarly, the incorporation of lubricating or tackifying additives into the extruder feed throat reduces the viscosity of the blend in a logarithmic fashion. Figure 1 below exemplifies this phenomena.

Fig. 1. Viscosity of a two-component system

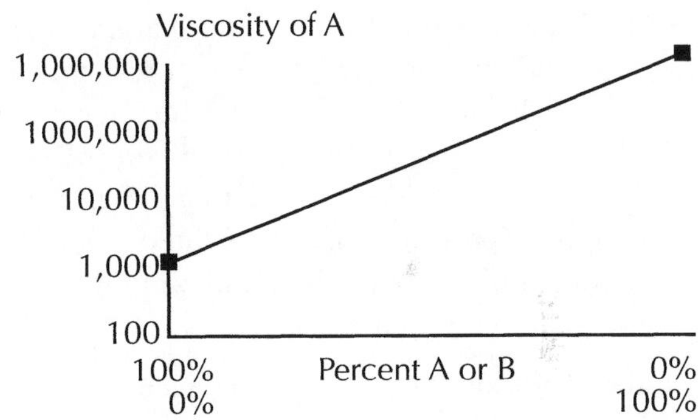

For example, a 50/50 blend of two polymeric fluids, one with a viscosity of 1,000,000 Pa·s and the other with a viscosity of 1000 Pa·s will have a viscosity of about 12,000 Pa·s, not the arithmetic mean of about 500,000.

In addition to this simple viscosity reduction, elevating the temperature of the additive-which is usually a much lower molecular weight and viscosity than the polymer being extruded-further reduces the viscosity of the blend, which leads to the slipping condition on the barrel previously described. When this happens, output is reduced and perhaps even becomes unstable. When the "melt pool" forms before the solid bed is fully compacted, bulk density can vary. When finally densified at some point down the barrel, output surges.

In addition to this surging phenomenon, improper melting entraps air, of which about 20% is oxygen. This entrapped oxygen is detrimental to extrudate quality and can lead to the generation of gels, carbon specs, etc.

These situations can exist or develop, for instance, when additives such as polyisobutene (PIB) are introduced directly into the feed throat in an attempt to make stretch cling film. When this is done, output reductions of 50% are not uncommon. The proper method of adding this additive is via an injection pump which delivers heated PIB about one-third the way down the length of the barrel. Similar situations can occur when slip additives, stabilizers, etc., are introduced to the extruder feed throat in a similar manner.

CHEMICAL, FLAME, AND CORONA TREATING METHODS

This section will explore the different priming methods available to the converter that are necessary when improved adhesion of the extrusion coating or laminating polymer to the substrate is needed (pre-priming) or when improved adhesion of an ink, adhesion, or coating to a finished web is required (post-priming). The three traditional methods available are chemical priming, flame treating, and corona treating.

Studies indicate that 60% of converters in the United States use corona treating, 50% use flame treating, and 50% use chemical priming *(3)*. This is an interesting finding indeed, indicating that most converters use a least two priming processes within their plants. The reason for this is that each system has its own merits and drawbacks, which are reviewed here.

The goal of any priming system is to remove impurities from the surface of a substrate or coating and to prepare that surface for subsequent converting operations, such as printing or laminating, so that the two mating surfaces are chemically compatible. In essence, all three systems accomplish the same end result, but use vastly different mechanisms to do so.

Chemical priming, as the name implies, utilizes a specially designed chemical primer to promote adhesion between a substrate and coating. Corona treating uses an electrically generated plasma to charge the surface of a substrate, polarizing it in the process. If the surface is a plastic, it also crosslinks the low molecular weight portions to form a skin layer of strong, high molecular weight material. Lastly, flame treating does the same thing as corona, but uses a combustion-generated plasma, namely a high-intensity gas flame. Each of these methods has its place, and their use will be determined by customer needs, converter equipment available, costs, and local environmental laws.

Chemical priming

Chemical priming is an established and proven technique that can work with every combination of paper, film, aluminum foil, and polymer. This technique generally provides the best adhesion at the lowest cost per unit area of all three methods. In addition, chemical priming can provide functionality such as barrier properties, where the other methods cannot. Though new aqueous primers are introduced almost daily to replace obsolete or outlawed solvent-based primers, they are not always as effective as their solvent predecessors.

On the negative side however, chemical VOCs are under scrutiny for air emissions, as are aqueous primers for dumping into ground water or city water systems. Equipment is quite expensive for this method, setup time excessive, and inventory space required for the raw material.

Chemical priming or coating is best suited for converters that produce high-quality films or laminations where the use of other priming methods can compromise product integrity. Some examples where chemical priming is most appropriate are listed in the table below.

Table 2. Uses for chemical priming

1. Treating a PP film for subsequent use by a printer or converter that has no in-line corona treating ability.
2. For product constructions where there is no single extrusion resin available to join to dissimilar materials and coextrusion is not an option, e.g., extrusion laminating an OPET film to a linear low density polyethylene LLDPE without the use of coextrusion.

3. As a primer for agricultural or chemical use multiwall sacks for which an oxygen or aroma barrier are required.

Corona treating

Corona treating has become very popular due to its good effectiveness on polyolefin films, ease of installation and use, and freedom from volatile organic compounds (VOCs). It is a versatile tool in that it can be used in primary operations (such as extrusion) or in secondary operations (such as printing), and can handle paper, paperboard, film, or metal aluminum foils. Other than threading through the unit, there is no setup time involved with corona treating. Adjustment of treatment level is as simple as turning a dial, and sophisticated units have watt meters to indicate exact treatment levels.

On the negative side, the testing of treatment level does involve the use of hazardous chemicals that require special handling. Also, polypropylene (PP) is known to be difficult to treat with corona. Corona treatment is not effective on paper substrates that are to be PP coated. Also, corona treating can cause pinholes and is known to decay over time on PP films and coatings, making this treatment technique inferior to the other two with this material.

Corona treating is best suited for situations where a chemical priming unit is not practical, feasible, or affordable. Some examples of where corona treating is most appropriate are listed below.

Table 3. Uses for corona treating

1. For converters that have many short orders that would require substantial setup time if chemical priming were used.
2. For small converters where the initial cost of coating stations and dryers or environmental laws are prohibitive.
3. Where the product line is limited and corona treating provides sufficient adhesion for the product mix, or when the drying of the sheet with flame treating might be undesirable.

Flame treating

Flame treating has been a popular traditional treating method for extrusion coating of paper and paperboard. New computer technology has vastly improved safety and treatment regulation, with gas composition and humidity analyzed and controlled to yield a consistent treatment level. Today, flame treaters in combination with a chilled backing roll are used to treat unsupported film.

On paper substrates, the high-intensity flame, which also can be thought of as plasma, oxidizes and polarizes the paper surface while simultaneously preheating it. On polymeric substrates, the flame boils off low molecular weight materials on the surface and provides a much deeper surface treatment than corona treating on polymer films. This extra depth of treatment provides a more effective and longer lasting treatment. Because of the nature of flame treating, their is no back-side treating and pinholes are not a problem with this technique-both of which are potential problems with corona treating.

The initial cost of a flame treating unit can be slightly higher than corona, but operating costs are lower, productivity is higher, and in general quality is superior to corona treating.

Flames treating is best suited for very demanding applications in which the power required for treatment can exceed the capability of corona treating or when more than one treatment technique is required.

Table 4. Uses for flame treating

1. When coating LDPE onto clay coated SBS board for which chemical priming or corona treating alone are insufficient.
2. When a product requires a long-lasting, high dyne level treatment, such as for overseas shipments.
3 When dealing mostly with paper and paperboard, and substantial preheating of the substrate is required.

LAMINATING VS. COATING

Die streaks

Question: Why have we experienced problems with carbon buildup, poor gauge control, and excessive neck-in and edge weave with polypropylene?

Answer: New users of polypropylene (PP) resins find many processing differences compared to standard LDPE resins. These differences are almost always negative, such as increased carbon generation, poorer gauge control, and excessive neck-in and edge weave, and are due to the structural differences in the polymer backbone.

Polypropylene is similar to polyethylene, with the exception that it has a methyl (CH_3) group hanging from one of the carbon atoms. This simple change makes all the difference in the world. It forces the polymer to line up in a particular order when frozen, thus making it highly crystalline. This is positive in terms of physical properties, but negative in terms of processing characteristics. This methyl group (called a tertiary carbon atom) is big and bulky when compared to the hydrogen atom it replaces, and is thus prone to being torn (sheared) off during extrusion. The action of shearing off the tertiary carbon degrades the polymer, causing the carbon buildup seen in the converter's screen packs. Despite the heavy stabilization of PP resins, carbon buildup remains problematic and causes more screen changes than desired. A modified screw design will help somewhat. Lower RPMs might be necessary to minimize shear rate in the extruder barrel.

Rheological properties of PP resins are responsible for the poor gauge control, neck-in, and edge weave properties cited by the questioner. In relative terms, PP is a low-viscosity, narrow molecular weight distribution resin. Most extrusion coating grades of PP are in the 40-70 MFI range. These attributes, combined with the absence of chain branching, account for all the difficulties cited above.

The poor gauge control this converter is experiencing is probably due to low die pressure as compared to standard conditions with LDPE. Steps therefore must be taken to maximize pressure in the die. These include backing out the adapter valve and/or reducing the number of screen packs used, decreasing the die gap, and increasing output.

People quite often confuse head pressure with pressure in the die, but the opposite is really true. Energy taken out of the polymer flow in the form of increasing head pressure only reduces the energy left to push it from the adapter through to the die. Screen pack usage is quite often quality related, so I am hesitant to recommend a particular screen pack pattern. Reducing die gap will increase pressure in the die dramatically, since pressure rises with the cube of die gap. For PP, I recommend starting at a 40:1 drawdown ratio of die gap to finished thickness, and modifying as necessary. Increasing output causes a linear increase in die pressure, and also serves to pressurize the die lips and provide more uniform coating weights. One should strive for at least 10 lb/h per inch of die width. Eastman did some work many years ago showing that short die lands also improves gauge control with PP resins.

Again, the chemical structure of the PP polymer backbone is conducive to excessive neck-in, and in severe cases, edge weave. To minimize this, melt temperature and air gap should be minimized within the constraints of the system and quality requirements. Unless you have the option of coextrusion, what you see is what you get. If coextrusion is an option, try coextruding a LDPE/PP blend next to the substrate. The addition of LDPE, with its chain branching and good melt strength, reduces neck-in, and stabilizes the melt curtain to minimize edge weave. Some converters successfully use air jets on the edge of the web to freeze or pin it in position.

Slitter dust

When converting paper-based products, the use of slitters is almost unavoidable. In the course of time, invariably slitter knives become dull, are improperly ground, or misaligned. Either one of these cases will be a cause of slitter dust. In non-demanding applications, this will not be a problem; but, in any type of printing or consumer application, slitter dust is not tolerable and must be eliminated somehow.

Question: Why do we experience slitter dust, and how do we eliminate it?

Answer: This producer of film and extrusion coated products is experiencing a very common phenomena in the converting industry, namely surface contamination. This occurs when less than optimal slitting and operating conditions exist in the process. The result is the generation of physical dust when the paper is slit.

I generated an Ishikawa diagram to determine the possible causes of this problem. Basically, the causes can be either process and material related, or both. In the basic poly coated paper scenario in which shear knife slitters are used, two knives come into close proximity of each other, between which is the product being slit. Slitter manufactures have specific instructions as to the location of slitter knifes, the angle to which the knifes should be sharpened, the distance between knifes, etc. Deviation from these specifications causes imperfect slitting and is a cause for slitting dust. As a general rule, more dust can be expected from score slitting than shear slitting.

Another source of slitter dust is the paper itself. Papers that contain a high percentage of fillers or are made from short, stiff fibers tend to generate more slitter dust under the same conditions than papers with different compositions, especially with dull or misaligned slitter knives.

As a matter of principle, I always recommend eliminating vs. masking the problem. The first cause (out-of-specification slitters) is easy to fix. Grinders must be properly set up to sharpen knives correctly and slitter positioning is critical. Because of the extremely wide product range made by the readership, a single concrete set up will not be given. Rather, simply follow manufacturers' specifications. Operator training is critical, as is the ability to make very small adjustments to the slitter knifes. Capability for on-line slitter adjustment is preferred. Razor knifes are recommended for film-only slitting applications.

With regard to substrate caused slitter dust, the solution can be difficult. Quite often papers are specified by the customer or purchasing and not subject to being changed easily. If the paper cannot be changed, consider requesting the addition of binding aids that bind the filler more tightly to the paper fiber.

Unfortunately, there are some cases where the paper cannot be modified in any way, and slighter dust is a fact of life. Some operations find that their products are too abrasive and slitter life is simply shorter than desired, and changing out knifes too frequently has an adverse impact on productivity. This causes management to make a productivity/quality trade-off decision. Fortunately, there are other options.

If the dust creation cannot be eliminated, then alternatives exist to remove the dust and debris from the surface of the web. These include the use of ionized air, vacuum systems, and contact cleaning rolls.

Ionized air blowers simply overcome the effect of static electricity and remove much of the larger surface contamination by the physical action of air moving past the web. Vacuum systems work in much the same way, but utilize negative rather than positive pressure, and can even be used in conjunction with ionized air blowers. The main drawback with these systems from a technical point of view is that they cannot overcome the very thin boundary layer of air on the surface of the substrate to be cleaned, and thus might not clean off the very small particles.

Contact cleaning rolls are simple devices that employ an elastomeric roll covering with a specifically designed tack that allows surface contaminates to be removed without affecting web handling. Their efficiency is about 90% removal vs. about 60% for noncontact systems. They are simple, inexpensive, efficient devices that offer a solution to the toughest dirt, dust, and contamination removal that converters encounter.

REFERENCES

1. Darnell, W.H. and Mol, E.A.J., SPE J. 1956.
2. Gregory, R.B., "Friction Coefficients of Plastic and Steel," SPE 1969 ANTEC Conference Proceedings, Society of Plastics Engineering.
3. American Paper Institute, confidential user survey, 1989.

CHAPTER 35

Troubleshooting the Extrusion Coater

Robert F. Bernacki, Michael R. Fitzsimmons, and Frances M. Hammond

EXTRUSION COATING/LAMINATING CONTROL OPERATION

Despite its apparent simplicity, successful operation of an extrusion coating/laminating operation depends upon painstaking attention to a multitude of details, 24 hours a day.

A partial list of these details must include:

1. **Material handling practices**. Materials used must be kept clean and dry.
2. **Extruder design.** The extrusion equipment used for a specific job must be able to handle the specific materials necessary for that job.
3. **Extruder profiles.** Different polymers require different processing conditions, hence different extruder temperature profiles, for successful processing. Extruder feeding, surging, melt uniformity, and the temperature of the melt are all dependent upon the profiles selected.
4. **Back pressure controls.** Back pressure influences extruder output, the temperature of the melt, melt uniformity, and sometimes adhesion; therefore, it must be carefully controlled. The back pressure is commonly controlled by means of a back pressure valve and/or changes in screen pack configuration.
5. **Transfer pipe and die temperatures.** Ideally, all metal surfaces in contact with the polymer melt after it leaves the extruder barrel should be the same temperature as the melt itself. Melted polymer flows without mixing and is a good thermal insulator, so trying to heat or cool the melt only results in temperature striations.
6. **Die gap.** The proper die gap setting ensures uniform gauge and can also influence the stability of the molten web.
7. **Air gap.** The amount of time that the molten polymer remains in contact with air has an impact on adhesion. A long gap cools the polymer and, if the melt temperature is over approximately 600°F, increases the amount of oxidation of the surfaces. Since cooling decreases adhesion while increased oxidation improves adhesion, an optimum air gap exists for each set of circumstances that allow maximum adhesion.
8. **Substrate preparation.** Occasionally, the adhesion obtained with certain polymer/substrate pairs is not sufficient. Methods used to increase adhesion include chemical priming, corona treatment, flame treatment, ozone, or a combination of the four.
9. **Chill roll temperature.** The optimum chill roll temperature is cool enough so that the coating releases cleanly, yet warm enough so that the polymer does not freeze too rapidly before having a chance to wet the substrate well.

EXTRUSION COATING LAMINATING DEFECTS

Control of the previous parameters is essential to the profitable operation of lines incorporating extrusion coating and laminating equipment. In the event of defects, rapid diagnosis of the problem reduces the amount of poor quality material produced. The following sections of this troubleshooting paper have been written to aid the rapid solving of common problems in this industry:

1. Poor adhesion
2. Voids
3. Pinholes
4. Gauge problems
5. Gels and oxidation
6. "Applesauce"
7. Molten curtain breaks
8. Poor heat sealing.

ADHESION

Adhesion of polymer coatings to various substrates is caused by mechanical and/or chemical effects that are primarily related to polymer temperature.

Mechanical effects

The adhesion of polymers to fibrous substrates is largely attributed to mechanical adhesion, although chemical attraction can have some influence. The polymer temperature controls the viscosity, which in turn controls the degree of wetting or penetration, and mechanical gripping of the polymer to the fibrous substrate. Typically, but not always, poor mechanical adhesion is caused by a melt temperature that is too low. This can be attributed to several factors-the most obvious being low temperature settings which can be corrected by increasing the temperature profile. If the temperature profile is correct, look for equipment malfunctions such as burned out heater elements, blown fuses, poorly tuned or dirty control instruments, or temperature controllers out of calibration. If there are no equipment malfunctions, try increasing the extruder back pressure. The majority of heat being supplied in the extruder is frictional with the remainder being supplied by the heaters. If the back pressure is too low, not enough frictional heat can be generated to achieve the desired melt temperature.

There are two factors that can affect the temperature of the polymer (and therefore its wettability) after it leaves the die. The first is air gap, or the vertical distance between the die lips and the nip. A certain minimum air gap is required to allow oxidation to occur on the polymer surface aiding in chemical attraction to the substrate. If the air gap is lengthened too much, the polymer will begin to cool prematurely before it contacts the substrate, thus reducing its wettability. This excessive air gap will also result in high neck-in and edge bead. The other factor affecting wettability is chill-roll temperature. If the chill-roll temperature is too cold, the polymer can cool too quickly before it gets an opportunity to penetrate the substrate.

In some instances, increased polymer viscosity, leading to poor mechanical adhesion, can be caused by extremely high melt temperatures. At very high temperatures, molecular changes such as cross-linking occur due to thermal degradation. Cross-linking, or joining together of molecules, causes an increase in viscosity and can eventually lead to gel formation.

Chemical effects

The other mechanism for adhesion is chemical attraction. The chemical effects are influenced by polymer temperature and oxidation, but are also dependant on comonomer content in the case of copolymers such as EAA, EVA, and ionomers. Chemical adhesion is usually more important for substrates with smooth, nonporous surfaces such as films or aluminum foils, although wetting of the surface is still important.

Most of the chemical bonding that takes place in homopolymers is attributed to oxidation of the molten film in the air gap, and/or oxidation of the substrate surface. The two factors that control the degree of oxidation of the melt are polymer melt temperature and length of the air gap. As melt temperature is increased, thermal oxidation degradation increases, forming polar groups in the polyethylene that are attracted to the substrate. As the length of the air gap is increased, the exposure to oxygen in the air is increased, which results in greater oxidation of the web. There are limits, however, on how high one can raise either the melt temperature or the air gap without adversely affecting other properties. Over-oxidation of the polymer web due to high temperature and/or a long air gap will result in poor heat sealibility and increased taste and odor contributions. Excessive melt temperature can also lead to gel formation. An excessive air gap will result in increased neck-in and edge bead. Also, premature cooling can result in decreased mechanical adhesion as discussed earlier. Therefore, the best compromise of melt temperature and air gap must be chosen with consideration of these conflicting effects.

Most ethylene copolymers such as EAA, EVA, and ionomers differ greatly from homopolymers with respect to chemical adhesion. Carboxylic acid groups distributed throughout the polymer are polar and provide sites for strong chemical bonds. The degree of adhesion is directly related to the percent comonomer, and oxidation of the molten web is not necessary to obtain adhesion. This is why most copolymers can achieve excellent adhesion to polar substrates at melt temperatures lower than a typical LDPE. It is also recommended that the air gap be kept at a minimum with these copolymers to minimize neck-in and odor effects and maximize mechanical adhesion.

Treatment of the substrate can also improve adhesion. In the case of paper or paperboard, flame treatment is commonly used to oxidize the surfaces and to heat the web, allowing better penetration of the molten polymer. In the case of aluminum foils and films, corona treatment can be used to oxidize the surface of the substrate and burn off impurities. A chemical primer or wash-coat can be applied to provide sites for chemical bonding.

Other effects

There are a few other factors that can affect adhesion that have not yet been covered. Increasing the nip pressure can improve adhesion by allowing better polymer penetration. The position of the die relative to the nip is also important, especially for laminations. Frequently, the die is positioned so that the molten polymer contacts one substrate earlier, enhancing adhesion to that substrate. It was previously mentioned that a chill roll temperature that is too cool could result in poor mechanical adhesion; yet, if the chill roll temperature is too high (causing the polymer to stick to it), the adhesion of the polymer to the substrate is sacrificed.

Line speed and/or coating thickness are other factors affecting adhesion. As line speed increases, the amount of time the polymer remains in the air gap or in the pressure nip decreases, which can affect adhesion. Typically thicker coating weights provide better adhesion because they do not cool as quickly.

Adhesion is also affected by the presence of additives. Some additives bloom to the surface and interfere with chemical bonding taking place. Antioxidant, if used in excess, can prevent oxidation of the polymer web, thus preventing chemical adhesion. If additives are blended into the product at the extrusion coater's facilities, extra care should be taken in choosing the proper additives and carrier resins and measuring the amount to be added.

VOIDS

Most voids are caused by volatiles contained either in the molten web or on the substrate. Occasionally, however, voids might be caused by large gels or particles of oxidized polymer and will be discussed later. If no particles are visible, the culprit is almost sure to be volatiles from either moisture or polymer degradation.

Moisture

Very small amounts of water can cause significant problems. For instance, subjecting a droplet of water to 620°F raises its pressure to 1800 psi if contained, and causes it to expand explosively at a 1000:1 volume ratio as the droplet emerges from the die. Porous substrates like paper, and certain polymers like ionomers, polyester, or nylon have a tendency to absorb enough moisture from the air to cause voiding. Humidity can also cause problems with other substrates and resins if abrupt temperature changes cause surface condensation. Obviously, gross moisture contamination from conditions like leaky roofs, silo leaks, or unloading bulk containers in inclement weather must also be avoided.

Polymer degradation

Subjecting polymers to high temperatures causes a series of complex chemical changes to occur, some of which lead to the formation of volatile materials. Different polymers exhibit different degrees of stability; therefore, it is important not to exceed the manufacturer's recommended maximum melt temperatures for a specific polymer.

PINHOLES

"Pinholes" can refer to smaller holes which appear anywhere in the packaging structure, either in substrates such as aluminum foil or in coating or tie layers. Substrate pinholes are typically caused by poor substrate quality or improper handling practices not directly related to the extrusion or laminating process and will not be addressed here.

Pinholes in coatings are often caused by penetration of fibers through the coating, which allows wicking of liquids or reduced vapor barrier of the finished product. Excessive substrate roughness can cause the problem in the case of paper or paperboard and is usually corrected by flame treating to burn off loose surface fibers. Even small protuberances can stick through very thin coatings. Occasionally, low coating weights cause scattered pinholes. The appearance of pinholes in a machine direction band can indicate a thin streak that can possibly be too narrow to pick up with some gauge profiling techniques.

GAUGE PROBLEMS

Product performance usually dictates the minimum allowable coating thickness, while any area where this minimum is exceeded represents lost profits. Therefore, maintaining uniform gauge at the desired thickness is important for both product performance and economics. Several obstacles will be discussed which can interfere with the gauge of the coating.

GAUGE BANDS

Gauge bands are strips of varying thickness in the polymer web. The first item to check when gauge bands occur is the die. The functions of the die are: (a) to distribute the melt with a uniform coating thickness, (b) to release the melt at the desired width for the substrate, and (c) to maintain the melt at a constant temperature.

To perform these functions, the die must first be adjusted properly to obtain a uniform coating thickness across its width. It should also be free of any scratches which could cause gauge bands or die lines. In addition, it should be cleaned periodically to make sure foreign material has not lodged in the die which could be causing thin bands. If the die and die lips are clean and properly adjusted but the gauge bands still occur, varying temperatures could be the cause. The molten polymer must be at the desired uniform temperature and viscosity as it leaves the die lips. Extrusion coating requires high temperatures for which heating elements in the extruder barrel and die are needed. Malfunctioning or loose heaters can cause temperature striations that create viscosity variations resulting in gauge bands.

SURGING

Surging is the term that refers to an unsteady flow of polymer through the die. This could be caused by an uneven flow of resin into the extruder feed zone due to bridging in the hopper or an improperly designed screw. Proper screw design takes into consideration length-to-diameter ratio, feed, compression and metering section lengths, and the correct pitch and depth of the flights tailored to the specific polymers to be processed.

EDGE TEAR

Edge tearing is occasionally exhibited during extrusion and is a kind of "warning sign" before the web completely breaks. Edge tearing can affect the width of the coating and can be caused by cold die ends. Another cause could be a draw ratio that is too high. The draw ratio is the die gap divided by the final coating thickness. By decreasing the die gap or increasing the coating weight, the draw ratio can be reduced, thereby reducing the stress on the molten polymer. Also, if the melt temperatures deviate from the optimum melt temperatures-by being either too hot or too cold-edge tears might be noticeable. If the melt is too cold, the viscosity is too high to allow the resin to flow properly. If the melt is too hot, the viscosity can increase due to cross-linking, inhibiting proper flow. Improper deckle positioning in the deckle-holder can cause the polymer to squeeze over the Teflon deckle and initiate edge surging.

DRAW RESONANCE

Draw resonance is the term used to describe a thickening and thinning of the web in the machine direction as well as an oscillation in the width of the coating. Draw resonance is sometimes mistaken for surging, but differs in that the amount of polymer flowing from the die remains constant. Draw resonance tends to be characteristic of linear low-density polyethylene, and some of the solutions are the same as for edge tears. First, the drawdown ratio could be too high. To lower this, either decrease the die gap or increase the coating thickness. Secondly, the melt temperature could be too hot, causing the polymer to degrade and crosslink. This can occur unevenly across the web, producing variances in the thickness of the coating.

GELS AND OXIDATION

Imperfections such as gels or oxidation in the molten web not only have a bad appearance, but can adversely affect important end-use properties. Gels are clear, raised imperfections that vary in size from 1/64 in. to as large as 1/16 in. or more in diameter. Degraded or oxidized polymer often appears as a gel, but is usually distinguished by its brown or yellowish appearance.

Gels can be generated in the melt stream of the extrusion coating process or can be caused by resin contamination during transfer and handling. Care must be taken in keeping the resin transfer and storage systems clean and free of debris. The hopper on the extruder should remain covered, and boxes should not be left open for extended periods of time. Gels formed in the melt stream are a result of thermal degradation that is related to temperature and residence time. A certain amount of degradation is necessary for oxidation and good adhesion, but if the melt temperature is too high (bulk temperature or hot spots) or the polymer residence time in the extruder is too long (low rpm), excessive degradation will occur leading to cross-linking and gel formation. If gel formation becomes a problem, lower the melt temperature, increase screw speed, and check for localized hot spots.

Gels can also be caused by incompatibility of resin blends, poor dispersion of pigments or additives, or unmelted pellets. These problems can be inherent with the combination of resins and pigments involved, as well as with improper screw design. However, these problems can be solved by increasing the back pressure using the back pressure valve or by changing the number of density of screen packs to allow better mixing.

Sloughing off of degraded, oxidized polymer can be the result of poor streamlining or improper purging and shutdown procedures. If there are non-streamlined areas in the extruder, adapter, or die where the polymer can get hung up and remain at high temperatures for long periods of time, degradation and eventual slough-off will occur. One of the most common areas for degraded polymer to build up is behind the deckles where a dead spot is formed. The problem is usually more severe with copolymers because they have a tendency to adhere to metal surfaces. It is helpful to clean all sections of the extrusion coater on a regular basis as part of a preventive maintenance program. It is also very important to use proper purging procedures as well as shutdown and startup procedures to eliminate gels and degraded polymer. Deckles should be removed for purging purposes and temperatures should be reduced. It is important that the extruder not be shut off for extended time periods until all temperatures are below 400°F for most homopolymers such as LDPE and LLDPE, and the extruder should never be shut down with acid copolymers or ionomers in it. It is recommended that a drool stick be placed underneath the die lips to seal off the polymer from possible oxidation.

APPLESAUCE

"Applesauce" is a term that refers to a rough, wavy appearance of the web. It can also be referred to as "orange peel" or flow patterns. This wavy appearance could be due to several factors. If blending two or more resins, they could be incompatible, either rheologically or chemically. Rheological mismatch occurs if their melt indices are too far apart. If the resins' chemistries do not closely coincide, web patterns could also result. Chemical incompatibility often occurs when attempting to blend a copolymer with a homopolymer. Poor purging procedures could also result in applesauce if the residual resin is incompatible with the subsequent resin. Therefore, the extruder should be completely and thoroughly purged before changing resins.

Melt temperature can also be the reason for a wavy web appearance. If the temperature is too high for the resin, degradation and cross-linking will occur. If the temperatures are too low, increased viscosity can result in a slip stick phenomenon called melt fracture.

A check should also be made to assure the temperature across the web is uniform since temperature striations could cause flow patterns and an "applesauce" appearance.

MOLTEN CURTAIN BREAKS

Extrusion coating or laminating requires a polymer to be drawn down from a thickness determined by the die gap (typically 20-40 mils) to a thickness determined by the substrate speed (often >1500 ft/min) and the amount of polymer being supplied to the die (up to 40 lb/h per inch of die width). This requires the melt to accelerate between the die and the nip at a rate referred to as the draw ratio. For instance, a 30-mil die gap being used for a 0.75-mil coating results in a draw ratio of 30 divided by 0.75, or 40. In other words, the melt has to be traveling 40 times as fast when it arrives at the nip as when it left the die lips. This acceleration builds up stresses in the molten web which, if they become too great, cause the curtain to break. Ways to reduce the draw ratio include increasing the coating weight (usually not an economical solution) or decreasing the die gap. The minimum die gap is controlled by back pressure and die design considerations.

Each polymer designed for extrusion coating has a limited draw ratio for a given set of extrusion conditions. Temperature is a very important determinant of the maximum draw ratio of a specific polymer. Generally, an extrusion coating resin is designed so that optimum drawdown characteristics are experienced at the recommended melt temperatures.

POOR HEAT SEALABILITY

Poor heat sealability is usually caused by over-oxidation or degradation of the polymer coating. This can be a result of excessive melt temperatures, air gap, or a combination of both. It is recommended to use the minimum melt temperature and air gap at which adequate adhesion is still achieved. If the temperature of the polymer web exceeds the setpoints, reducing back pressure (or screw speed), or increasing barrel cooling can help. If these steps are unsuccessful, inspect the heaters and control instrumentation for malfunctions as discussed earlier in the adhesion section.

Over-oxidation of the coating can also be caused by excessive corona treatment after coating. Many extrusion coated surfaces are corona treated to improve their adhesion to printing inks. If the level of corona treatment is too high, over-oxidation and therefore heat sealing problems will result. The operation of the corona treater should be checked periodically, and the amount of corona treatment being applied can be checked by measuring the surface tension of the coating.

Poor heat sealability can also be caused by contamination of the coated surface. This contamination can be from: (a) additives within the resin that bloom to the surface; (b) dirt, dust, or other impurities that come in contact with the polymer surface due to improper storage or handling; or (c) the product being packaged remaining in the seal area due to static buildup or splashing. If using polymers with additives, be sure that the additive does not adversely affect sealability and make sure the proper amount is used. The coated substrate should be kept in a clean environment and the amount of handling should be minimized.

Different heat-seal strengths can be expected from different resins. LLDPE resins offer improved heat-seal strengths at conventional LDPE sealing temperatures. Specialty copolymers such as EAA, EMAA, EVA, EnBA, and ionomers offer not only improved seal strengths, but they initiate at considerably lower temperatures than LDPE or LLDPE.

TROUBLESHOOTING REFERENCE GUIDE

Problem	Cause (reason)	Solution
Poor adhesion	a. Melt temperature (increased viscosity)	1. Increase temperature too low settings. 2. Check for heater and/or controller malfunctions. 3. Increase back pressure.
(Increased viscosity or degradation)	b. Melt temperature	1. Decrease temperature too high settings 2. Check for heater/controller malfunctions. 3. Decrease back pressure.
	c. Inadequate oxidation	1. Increase air gap. 2. Lower line speed.
	d. Premature cooling	1. Decrease air gap. 2. Increase chill-roll temperature.
	e. Chill-roll tack	1. Decrease chill-roll temperature.
	f. Non-optimum substrate	1. Pretreat substrate surface. 2. Change substrate.
	g. Additive effects of additive are used.	1. Confirm that proper type and level
Excessive odor	a. Over-oxidation	1. Decrease melt temperature. 2. Decrease air gap.
Edge tear	a. Draw ratio too high	1. Decrease die gap. 2. Increase coating weight.
	b. Die end temperature	1. Increase temperature of die too low ends.
	c. Improper polymer	1. Adjust temperature temperature to optimum.
	d. Improper deckle setting	1. Adjust deckle.
Draw resonance	a. Draw ratio too high	1. (See edge tear.)
	b. Temperature too high	1. Decrease polymer temperature. 2. Repair runaway heaters.
Gels	a. Resin contamination	1. Keep extruder and storage areas clean. 2. Keep resin boxes covered when not in use.
	b. Thermal degradation	1. Decrease polymer (cross-linking) temperature. 2. Increase screw speed. 3. Check for hot spots.
	c. Poor purging	1. Use proper purging techniques.
	d. Poor mixing	1. Increase back pressure. 2. Increase number or density of screen packs.
Oxidation	a. Degradation	1. Decrease temperature. 2. Increase screw speed. 3. Clean non-streamlined areas often. 4. Reduce back pressure.
	b. Poor purging	1. Use proper techniques, usually lower temperatures. 2. Do not stop extruder when still above 400°F. 3. Always purge out copolymer with a homopolymer before turning off extruder.

Voids	a. Moisture	1. Avoid abrupt temperature change in prod uct storage. 2. Check for leaks in resin handling system. 3. Protect hygroscopic resins from humidity.
	b. Polymer degradation	1. Do not exceed recommended maximum melt temperatures.
Pinholes	a. Substrate roughness	1. Flame treat substrate. 2. Increase polymer thickness.
Gauge bands	a. Unclean die	1. Clean die interior periodically. 2. Scrape inside of die lips to dislodge foreign material.
	b. Improperly adjust die.	1. Adjust die bolts. 2. Measure die gap periodically.
	c. Temperature variations	1. Repair heater malfunctions. 2. Keep adapter, transfer pipe, and die temperature close to extruder melt temperature.
	d. Temperature too high	1. Lower temperature settings. 2. Repair malfunctioning heaters or controllers.
Surging	a. Hopper bridging temperature.	1. Lower feed zone (improper feeding)
	b. Improper screw design	1. Use different resin or screw.
Applesauce	a. Resin incompatibility	1. Check on resin compatibility (chemical and rheological). 2. Make sure previous resin is completely purged.
	b. Temperature too high	1. Decrease temperature.
	c. Temperature too low	1. Increase temperature.
Molten curtain breaks	a. Draw ratio too high	1. Decrease die gap. 2. Increase coating weight.
	b. Improper melt	1. Adjust to optimum temperature
Poor heat sealability	a. Over-oxidation	1. Decrease melt temperature. 2. Decrease air gap. 3. Check corona treatment.
	b. Contamination	1. Check on additives used. 2. Keep handling of sealant surface to minimum and keep in clean dry area.

GLOSSARY

Air gap

The vertical distance between the die lips and the nip. Air gap is less critical for copolymers than for LDPE.

Applesauce

Rough wavy appearance of the web similar to orange peel.

Coating vs. lamination

Coating is the process of applying a molten polymer web to a moving substrate, whereas lamination uses the molten polymer to adhere two substrates to each other.

Copolymer

Polymer that has a functional chemical group added for improved properties. Examples are EAA adhesive polymers, EVA, or EMAA.

Cross-linking

A mechanism of degradation by which molecules join together resulting in increased high-molecular-weight portions. Cross-linking will increase polymer viscosity. In fact, many gels are actually highly cross-linked material.

Deckle rods

Adjust devices used to control the width of the polymer as it leaves the die lips.

Die gap

The distance between the die lips through which the polymer flows (typically 20-40 mils).

Drawdown

The complete and instantaneous tearing or breaking of the molten film across its entire width.

Draw ratio

The die gap divided by the coating thickness.

Draw resonance

A different form of drawdown failure characterized by wandering or wavering of the edge of the molten film or loss of thickness uniformity across the width of the coating.

Edge bead

A buildup of polymer along the edges of the web due to neck-in. A cross section of this beaded edge is teardrop-shaped tapering into the film. Internal reduction rods are used to minimize edge bead.

Edge tear

Partial tearing of the web along its edges. The line speed (nip-in) at which edge tear occurs, along with extruder output rate determines minimum coating thickness.

Neck-in

Contraction of the molten polymer as it leaves the die. It is measured as the difference between the die width and the coating width.

Nip

The point of contact between the pressure roll and the chill roll where the molten polymer first contacts the moving substrate.

Processability

The ease with which a polymer can be extruded and coated. At a given screw speed, reduced extruder amps and reduced back pressure characterize improved processability.

Single heat

The use of one, two, or three extruders in series to tandem/triplex produce a multi-layer structure.

Wettability

The relative tendency of a fluid to flow out onto a surface. It is dependent on surface tension, viscosity, and chemical compatibility.

CHAPTER 36

Process Troubleshooting Guide for Coextrusion Coatings

Paul H. Jackson, Equistar Chemicals LP

Coextrusion processing is the conversion of multiple polymers (flowing through separate streams) that are combined into a common primary passage and then shaped by a die. The manufacture of coextrusion film has been in existence for more than 30 years and continues to flourish because multiple layers of two or more material types provide properties that cannot be provided by a single material in an economical manner. Coextrusion is quite efficient due to its flexibility with regard to processing parameters, material selections, layer construction, and ease of operation. Multiple layer constructions using EVOH, nylon, or PVDC high barrier materials (as well as the coming of age for recycle content packages) have contributed significantly to the growth of coextrusion. Coextrusion is expected to continue its expansion as polymer and equipment technologies develop and as the understanding of processing progresses.

Troubleshooting methods for coextrusion are more complex than those for monolayer film extrusion. Coextrusion processing becomes increasingly complex as the number of layers in the structure increase, as the asymmetry of multilayer construction grows, and because processing and rheological characteristics of coextruded materials differ greatly from one another. Coextrusion processing obstacles can be overcome by being aware of the causes for difficulty and following basic guidelines for resolution.

Successful production of coextruded products depends on several key factors which include polymer selections, hardware design (screw, feedblock/die, handling equipment), layer construction, and optimal processing conditions. Proper selection and adjustment of each key factor will minimize difficulties and assure high-quality coextrusion results.

Troubleshooting coextrusion films requires emphasis on the explanation of problems associated with non-uniform layer distribution and interfacial instability between layers or on film surfaces. Non-uniform layer distribution can appear in the direction of extrusion or transverse to the direction of film production.

Layer uniformity in the machine direction can be influenced by die imperfections, poor die design or adjustment, excessive extruder pressure variation, variable film tension, or film instability. Non-uniform layer distribution between layers in the machine direction are equally common during monolayer extrusion and require less detailed discussion.

Layer uniformity transverse to machine direction can be influenced by poor melt temperature uniformity, viscosity mismatch between layers, poor hardware design, or viscoelastic flow characteristics induced by excessive shear stress. Coextrusion requires emphasis on the explanation of problems associated with non-uniformity of layers transverse to the machine direction.

Non-uniform melt temperature across a melt pipe, feedblock, and/or die-as well as poor melting in an extruder-cause poor layer uniformity transverse to machine direction. Melt temperature variance alters viscosity uniformity which exhibits a change in flow characteristics and layer distribution. Homogenous melt temperatures can be achieved through installation of a static mixer in the melt pipe, a dynamic mixer on the extrusion screw, installation of a more efficient screw design, or adjustment of pipe, feedblock, and/or die temperatures. Melt temperature of a single polymer stream can often vary by as much as 30°F. General rule of thumb is to achieve +/- 2°F or less variation in melt temperature for each extruder.

Viscosity mismatch between layers can cause variation in the thickness of a film that eventually reaches a steady-state condition of non-uniformity (assuming homogenous melt temperature conditions for each polymer). In a coextrusion system, lower viscosity polymers migrate to the region of highest shear stress (nearest the die wall) and tend to encapsulate higher viscosity polymers. The amount of migration is dependent on the degree of viscosity mismatch, the length of the flow path, and the shear stress in the system. Improvements in layer variations caused by viscosity induced flow behavior can be achieved through adjustment of melt temperature, modification of distribution channels in the feedblock or die, or selection of a polymer with different viscosity characteristics which is most commonly measured by melt index. Annular dies are typically more tolerant of viscosity mismatch than flat die systems.

Poor hardware design can also cause non-uniform layer distribution in the direction transverse to extrusion. Improperly designed flow channels of the feedblock or die can cause poor steady-state layer distribution of materials even with the most closely matched viscosities.

Interfacial instability between layers or on film surfaces can cause non-uniform distribution of layers in the form of parabolic flow lines, intermixing of layers, roughness between polymer boundaries, melt fracture, or uncharacteristically high haze. The instabilities are believed to be a result of the viscoelastic behavior of polymers at the die land or region of highest shear stress. Improvements in layer instabilities can be achieved by reducing the shear stress between coextrusion layers and/or the die land surfaces. Shear stress can be reduced by decreasing total output rate, increasing skin layer melt temperature (decrease in viscosity), increasing the die gap, adding a process lubricant to the skin layer material, or selecting a lower viscosity skin layer material. An increase in the thickness of the skin layer can also reduce instability between polymer layers by moving the interface further from the high shear stress die wall. This is especially significant for asymmetric coextrusion constructions. Finally, if coextrusion layers exhibit dramatic differences in melt elasticity, then choose materials that match more closely in extrudate elasticity as measured by extrudate swell.

Film\Web Defect	Cause	Symptom	Solution
Surface flaws that appear as lines	Non-uniform layer distribution in direction of extrusion	Die imperfections	Clean die buildup. Remove contaminant from polymer melt channel. Repair die nicks or burrs.
Gauge band(s) on roll		Poor die design	Install improved die design to eliminate distribution of polymer at exit.
		Poor die adjustment	Adjust die gap for uniform web distribution.
Repeating pattern of variation in thickness of layer(s)		Excessive extruder pressure variation (surging)	Achieve +/- 1% or less variation in total head pressure for each extruder.

Film\Web Defect	Cause	Symptom	Solution
			Adjust extruder temperature profile(feed and transition zones).
			Increase back pressure with restrictor flow plug.
			Increase back pressure by installing fine mesh screen pack.
			Change screw design of surging extruder(s).
			Check for worn screw(s) and replace if necessary.
			Check extruder feed throat for bridging and correct if necessary.
Variable film/web tension			Eliminate variability in drive speed.
Film/web instability			Protect film/web from atmospheric air turbulence.
Intermittent and somewhat random variation in thickness of layer(s)	Non-uniform layer distribution in direction transverse to extrusion	Poor melt temperature uniformity	Achieve +/- 2 °F or less variation in melt temperature for each extruder
			Adjust extruder temperature profile to assure complete melting of extrudate.
			Install static mixer in feed channel(s).
			Install new screw design with dynamic mixer or more efficient melting capacity.
			Reduce screw speed for increased residence time to complete melting of polymer.
			Adjust temperature of feed channels, die, and/or feedblock.
Variation in thickness of layer(s) that reaches steady-state distribution		Viscosity mismatch of polymer layers	Select materials that match in viscosity.

Film\Web Defect	Cause	Symptom	Solution
			Adjust temperature of material(s) to aid in viscosity matching.
		Poor hardware design	Change die and/or feedblock design.
Uncharacteristically high film haze	Interfacial flow instability between layers and/or die surfaces	Viscoelastic flow characteristics induced by excessive shear stress	Select lower viscosity skin layer material(s).
Parabolic shaped flow lines in direction of extrusion			Increase melt temperature of skin layer material(s). Increase die temperature(s).
Intermixing of polymer layers			Reduce total extrusion output.
Roughness between polymer layer boundaries			Increase die gap opening.
Melt fracture of film\web surface			Add process lubricant to skin layer material(s). Increase thickness of skin layer(s).
			Select material(s) that exhibit similar melt elasticity behavior (extrudate swell).